MECHANICS

MECHANICS

George Kofi Collins

Oxford Book Company
Jaipur ▪ New Delhi

Published By

Oxford Book Company

Regd. Off. "Bony Residency", Gate No. 2
Opp. Tilak Public School, Vishweswariya Nagar,
Gopalpura Road, Jaipur - 302018, India
Ph: 0141 - 2761280, Telefax: 2761381
email - oxfordbookjpr@gmail.com
website - oxfordbookcompany.com

Delhi Office: 102, 1st Floor, Satyam House,
4327/3, Ansari Road, Darya Ganj,
New Delhi - 110002, India
Ph: 011-45652440

Published in India, 2025

ISBN : 9788189473990

Price : ₹ 2295/-

Typeset by Shivangi Computers, Jaipur

Printed and bound in India by Thomson Press Ltd.,
New Delhi

Preface

There are many excellent physics textbooks on mechanics, this book covers such topics as Newtonian Mechanics, Gravitation Thermodynamics, Surface tension and Energy. The goal of this book is to introduce graduate students with no prior knowledge of physics to the basic concepts and methods of mechanics. For the last few decades mechanics has been a driving force behind many dramatic achievements in mathematics, similar to the role played by classical physics in the seventeenth to nineteenth centuries. Classical physics, especially classical mechanics, was an integral part of mathematical education up to the early twentieth century. Surprisingly, quantum physics, especially quantum mechanics, with its intrinsic beauty and connections with many branches of mathematics, has never been a part of a graduate mathematics curriculum.

This book is developed to partially fill this gap and to make mechanics accessible to graduate students and research mathematicians. Research in mechanics has grown dramatically in recent years, to the extent that students, researchers, and practitioners in mechanics now outnumber those working in the underlying discipline of mechanics itself. A courageous reader can actually learn by studying the main text, and with enough sophistication, could "translate" corresponding parts in physics textbooks. It is worth mentioning that development of many mathematics disciplines has been stimulated by mechanics. The book satisfies the needs of many foundation year engineering courses as well as many other engineering courses.

George Kofi Collins

Contents

Chapter 1

History of Mechanics

The study of motion of physical bodies is called mechanics. This subject dates back to ancient times. Many wise and curious people have wondered about the motion of heavenly bodies, often in preference to terrestrial bodies. The debate whether the Sun goes around the Earth, or the Earth goes around the Sun was one of the most heated issues in the history of science.

A study of history reveals that Aristotle formulated the first theory of mechanics. Aristotle's theory of motion was rather philosophical and relied heavily on logic.

Many centuries later it was observed that Aristotle's predictions did not match with experimental (or empirical) observations. Around the sixteen century, Galileo, Newton, and other physicists formulated a completely new theory of mechanics which forms the basis for modern science. In this new theory, observations and experiments have a very important role as they are used to verify or reject a physical theory.

At the turn of the twentieth century it was found that the predictions of Newton's laws do not hold for particles moving with speeds close to the speed of light, and for microscopic particles, e.g. electrons. To understand these phenomena, two new theories were proposed: Einstein's theory of relativity and quantum mechanics.

The present book will focus on Newton's laws, and it will only touch slightly on the above modern theories. While studying the history of mechanics, it will become evident that mechanics was the front-runner in founding modern scientific ideas.

ANCIENT THOUGHTS

Among ancient thinkers, Greeks philosophers contributed the most in the field of science. They discovered that the Earth is round. They estimated the radius of the Earth fairly well. They also attempted to measure the Earth-Moon and Earth-Sun distances.

They thought up many interesting ideas about the planets, the stars, and the universe as a whole. Due to lack of time and space, we cannot describe all these ideas here. In the following paragraphs we will only discuss the ideas of Eudoxus, Aristotle, and Ptolemy. Interested readers can refer to Ferris (1988), Spangenburg and Moser (1993), and Wikipedia for details.

Many Greek thinkers addressed the movement of stars and planets. The most prominent among them was Eudoxus (410 or 408 bce to 355 or 347 bce). He belonged to Plato's academy in Athens. Deeply influenced by Plato's appeal for perfect geometrical forms (sphere, regular solids etc.), Eudoxus proposed that the universe is composed of 27 concentric spheres surrounding the Earth.

These spheres rotate with constant velocity, with the axes of some of the spheres inclined to each other.

The stars were carried by the outermost sphere. Each planet moves in four spheres (20 for 5 planets), and the Sun and the Moon move in three spheres each.

By adjusting the rates of rotation and the inclination of the axes of the spheres, Euxodus could explain the movements of heavenly bodies. Since the Earth is at the centre of the universe in this model, it is called the geocentric model of the universe.

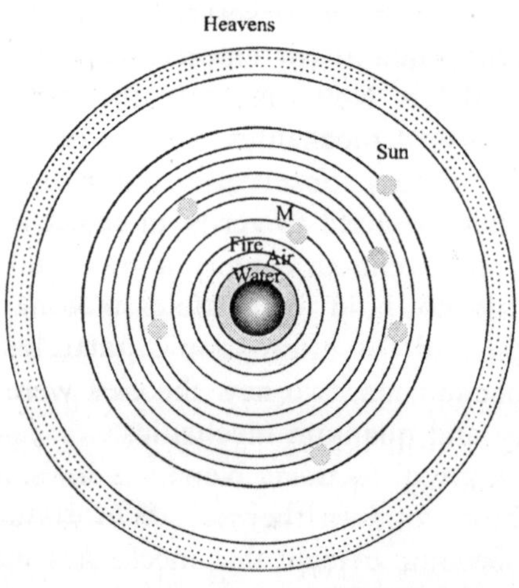

Fig. The Cosmos According to Eudoxus

The most complex motion in the sky is that of the outer planets. When we observe these planets from the Earth at a fixed time in the night, they move westward, then eastward, and again westward. This is called retrograde motion of the planets.

Eudoxus explained this phenomena by introducing epicycles. The planet moves on a smaller sphere that itself rotates around a larger sphere. The apparent motion of the planet from the Earth in the epicycle model. Note that a single epicycle produced the trajectory reproduced; more than a single epicycle is required to generate the retrograde motion seen in the sky.

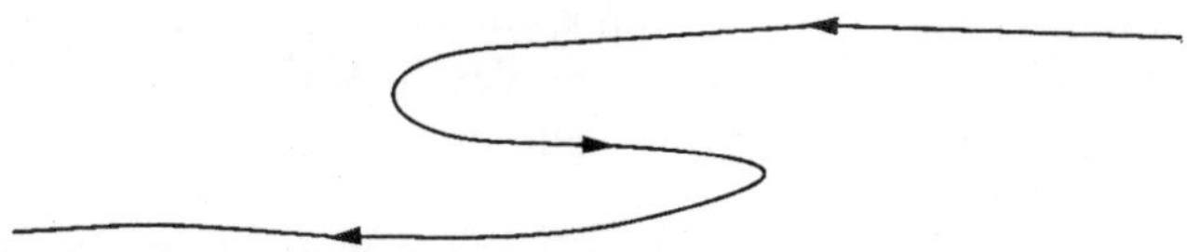

Fig. Observed Motion of Mars and Saturn. These Planets move Westward, Eastward, and Again Westward. This Phenomenon is called Retrograde Motion

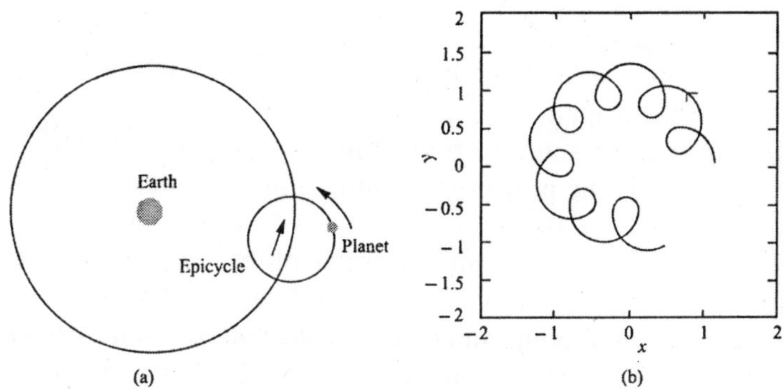

Fig. (a) The Retrograde Motion of an outer Planet Explained using Epicycles. The planet moves on a Sphere that itself Rotates around a Sphere whose Centre is the Earth. (b) The Apparent Motion of the Planet in Earth's Frame of Reference with a Single Epicycle.

Aristotle ntroduced more spheres to take the count up to 55. The results from later refined observations however did not match with the Aristotelian theory. To keep the geocentric picture intact, more complex set of spheres were introduced.

Claudius Ptolemy constructed the most refined geocentric model that lasted for nearly 1400 years. Note that the heliocentric picture, much simpler than the geocentric one, can explain all the observations. Unfortunately, prevalent prejudices typically hindered thinkers from attempting simple alternate ideas.

Aristotle set out to explain why the heavenly spheres rotated. He postulated that the prime mover rotated the spheres containing the heavenly bodies. The planets, the Moon, and the Sun are regulated by several spheres.

For Aristotle, the heavenly cosmos was imperishable and perfect, and the heavenly spheres moved in circular orbits .which is the most perfect trajectory. On the contrary, everything on the Earth was changeable and corrupt. The motion on the Earth is linear, not circular.

To explain the motion in the terrestrial environment, Aristotle argued that each object is made up of four elements—earth, water, air, and fire in various proportions. Each object moves in such a way as to return to its natural state, that is, one of the spheres on the Earth (sphere of earth, water, air, and fire).

Solid objects like stones would tend to fall to the Earth, their natural place. Air and fire are above the ground; objects composed mainly of air or fire would go upward towards their natural place. A consequence of this theory is that heavier bodies containing more earth will fall to the Earth faster than lighter ones.

Aristotle's arguments for the vertical motion on the Earth are logically consistent, but they do not work for the projectile motion. He postulated a rather awkward argument to explain projectile motion, theorising that air rushes into the vacuum created by the projectile's trail, and the incoming air pushes the projectile. Compare the complexity of Aristotelian theory with the relatively simpler Newton's laws of motion.

According to Aristotle, an unforced object comes to rest when it reaches its natural place. Hence a block sliding on a horizontal plank will come to rest unless forced, since rest is the natural state of the block.[2] Aristotle also gave a theory for the origin of force. For a projectile motion, thrower's arm provides force to the projectile; the thrower to his/her arm; and so on, forming an infinite chain of causes. Aristotle argued that there must be an unmoved mover, something which can initiate motion without itself being set in motion.

This view was preserved by the medieval Church during the Dark Ages, and it became a ruling paradigm.

In hindsight, Aristotle's theory of cosmos and causation were constructed by pure logic without resorting to proper observational and experimental tests. So they tended to be metaphysical and philosophical. Around two thousand years later, many inconsistencies were found in the Aristotelian theory, the resolution of which led to the birth of modern science.

RENAISSANCE—HELIOCENTRIC THEORY

Nicolaus Copernicus

In medieval times, astronomers performed very accurate observations of planetary orbit. Ptolemy's system of planetary motion could not explain the refined observations on the retrograde motion of Mars and Saturn.

Copernicus (1473-1543), a Polish astronomer, discovered that the retrograde motion of the planets could be explained quite easily if one assumes that the planets go around a stationary Sun. The time period of Earth's revolution around the Sun is smaller than that of Mars. In the heliocentric picture, Mars exhibits retrograde motion when seen from the Earth. At instants 1,2,3 and 6,7 Mars is moving westward, and in between at instants 4 and 5, it moves eastward. These observations led Copernicus to propose the heliocentric model.

This was the beginning of heliocentric theory.

Example: In the heliocentric picture, describe the motion of Mars in Earth's sky. Assume the orbits of Mars and Earth around the Sun to be circular.

Solution: In the heliocentric model, the motion of Mars and Earth. We assume circular orbits for both these planets. We also assume that the Earth, Mars, and the Sun are collinear at $t = 0$. In a coordinate system fixed to the Sun, the coordinates of Mars and Earth are

$$x_{Mars} = R_{Mars}\cos(\omega\, M_{ars}t);\ y_{Mars} = R_{Mars}\sin(\omega\, M_{ars}t),$$
$$x_E = R_E\cos(\omega_E t);\ y_E = R_E\sin(\omega_E t);$$

where R_{Mars} and R_E are the distances of Mars and the Earth from the Sun respectively, and ω_{Mars} and ω_E are the angular velocities of Mars and the Earth respectively around the Sun. The relative coordinates of Mars with respect to the Earth is

$$x_r = x_{Mars} - x_E = R_{Mars}\cos(\omega_{Mars}t) - R_E\cos(\omega_E t),$$
$$y_r = y_{Mars} - y_E = R_{Mars}\sin(\omega_{Mars}t) - R_E\sin(\omega_E t).$$

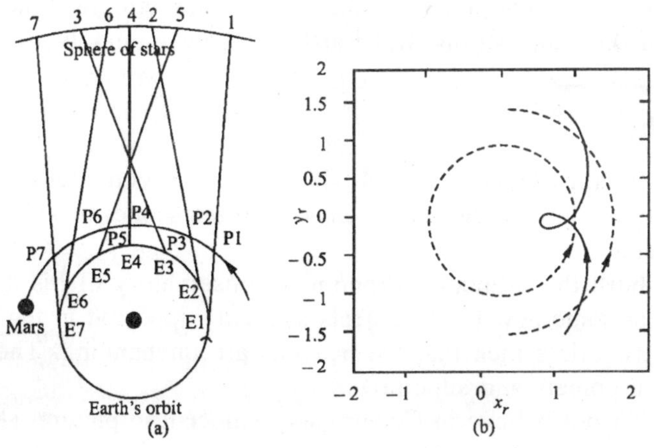

Fig. (a) The Apparent Retrograde Motion of Mars as seen from the Earth in the Heliocentric Model. (b) In the Heliocentric Model, the Apparent path of Mars as seen from the Earth is Depicted as a Solid Line.

The path of the Earth and Mars are also shown as inner and outer dashed lines respectively. The trajectories shown here cover $t' = t/T_E = \omega$ 0.5: 0: 0.5. The retrograde motion takes place when the Sun, Earth, and Mars are in an approximate straight line.

If we measure time in units of one terrestrial year, and the distance in the units of R_E, then

$$t = t'T_E = t'\frac{2\pi}{\omega_E}; \quad r = R_E r'$$

where t', r' are the non-dimensional time and distance respectively. In these units

$$x'_r = \left(\frac{R_{\text{Mars}}}{R_E}\right)\cos\left[\left(\frac{T_E}{T_{\text{Mars}}}\right)2\pi t'\right] - \cos(2\pi t')$$

$$y'_r = \left(\frac{R_{\text{Mars}}}{R_E}\right)\sin\left[\left(\frac{T_E}{T_{\text{Mars}}}\right)2\pi t'\right] - \sin(2\pi t')$$

The ratios are $\frac{R_{\text{Mars}}}{R_E} = \frac{2.27\times10^{11}}{1.5\times10^{11}} = 1.51$ and $\frac{T_E}{T_{\text{Mars}}} = \frac{365.25}{686.93} = 0.53$.

The angle made by Mars will be

$$\tan\theta = \frac{y'_r}{x'_r}$$

The plot of y'_r vs. x'_r for $t' = -0.5$: 0: 0.5. The retrograde motion takes place when the Sun, Moon, and Earth are approximately in a straight line.

Tycho Brahe

Tycho Brahe (Denmark, 1546-1601) can be considered to be the greatest astronomer before the advent of the telescope. In an island near Copenhagen,

Brahe built the ultimate astronomical observatory of his day. Brahe observed that some new bright objects suddenly appeared in the sky. This dispelled Aristotle's idea that the heavens are unchanging. These bright objects were comets and supernova.

Brahe did not believe in Copernicus' heliocentric picture. He thought of a composite model in which the planets moved around the Sun, but the Sun itself moved around the Earth.

To get observational support for his model, Brahe made very precise astronomical observations of planets (up to 1 minute of an arc) from 1576 to 1597, which provided him a huge data bank. Brahe was a brilliant astronomer, but not a very good theorist. He could not analyse his data; this task was done by his student Kepler after his death.

Johannes Kepler

Kepler (Germany, 1571-1630) inherited Tycho Brahe's data and analysed them with great care. The analysis of the Mars data itself took 8 years. He summarised his findings in the following three laws of planetary motion (1680):

- The orbit of a planet lies in a fixed plane containing the Sun. The planet moves in the plane in an elliptical orbit with the Sun as one focus.
- The radius vector from the Sun to the planet sweeps out equal areas in equal amount of time.
- $T^2 = kR^3$, where T is the period of revolution, R is the semi-axis of the elliptical path of the planet, and k is a constant.

Recall that all the earlier thinkers believed that the planetary orbits were circular. This prejudice came from Plato's fondness for symmetry in nature. Kepler claimed Mars' orbit to be elliptical.

This claim was quite courageous because the eccentricity of Mars' orbit ($e = (r_{max} - r_{min})/(r_{max} + r_{min})$) is 0.093, that is very close to zero (eccentricity of a circle). Kepler strongly believed in Brahe's data and his own calculations.

Kepler's laws played a very important role in the development of mechanics. Newton used Kepler's laws to arrive at the law of universal gravitation.

BIRTH OF MODERN SCIENCE—GALILEO AND NEWTON

Around the fifteen century, the Renaissance brought in many new ideas to Europe. In this vibrant period, modern science was born due to the contributions of many thinkers like Descartes, Galileo, Newton, Laplace, etc. Here we will focus on Galileo and Newton who formulated the theory of mechanics.

Galileo Galilei

Galileo Galilei (Italy, 1564-1642) is considered to be the father of modern science. He placed a firm emphasis on observation. For example, he rolled balls down an inclined plane and measured the time elapsed using a water clock (similar to an hour glass).

He recognised this motion to be equivalent to free fall. He observed that all the balls, independent of its constituents, took the same time. This observation clearly disproved Aristotle's idea that heavier bodies fall faster than lighter ones.

Galileo made numerous important observations. For the first time he showed that the time period of a pendulum is independent of its amplitude, which led to the invention of clocks. He also got hold of a then-recently invented instrument called the telescope and started observing the sky.

He discovered the moons of Jupiter. The moons revolving around Jupiter rather than the Earth provide further credence to Copernicus' heliocentric theory.

One of the most important contribution of Galileo is the law of inertia according to which a body retains its velocity unless a force acts on the body. This law contradicts Aristotle's theory that an object slows down and stops unless some force acts on it.

Galileo's law of inertia is valid only for a free particle, which is an abstraction. Galileo also gave a law of relativity according to which the laws of nature are the same in all reference frames moving with constant velocity.

Galileo provided a new paradigm for science in which the laws are constructed and refuted on the basis of observations. Einstein said the following in his homage to Galileo, "Pure logical thinking cannot yield us any knowledge of the empirical world; all knowledge of reality starts from experience and ends in it... Because Galileo saw this, and particularly because he drummed it into the scientific world, he is the father of modern physics—indeed, of modern science altogether."

Isaac Newton

Isaac Newton (England, 1642-1727) was born the year Galileo died. Newton synthesised the ideas of Copernicus, Kepler, and Galileo, and produced compact and abstract laws of motion and the law of universal gravitation.

These theories had an enormous impact on science. Newton was very strong in mathematics; he is the first physicist who wrote the laws of physics in precise mathematical forms. Note that Newton is also the discoverer of calculus.

You would have been aware of Newton's laws of motion from your school days onwards.

So we will not delve into it except to remind you that Newton's laws are universal, i.e., they are applicable to all systems except those that move with speeds comparable to the speed of light.

There is a story that Newton discovered the idea of gravitation when an apple fell on his head.

This is a story for children and definitely untrue. Newton used Kepler's laws to discover the law of universal gravitation. Huygens

(Netherlands, 1629-1695) first derived an expression for the acceleration of an object going around in a circular path. Using Huygens' formula, the acceleration of a planet going around the Sun is

$$a = \omega^2 R = \frac{4\pi^2}{T^2} R ,$$

where ω is the frequency of the circular motion, and R is the distance of the planet from the Sun. According to Kepler's third law, $T^2 = kR^3$ (k is a constant), substitution of which in equation yields

$$a = \frac{4\pi^2}{k} \frac{1}{R^2} .$$

From Newton's second law of motion, the force acting on the planet is

$$F = ma = m \frac{4\pi^2}{k} \frac{1}{R^2} ,$$

which is inversely proportional to R^2. So Newton concluded that the Sun attracts planets by a gravitational force that varies as $1/R^2$.

Newton felt that the same force must be making the Moon go around the Earth. He also noted that the moon falls towards the Earth all the time while still following a circular path. Newton cleverly used Kepler's third law for the Earth-Moon system. The acceleration of the Moon towards the Earth is

$$a_M = \frac{4\pi^2}{T_M^2 R_{EM}} ,$$

where R_{EM} is the distance between the Earth and the Moon, and T_M is the time period of the Moon's orbit around the Earth.

After this Newton used equation to deduce the acceleration of an object on the surface of the Earth:

$$a_E = \left(\frac{R_{EM}}{R_E}\right)^2 = \frac{4\pi^2}{T_M^2} \left(\frac{R_{EM}}{R_E}\right)^3 R_E$$

where R_E is the radius of the Earth. Substitution of T_M = 27 days 7 hours and 43 minutes, $R_{EM}/R_E = 60.1$, and R_E = 6400 km yields

$$a_E = 9.84 \text{m/s}^2.$$

Newton measured the acceleration due to gravity by rolling various masses on an inclined plane, and found the experimental value to be in general agreement with the above theoretical prediction.

Using this result, Newton concluded that the forces acting between the Sun and planets, the Earth and Moon, and the Earth and objects on the Earth have the same origin, i.e. gravitational force.

A remark is in order here. Newton proved the inverse square law of gravitational force for elliptical orbits. The proof is geometrical, and somewhat involved.

From the above calculations, it is also clear that the acceleration due to gravity on the surface of the Earth is independent of the mass of the object. Newton verified this theoretical observation by measuring the acceleration of objects made of various materials on an inclined plane. Thus Aristotle's theory of motion was proven to be incorrect.

This is the genius of Newton. He used Kepler's third law of planetary motion, and some mathematics to deduce the universal gravitational force between objects at disparate scales.

The universality of gravitation indicates that the laws of physics on the Earth and in the heavens is the same, contrary to Aristotle's belief. Newton deduced the universal law of gravity according to which any two masses m_1 and m_2 separated by a distance r exert an attractive force on each other whose magnitude is

$$f = G\frac{m_1 m_2}{r^2},$$

where G is Newton's constant of gravitation, and whose direction is towards each other.

Although Newton constructed general abstract laws in the language of mathematics, these laws are based on observations. Newton made various predictions based on his laws, e.g., the orbits of comets, tidal eZωects, the oblate nature of the Earth, perturbations of planetary orbits from the ellipse due to other astronomical objects etc.

Many of these predictions could be tested from observations. This method, started by Newton and Galileo, is the method of modern science. Newton's ideas had far-reaching consequences in the field of science. For his wide-ranging contributions, many people consider Newton to be the greatest scientist of all time.

Newton wondered very deeply about space and time; he conjectured them to be absolute, i.e. the same for all observers. Many of Newton's predictions based on absolute space and time proved to be correct. But problems surfaced when the speed of the particle is close to the speed of light. That's where Einstein comes to rescue.

FUNDAMENTAL REVISION OF SPACETIME—EINSTEIN

At the turn of the twentieth century, Michelson and Morley observed a very puzzling experimental result. They measured the speed of light to be the same while an observer is moving towards the source or away from the source. In addition, the Maxwell's equations too indicated that electromagnetic waves move with a constant velocity independent of the motion of the source or the observer. These results contradict the addition of velocities given by Newton and Galileo.

Many scientists tried to explain the constancy of the speed of light using various tricks within Newton's framework. However iconoclast Einstein (Germany and USA, 1879-1955) had another idea.

He noticed that the constancy of the speed of light and the postulate of absolute space and absolute time are contradictory, and abandoned the concepts of absolute space and time.

He proposed a new theory called the special theory of relativity in which the time dibωerence between any two events could be different for different observers.

The spacetime structure of special relativity differs completely from that of Newton. The time difference between two events could be measured differently in two inertial frames; the length of a metre stick also is different in two inertial frames.

The theory of relativity has many other startling predictions such as it is possible to convert mass to energy and vice versa. All the predictions of relativity have been found to be consistent with experiments, and hence it is considered to be a correct theory. For particles moving with small speeds, the kinematics and dynamics of the special theory are consistent with Newton's theory of motion.

Einstein went further and discovered an equivalence between accelerating frames and gravity.

Using this idea, he proposed a new theory of gravity that is more general than Newton's theory of gravity. For example, Einstein predicted that light bends near a star, but no such conclusion can be drawn from Newton's theory of gravity. Einstein's theory of gravity is called the general theory of relativity. Einstein's theory of relativity has had a revolutionary impact in physics.

MECHANICS AND MODERN PHYSICS

Modern science essentially starts with Galileo and Newton's theory of motion. New areas like fluid mechanics, elasticity and wave theory were directly inspired by Newton's theory.

Electrodynamics and statistical mechanics took birth around mid-nineteenth century. Two revolutionary theories, relativity and quantum mechanics, were born in the twentieth century when Newton's theory was observed to fail at high speeds and at small length scales. These theories form modern physics.

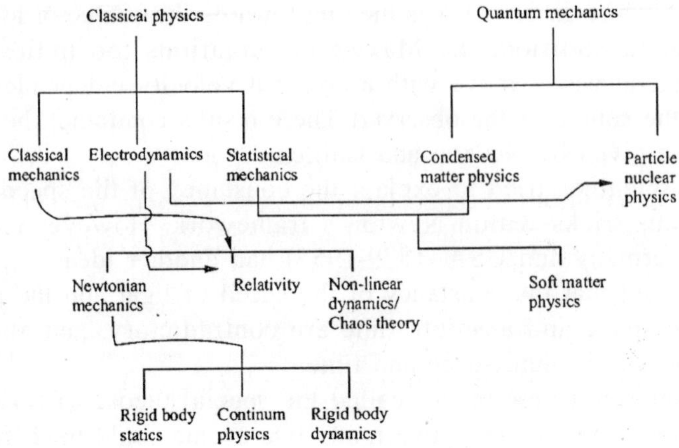

Quantum mechanics differs fundamentally from Newton's and Einstein's conception of dynamics. For example, in quantum mechanics we cannot simultaneously specify the position and the velocity of a particle precisely—unlike in Newton's formalism. The ideas of force and trajectory are abandoned in quantum mechanics.

In the major part of the twentieth century, the study of elementary particles was one of the main areas of research. A successful theory of elementary particles called quantum field theory, was discovered by marrying electrodynamics, relativity and quantum mechanics.

The quantum field theory, which is still incomplete, is presently the best known physical theory of nature. One of the major unsolved problems in quantum field theory is the quantum theory of gravity. Quantum field theory is also used to study properties of matter like metals, semiconductors etc. Another connected area of research which has come up in recent times is nonlinear dynamics and chaos.

We will explain this field using an example. Newton solved the twobody problem exactly—bound particles under gravitational potential have elliptical orbits. Way back in 1888, Poincar´e showed that no such solution exists for a three-body problem.

In fact, if a test particle starts from two nearby initial conditions, the future for these two initial conditions could be very different. This kind of sensitivity to initial conditions is called *chaos*. A similar feature has been found for weather systems—this property of the weather makes long-term prediction of the weather impossible. Chaos theory and related fields are active areas of physics research.

Presently, it appears that we know all the basic equations needed to describe nature.

For example, physicists are quite confident that quantum chromodynamics describes the physics of the nucleus; Navier-Stokes equation describes turbulent flows; certain quantum mechanical equations describe the physics of superconductors and other complex materials. The major difficulty in physics, however, is how to solve these equations. Physicists have been trying very hard to solve them using analytical methods and powerful computers, yet complete success has eluded them so far.

DYNAMICS

We begin to develop "participator dynamical systems" on environments supported by reflexive frameworks. We introduce the notions of action kernel and participator. For the cases of one and two participator systems, we give a description of the participator dynamics in the language of Markov chains. Motivational deals intuitively with very restricted cases.

MATHEMATICAL TERMINOLOGY

The dynamics developed makes use of several mathematical concepts from the theory of Markov chains. We collect basic terminology and notation for the convenience of the reader.[1] We assume a familiarity with the notions of conditional probability and expectation.

Let (E, ε) be a measurable space. The set of measurable functions $f: E \to R$ that are bounded is denoted by $b\varepsilon$, and the set of measurable nonnegative functions by ε_+.

A kernel P on E is said to be positive if its range is in $[0, \infty]$. It is called a transition probability or a submarkovian kernel if $P(e, E) \leq 1$ for all $e \in E$. It is called markovian if $P(e, E) = 1$ for all $e \in \varepsilon$. The abbreviation T.P. is sometimes used for transition probability. If P is a positive kernel and $f \in \varepsilon+$, Pf defined by $P f(e) = \int_E P(e, dh) f(h)$.

Similarly, if ν is a positive measure on E, then P can be viewed as the operator on measures $\nu P(A) = \int E\, \nu(dh)P(h, A)$ for $A \in \varepsilon$.

The composition or product of two positive kernels P and Q is the kernel $PQ(e, A) = \int E\, P(e,dh)Q(h, A)$. The n-fold product of a kernel P with itself is denoted P_n.

Let $(\Omega\ F, P_0)$ be a probability space and $Z = \{Z_n\}_{n\geq 0}$ a sequence of random variables $Z_{n:} \to E$. Such a structure is called a stochastic process with base space (Ω, F, P_0) and state space E. A sequence $\{G_n|_{n\geq 0}$ of sub σ-algebras of F, such that $G_n \subset G_{n+1}\ \forall n$, is called a filtration on (Ω, F).

Let $F_n = \sigma(Zm, m \leq n)$ and G_n be a filtration such that $G_n \supset F_n$ for every n. The sequence $Z = \{Z_n\}_{n\geq 0}$ is called a Markov chain with respect to the filtration $\{G_n\}_{n\geq 0}$ if, for every n, the σ-algebras G_n and $\sigma(Z_m, m \geq n)$ are conditionally independent with respect to Z_n, i.e., if for every $A \in G_n$ andB $\in \sigma(\text{Zm}, m \geq n)$, $P_0[A \cap B|Z_n] = P_0[A|Z_n]\ P_0[B|Z_n]$ a.s. The σ-algebras

Fn are referred to as "past" σ-algebras. When we say simply that Z is a Markov chain (with base space (Ω, F, P_0)) we mean that it is so with respect to the past algebras Fn. Intuitively, a sequence of random variables is a Markov chain if the probabilities for passing into the next state are completely determined by the current state of the system.

A sequence $Z = \{Z_n\}_{n\geq 0}$ of random variables is called a homogeneous Markov chain with respect to the filtration fGng with transition probability P if, for any integers m, n with $m < n$ and any function $f \in b\varepsilon$, we have $E_0[f(Z_n)| G_m] = P_{n-m} f(Z_m)$, P_0 a.s., where E_0 denotes the mathematical expectation operator with respect to P_0.

The probability measure ν defined by $\nu(A) = P_0[Z_0^{-1}(A)] \equiv P_0[Z_0 \in A]$, for $A \in \varepsilon$, is called the starting measure.

Let P be a T.P. on E. It is customary to extend the state space (E, ε) to the space $(E_\Delta, \varepsilon_\Delta)$, where Δ is a point not in E called the cemetery, $E_\Delta = E \cup \{\Delta\}$, and $\varepsilon_\Delta = \sigma(\varepsilon, \{\Delta\})$. P extends to a markovian kernel on $(E_\Delta, \varepsilon_\Delta)$ by setting $P(e, \{\Delta\}) = 1 - P(e, E)$ if $e \neq \Delta$, and $P(\Delta, \{\Delta\}) = 1$ A canonical probability space is the space $(\Omega\ F, P_0)$ where $\Omega\ \prod_{n=0}^{\infty} E_\Delta^{(n)}$, and $E_\Delta^{(n)}$ is a copy of E_Δ; where the σ-algebra F is generated by the semi-algebra of measurable cylinders of Ω (namely sets of the form $\prod_{n=0}^{\infty} A_n$, where $A_n \in \varepsilon_\Delta^{(n)}$, and A_n differs from $E_\Delta^{(n)}$ for only finitely many n); and where P_0 is a probability measure.

A point $\omega = \{\omega_n, n \geq 0\}$ of Ω is called a trajectory or path. The mapping $Z_n: \Omega \to E_\Delta^{(n)}$ taking $\omega = (\omega_0, \omega_1, \omega_2, ...) \in \Omega$ to its nth entry ω_n is called the nth coordinate mapping. If the sequence Z = fZng of coordinate mappings on the canonical probability space forms a homogeneous Markov chain with T.P. P, we call it the canonical Markov chain with T.P. P.,

The shift operator θ is the point transformation on Ω defined by $\theta(\omega_0, \omega_1, ..., \omega_n, ...) = (\omega_1, \omega_2, ..., \omega_{n+1}, ...)$. We write θ_n for the n-fold iteration of θ: $\theta(\omega_0, \omega_1, ...) = (\omega_n, \omega_{n+1}, ..)$: A stopping time T of the canonical Markov chain Z is a random variable defined on $(\Omega\ F)$ with range in $N \cup \{\infty\}$ and such that for every integer n the event $\{T = n\}$ is in F_n. (N is the set of natural numbers including 0.)

The σ-algebra associated with T is the family F_T of events $A \in F$ such that for every n, $\{T = n\} \cap A \in F_n$. Notice that then the random variable $Z_T(\omega)(\omega)$ is F_T-measurable.

Let G be a group that is locally compact with countable basis (LCCB), and let G denote the σ-algebra of its Borel sets.

Given probability measures μ_1, μ_2 on G, their convolution $\mu_1 * \mu_2$ is defined to be the probability measure which assigns to $K \in G$ the measure $(\mu_1 * \mu_2)(K) = \iint 1_K(x+y)\mu_1(dx)\mu_2(dy)$. A right (left) random walk on G is a Markov chain with state space (G, G) and transition probability $\in_g * \mu$ $(\mu * \in_g)$, where μ is a probability measure on (G, G) which is called the law of the random walk, and $\in_g$ is Dirac measure supported at the point $g \in G$. On an abelian group there is only one random walk of law μ, and it is invariant under translations.

FUNDAMENTALS OF DYNAMICS

The conclusion of an observer O's perceptual inference is represented, as we have discussed, by a probability measure $\eta(s, \cdot)$. This conclusion is true in a given semantics, according to Definition, if $\eta(s, \cdot)$ is the actual regular conditional distribution, given s, of the measurable functions X_t. $\{X_t\}$ is a sequence of random variables indexed by a discrete time t, taking values in configuration space X, and whose domain is some unspecified probability space Ω.

In extended semantics there is a set B of objects of perception; for each t, a value of X_t is associated with an interaction of O with an element of B. These interactions are called channelings. In the case of an environment supported by a reflexive framework we have a set of observers β which is also the set of objects of perception for each of its members. At each instant of "reference" time (which, as we shall see, is not the time t of the random variables X_t) the totality of channeling interactions at that instant is described by a subset L of B and a relation $\tilde{x}$ on L.

We now begin to construct a class of models for environments supported by reflexive frameworks; these models are called "participator dynamical systems." We do this using entities called"participators"; a participator manifests as an observer in B_E at each instant of reference time. The subset D of B at reference time n always contains the set of participator manifestations at time n. The determination of D and B can be discussed in terms of participators. In the process of this development, an analytical viewpoint emerges in which the participators themselves are the centre of attention.

The Motivation for Dynamics

Consider two observers, A and B, in a reflexive framework $(X, Y, E, S, \pi.)$. This means there exist points $a, b \in E$ such that $A = (X, Y, E, S, \pi_a, \eta_A)$ and $B = (X, Y, E, S, \pi_b, \eta_B)$, where η_A, η_B are some conclusion kernels.

The observers A and B channel to each other. Each makes an inference about the perspective map of the other, i.e., about the point of E that represents the perspective of the other.

Figure shows the premise $s = \pi_a(b)$ of *A*'s perceptual inference, and the ray of configuration points x such that $\pi_a(x) = s$ (labelled in the figure as $\pi^{-1}_a\{s\}$). A's conclusion measure η_A is supported on the set $\xi = \pi^{-1}_a\{s\} \cap E$, which includes b. But ξ includes nfinitely many perspectives other than B's as well, some of which are indicated by the smaller dashed circles with numbers above.

Thus A is faced with perceptual uncertainty: What was the perspective of the observer that channeled? Was it 1, 2, b, 3,...? In general, A cannot pick just one perspective as the answer to this question. Instead A concludes that it is perspective 1 with probability P_1, perspective 2 with probability P_2, perspective b with probability Pb, and so on. This is the content of A's conclusion measure $\eta_A(s, \cdot)$.

How is A's conclusion measure $\eta_A(s, \cdot)$ to be chosen? On what basis can A conclude that the other observer's perspective was 1 with probability P_1, 2 with probability P_2, etc? The answer we give is roughly as follows. A markovian dynamics of perspectives naturally arises in the context of reflexive frameworks.

That observers in the framework perceive truly means that their η's should be related to the asymptotic behaviour of this dynamics. Intuitively, the probability assigned by $\eta(s, \cdot)$ to a point $e \in E$ should be a conditional probability derived from the frequency with which the perspective correspond- ing to that e is adopted by participators in the given dynamical context. In this sense, the given dynamics plays the role of the "environment" in which these observers are embedded. To make these ideas more precise, we begin by

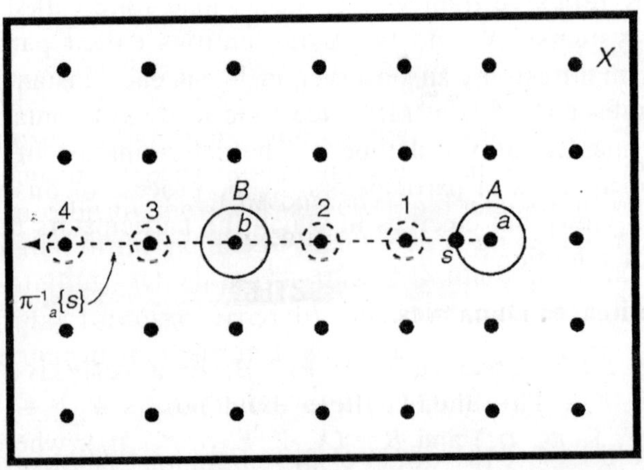

Fig. Perceptual Uncertainty in a Reflexive Framework.

discussing how a dynamics of perspectives arises on reflexive frameworks.

When A and B channel, the premise s of A's perceptual inference greatly restricts what A can conclude about B's perspective. Yet A has, in general, infinitely many choices remaining, for B's perspective could be any in $\pi_a^{-1}\{s\} \cap E$. Suppose A and B retain their perspectives after channeling.

Then if they channel again A has precisely the same set of choices and the same ambiguity|regarding B's perspective as before. In other words, if the observers do not alter their perspectives after an observation then there is no point to further observations. A can channel with B as many times as you like, but the same premise s will result every time, and with it the same ambiguity of interpretation.

Moreover, should A and B never change perspectives the whole question of how η_A is chosen would be trivial: the ideal η_A (s; ·) would be Dirac measure supported on b, and η_A $(s^{\prime} \cdot)$ for $s^0 \neq s$ would need not be defined. Indeed, the construction of reflexive frameworks would be pointless.

Let us, then, allow observers in a reflexive framework to change perspective following a channeling. That is, let us allow some kind of dynamics of perspectives on reflexive frameworks. Several questions immediately arise. How shall observers change perspective? Since the perspective of an observer in a reflexive framework (together with its conclusion measure) is its only means of individuation, does not a change in perspective actually mean a change in observer?

If so, then what is it that is manifesting itself as a different observer at each step of the dynamics? Furthermore, dynamics requires sequence. What is the formal structure of this sequence? What is the formal structure of the dynamics? How, precisely, shall η be related to the resulting dynamics?

Action Kernels

How should we allow observers to change perspective in reflexive frameworks? There are two basic issues. First, what information should be used to select the new perspective of an observer after a channeling? Second, should the new perspective be chosen deterministically or probabilistically? This figure shows observers A and B channeling to each other. In consequence of the channeling, A's premise is sA, and B's premise is sB.

The figure shows A changing its perspective from πa to πa^0, and B changing from πb to πb^0. Of course, after these changes A and B are no longer the same observers since they no longer have the same perspectives.

We denote the new observers A^0 and B^0. The only information available to choose A's new perspective is its current perspective and the premise s_A. Similarly, mutatis mutandis, for B.

Therefore, for maximum generality, we assume that an observer's next perspective is some function of its current perspective and current premise. Shall this function be deterministic or probabilistic? Again, for maximum generality, we assume that the next perspective is chosen probabilistically, according to some measure. Further, we assume this measure to be a probability measure; after we introduce the notion of participator, we will interpret this assumption.

In light of these considerations, we could propose that the change in per- spective of an observer A should be governed by a probability measure that is selected based on A's current perspective and current premise.

However, a sense of symmetry suggests that A's probability measure should depend not on its absolute perspective, but on the difference "between its perspective and that of B. Symmetry also suggests that the probability that A moves to A0 should depend only on the difference "in perspective between A and A^0. Now to talk about differences of perspective in E requires some structure on

E. For instance, E might be a principal homogeneous space for some group of "translations." More generally, the minimum structure necessary here is a

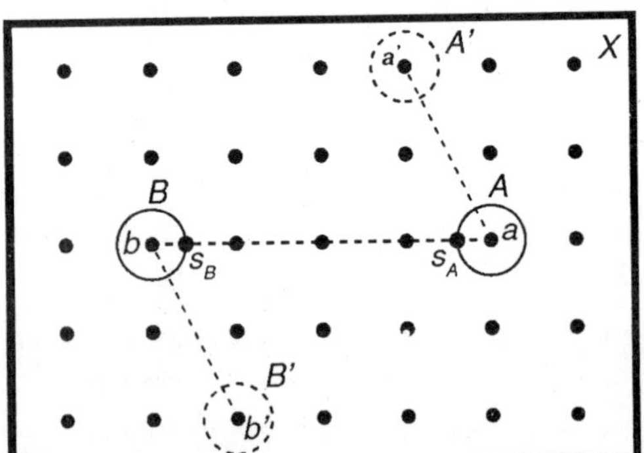

Fig. Changing Perspective on a Reflexive Framework.

symmetric framework.

Assumption: We are working in a symmetric framework (X, Y, E, S, G, J, π) in which $G = X$ is an abelian group written additively and $J = E$ is a subgroup. Equivalently, we can say that (X, Y, E, S, $\pi_.$) is a reflexive framework in which X is an abelian group, $E \subset X$ is a subgroup, and there exists a map $\pi: X \to Y$ such that for each $e \in E$, $\pi_e(x) = \pi(x - 4)$.

Thus we can speak of difference s in perspective" without thinking twice. The reader may rely for intuition one can think of X as R^n (with vector addition as the group operation) and E as a measure zero subgroup thereof. We return now to the question of the probability measure governing changes in perspective. In light of our assumptions, this is a measure on a group E, telling how probable are various translations from the current perspective.

We can capture the dependence of the measure on the observer's current premise by associating to each premise s of the observer a measure on the group of translations that acts on E. The appropriate mathematical device to do this is a kernel Q that we call an action kernel. For each premise s, the measure $Q(s, \cdot)$ is a probability measure on the group of translations that acts on E.

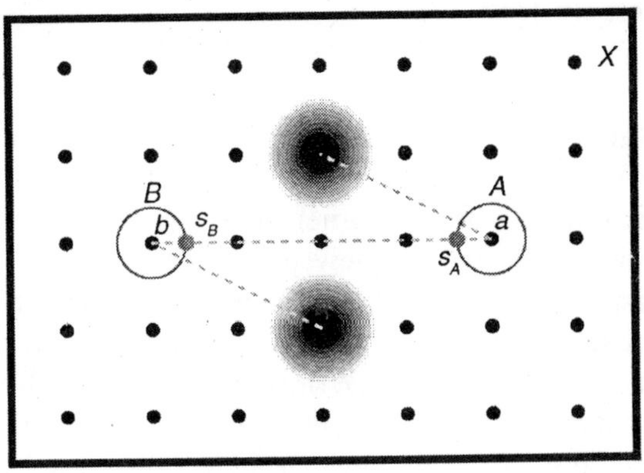

Fig. Action Kernels.

The shading of the upper circular region represents the density of the probability distribution Q_A (s_A, X). Similarly, the shading of the lower circular region represents the density of the probability distribution Q_B (s_B, X).

Once again, two observers A and B channel with each other. A's premise is s_A. The measure Q_A (s_A, X), derived from A's action kernel Q_A, is depicted by a shaded disk with a dashed line drawn from A to the centre of the disk. The darkness of a region within this disk encodes the probability that A will adopt a perspective in that region as its next perspective.

Darker regions are more probable than lighter ones. A's expected new perspective happens, to be the perspective represented by the centre of the disk. In general there will be some probability that an observer does not change its perspective after a channeling.

Definition: Under Assumption, an action kernel is a markovian kernel Q: $E \times E \to [0, 1]$ such that $Q(e, \cdot) = Q(e', \cdot)$ if $\pi(e) = \pi(e')$. Given Q, to each $e_1 \in E$ we can associate a kernel Q_{e1}: $E \times \varepsilon \to [0, 1]$ by

$Q_{e1}(e, \Gamma) = Q(e - e_1, \Gamma)$. Suppose Q is an action kernel. Since $Q(e, \cdot)$ depends only on $\pi(e)$ we could equally well define it as a kernel $s \times \varepsilon \rightarrow [0, 1]$. In fact we will sometimes write $Q(s, \cdot)$; this will mean $Q(e; \cdot)$ for any e such that} $\pi(e) = s$.

Similarly $Q_{e1}(e, \cdot)$ depends only on πe1 (e). The interpretation of the action kernel is as follows: $Q(e, \Gamma)$ is the probability that the observer will change perspective by an increment in the set Γ, given that it channeled with an observer whose perspective differed from its perspective by e. If the first observer is at e_1, then $Q_{e1}(e_1 + e, \Gamma)$ is an equivalent way to write this. The terminology action kernel," when used for a given kernel Q: $E \times E \rightarrow [0, 1]$, signals our intention to consider the family of kernels $\{Q_e\}_{e \in E}$.

Participators

In our discussion of action kernels we have spoken as though an observer in a reflexive framework could change its perspective map π. We said, for instance, that an action kernel gives the probabilities with which an observer might adopt various new perspectives.

Now this way of speaking, though convenient, cannot be correct, the Definition of observer does not permit a given observer to change its perspective. On the contrary, the Definition requires an observer to have a fixed perspective map $\pi: X \rightarrow Y$. Therefore, the formal entity that changes perspective according to the dictates of an action kernel is not itself an observer. Instead this entity manifests itself at each instant as an observer in the context of a reflexive framework. This new formal entity we call a "participator."

Definition: A participator on a reflexive framework $(X, Y, E, S, \pi_{\bullet})$ is a triple, $(\xi, \{Q(n)\}_n, \{\eta(n)\}_n)$, where n varies over the nonnegative integers, ξ is a probability measure on E, each Q(n) is an action kernel, and each η(n) is a family of interpretation kernels for the reflexive framework. (That is, $\eta(n) = f\eta_e(n)g_{e2E}$, where, for each e 2 E, $(X, Y, E, S, \pi_e, \eta_e(n))$ is an observer.)

If all the $Q(n)$ are equal to a fixed action kernel Q, we denote the participator simply by $(\xi, Q, \eta(n))$, and call it a kinematical participator with action kernel Q. If, for some n, a participator $A = (\xi, \{Q(n)\}_n, \{\eta(n)g_n)$ on a reflexive framework $(X, Y, E, S, \pi_{\cdot})$ has perspective π_n, then we call the observer $A_n = (X, Y, E, S, \pi_n, \eta(n))$ the manifestation of A at time n. We also say that A manifests as A_n. A preparticipator is a pair $(\xi, \{Q(n)g_n)$ with ξ and Q(n) as in a participator.

The formal Definition of participator is based upon the following intuitions. A participator must have a first perspective, this is the purpose of ξ.

The probability measure ξ on E, called the initial measure of the participator, governs the choice of the first perspective of the participator. When we say that a participator is initially "at e" or "has perspective e" we mean a participator for which ξ is Dirac measure at e, formally, we write $\xi = \in_e$. A participator must also have a means of changing perspective, this is the purpose of the action kernels $Q(n)$. The changes in perspective are discrete and sequential, with respect to a notion of time that we discuss shortly.

The notation means that the nth change of perspective in this sequence is governed by the action kernel $Q(n)$. Since the action kernels give probabilities for change of perspective conditioned by premises arising from channelings, the perspective changes of participators are probabilistic and are driven by channelings.

The terminology "kinematical participator," for the special case when all the $Q(n)$ are identical, indicates that this case gives rise to systems with a property analogous to constant velocity.

This does not mean that the motion of the participators is "linear" in the usual geometric sense of the word. Rather, it means that the instantaneous state-change data (in this case, given by the action kernel) is time invariant.

We discuss shortly a dynamics of perspectives that arises from the mutual observations of an ensemble of participators in a common reflexive framework.

This dynamics is a Markov chain whose state space is a product of copies of E, one for each participator in the ensemble.

A simplified version of the dynamics which is determined entirely by the action kernels and initial measures of the participators. To specify a (canonical) Markov chain on some space one need only give its initial measure and transition probability.

The initial measure of the markovian dynamics of perspectives is simply the product of the initial measures of the participators. In the special case of kinematical participators the resulting Markov chains are homogeneous. In this case we will sometimes use the word kinematics rather than dynamics.

A participator dynamics on a reflexive framework incorporates a nondualistic model of extended semantics. There is some set B of observers in the framework, B serves as the objects of perception for each observer in the framework.

In participator dynamics the set B has a special property, this property consists in a precise condition on the subset B_E. Let us begin with a fixed set K of participators, say it is this set of participators whose dynamical interaction constitutes the given participator dynamical system. B_E is then the set of all possible instantaneous manifestations for the participators in K. To be precise, suppose each j $\in K$ is represented (ξ_j, $\{Q_j(n)\}_n$, $\{\eta_j(n)\}_n$). Then

$$B_E = \bigcup_n \{(X, Y, E, S, \pi_e, \eta_j(n)) \mid e \in E, j \in K\}.$$

Taking a union over the various instants of time n implies that we do not distinguish the observers (X, Y, E, S, $\pi_\bullet$, η_j (n)) and (X, Y, E, S, $\pi_\bullet$, η_j (n')) (for e and j fixed) if it happens that the kernels η_j (n) and η_j (n^0) are equal for distinct times n, n: By contrast, for $j \neq j' \in K$ and for a given $e \in E$, the observers (X, Y, E, S, π_e, η_j (n)) and (X, Y, E, S, π_e, η_j0 (n)) are counted as distinct elements of B, even if the kernels η_j (n) and η_j0 (n) are the same.

For example we could have taken a disjoint union over n rather than ordinary union. This means that the manifestations of a given participator at distinct moments n, n' would always be viewed as distinct elements of B, even if the perspectives and conclusion kernels of the two observers were identical.

This would allow the present manifestation of a participator to interact with a previous manifestation of the same participator|let us call this a "memory interaction" in a manner which permits keeping track of the distinct times.

However, it is still possible for a present and a past manifestation of a single participator to interact. The difference is that now, if these two manifestations happen to be identical as observers, then they are also considered identical as objects of perception, they are represented by the same element b 2 B.

Thus the interaction in question is characterized by b channeling with itself. In other words, a memory interaction may be analytically indistinguishable from a self-channeling, whereas in the alternate (disjoint union) approach memory interaction and self-channeling are always distinct. Whether this difference is theoretically significant is an open question.

We can now interpret the requirement that action kernels are markovian, i.e., that if Q is an action kernel then for each $e \in E$ the measure Q–e, ·) is a probability measure on E.

This means that the set of participators which manifest themselves as observers is the same set at each instant of time: participators do not appear or disappear while a scenario is running. To see this, recall that if Q is the action kernel of a participator A, the measure Q(e, ·) assigns probabilities to A's perspective at time n + 1 (given that, at time n, A channeled with some participator whose perspective is e). And if Q is not markovian, i.e., if Q(e, E) < 1, then there is positive probability that A has no perspective at time n + 1, so that it is not manifested as an observer in the framework at that time.

However, though A must manifest as an observer at each time n, A's manifestation need not channel at each time n. In other words, the subset L of B which is the domain of the channeling relation at time n may be a proper subset of the set of all participator manifestations at time n. We see, then, that the markovian requirement on action kernels is a matter of convention, not a restrictive assumption: since we do not require the participators to channel at every instant, and since the dynamics is driven by channeling, the net effect on the dynamics is the same whether a participator does not manifest at time n, or manifests but does not channel at time n.

The only events in a reflexive framework with which to associate numbers are the discrete acts of observation and the consequent changes in perspective. To each participator, then, we assign a number, called the "proper time" of that participator, such that the number increases only when the participator makes an observation.

Every channeling that involves that participator increases its proper time. Thus discrete acts of observation constitute the units of subjective time in this framework. The examples we presenting are simplified (artificially) so that the proper time of each participator coincides with "reference time".

The setting of the dynamics described here is different from the spacetime setting assumed in physics. In place of physical space we have the space of possible observer perspectives, and in place of physical time we have the sequence of discrete observations of participators.

A particular channeling may not include the perspectives of some participators in the dynamics. In this case the proper times of the excluded participators are not increased, but the proper times of the others are increased.

Therefore, even if the proper times of all participators begin with the same value, say zero, their proper times will eventually differ due to channelings that exclude some participators. We cannot then, in general, take the proper time of any particular participator to be the time parameter of the markovian dynamics of the ensemble. For this we need a time parameter that increases for every channeling whether or not that channeling includes a particular par ticipator.

This time parameter we call "reference time." In a given dynamical setting in which we have a fixed set K of participators, we may take the reference time to be a copy of the nonnegative integers, called "R," which is the domain of the time index "n" of Definition for all the participators in K. Thus in speaking of reference time we are making the assumption that these indices have a common domain. The reference time in a given dynamical context (corresponding to a set K of participators) is not the same as the active time in the sense for the observers in the set B_E. In fact, reference time is associated to a set of participators, not to a set of observers. And the reference time need not include those instants when the participators channel only to non-distinguished objects of perception. It need only include those instants when participator observations occur, and by the term 'observation' we always mean a channeling which results in a distinguished premise (which causes the output of a conclusion, etc.). Now a channeling with a non-distinguished object of perception may result in a distinguished premise ("false targets"), and an instant of time in which this occurs (for the manifestation of a participator) would have to be included in reference time. But if no distinguished premises occurred at the given instant for any of the participators, then that instant would be excluded from reference time.

The active time of an observer indexes the Xt's, it consists by Definition precisely of those instants when the observer receives any channeling, from a distinguished object of perception or not.

Definition: (i) a participator channeling sequence is a function, ζ, from the natural numbers to the space of channelings, ζ: $N \to I$, with the following property. Let $\zeta\,(n) = (L_n, \tilde{x}_n)$, where $L_n \subset B$ and Bn is an involution of L_n.

Let $K_n = (D_n, x_n)$ denote of $\zeta(n)$. Then for each $n \in N$, D_n is not empty. N is called the reference time for the sequence. (ii) let $A_n \in B_E$ denote the manifestation of participator A at reference time n. To each participator A in the dynamics is associated its proper time, T_A: $N \to N$, defined inductively as follows:

$$T_A(n) = \begin{cases} 0 & \text{if } n = 0 \\ T_A(n-1)+1 & \text{if } A_n \in D_n \text{ and } \pi_{\Phi(A_n)}(\Phi(\chi_n(A_n))) \in S \\ T_A(n-1) & \text{otherwise} \end{cases}$$

At every instant of reference time, the proper time of at least one participator is increased. Definition says that the unit of subjective time for a participator is a single act of channeling, i.e., the performance of a single perceptual inference. Since at any step of reference time some participator manifestations may not channel, it follows that the proper times for different participators vary: proper time is relative to the participator.

The proper time of a participator A increases not only if its manifestation channels with a distinguished object of perception, but also if it channels with a false object. A false object is an object of perception $B_n \in B - B_E$ such that $\pi_{\Phi(An)}$ $(\Phi(B_n)) \in S$. If Bn is a false object then, $\Phi(B_n)$ is a false target. Channelings with false objects affect participator dynamics since participators, unable to distinguish false objects from true, change perspective according to their action kernels upon channeling with false objects. In this book we attempt no serious investigation of the role of such channelings in participator dynamics. In fact we ignore false objects and assume that, at each instant of reference time, participator manifestations channel only with other participator manifestations. This is the content of the following "closed system" assumption:

Assumption: Closed system. For each reference time n, D_n is contained in the set of participator manifestations at time n.

One further assumption should be noted. We conceive of the change of perspectives of participators on a reflexive framework as probabilistic. However, we have not given explicit details of the underlying probability spaces on which the dynamical mechanism depends.

Assumption: Independent action. At any instant of reference time, and given the current perspectives of all participators and the current channeling involution, the perspectives of the participators at the next instant of reference time are independent random variables.

For example, suppose we have three participators A, B and C with action kernels Q_A, Q_B and Q_C respectively, and with channeling involution $B = \{(A, B)\}$ (so that C is not channeled to). Then the probability that, at the next instant, $A \in \Gamma_A$, $B \in \Gamma_B$, and $C \in \Gamma_C$ is

$$Q_A, e_A\ (e_B, \Gamma_A)\ Q_{B, eB}\ (e_A, \Gamma_B) 1\ \Gamma_C(e_C).$$

That is, we need simply take a product of the appropriate probabilities for the individual participators.

KINEMATICS OF A SINGLE PARTICIPATOR

The kinematics of perspectives that arises in a system consisting of a single kinematical participator. We find that this kinematics is a random walk. In the next section we consider a kinematics of two participators.

Consider a single participator on a symmetric framework (X, Y, E, S, G, J, π) satisfying Assumption Let $\xi = \in_e$, $e \in E$. The first manifestation of this participator then has perspective map π_e, defined by $\pi_e(g) = \pi(g - e)$, $g \in X$.

The only channeling possible, since there is but one participator, is a "self channeling," viz., a channeling in which $\chi(e) = e$. The participator's premise is then π_e (e), i.e., $\pi(0)$, where 0 denotes the identity element of our additive abelian group E.

This applies to each instant of the participator's proper time and, since there are no other participators, the system is inert at all other instants. It follows from this that the same perceptual premise $s_0 = \pi(0) \in S$ obtains at each step of the kinematics. And, denoting by Q the action kernel of the participator, this implies that the same probability measure $Q(s_0, \cdot)$ for the next perspective obtains at each step of the kinematics. This implies that the kinematics is a random walk of law $Q(s_0, \cdot)$ with respect to the discrete time which is the participator's proper time and, in this special case, the reference time.

KINEMATICS OF PAIRS

We now consider a system involving two kinematical participators. In such a system each participator might channel with itself, with the other participator, or not at all, at each step of reference time.

Again we are in the situation of Assumption. When two participators, A and B, observe each other, each changes its perspective according to its action kernel. This leads to a new difference in their perspectives.

This change in the relationship between their perspectives is governed by a kernel P which we can define as follows: for each $e \in E$ and $\Gamma \in \varepsilon$, $P(e, \Gamma)$ is the probability that, as the result of a change in their perspectives, the new perspective of B relative to A (i.e., the difference of their new perspectives) will lie in the set Γ, given that the present difference in their perspectives is e.

We can compute P from the action kernels of the individual participators. The figure shows two participators with initial perspectives a and b. The perspective of B relative to A is e (that of A relative to B is $-e$). After observing, A changes perspective by an amount dk and B changes perspective by an amount dh. This leads to a new difference in perspective $e - dk + dh$ (or $-(e - dk + dh)$).

Let Q and R denote the action kernels of the participators whose current perspectives are a and b respectively. Then the probability that A changes perspective by an amount dk given that B's perspective differs from A's by an amount e is $Q(e, dk)$. Similarly, the probability that B changes perspective by an amount dh given that A's perspective differs from B's by an amount Γe is R(Γe, dh).

The probability of the joint event that A changes by dk and B by dh. $Q(e, dk)\mathrm{R}(-e, dh)$. That is, the probability that the new difference in perspective is $e - dk + dh$, given that the old difference in perspective was e, is given by $Q(e, dk)\ \mathrm{R}(-e, dh)$.

Thus, to determine what is the probability that the new difference in perspective lies within a region $\Gamma \in \varepsilon$, we simply find the measure of the region $\{(k, h) \in E \times E |\ e - k + h \in \Gamma\}$ with respect to the product measure $Q(e, dk) \otimes R(-e, dh)$ on $E \times E$. This is the same as the integral

$$\int_{E\times E} 1_\Gamma(e-k+h)Q(e,dk)R(-e,dh);$$

we conclude that $P(e, \Gamma)$ is this integral.

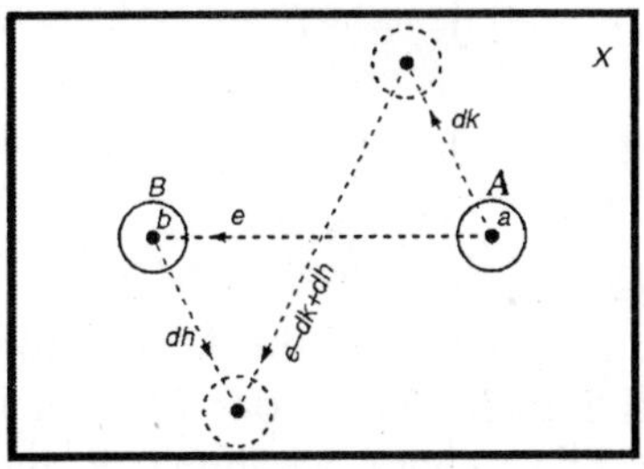

Fig. Two Participators Change Perspective.

Note that P is time independent (assuming, as we do, that Q and R are) and is also independent of the absolute perspective. Thus we can summarize:

Suppose A and B are participators with action kernels Q and R respectively. Assume a channeling sequence where A channels only to B and vice versa. Then the proper times of A and B are the same.

With respect to this proper time the successive perspectives of B relative to A (i.e., the successive differences in their perspectives) form a homogeneous Markov chain with state space E and transition probability P given by $P(e, \Gamma) = \int_{E\times E} 1_\Gamma(e-k+h)Q(e,dk)R(-e,dh);$.

The dependence of P on the action kernels Q and R can be conveniently and suggestively expressed in terms of a natural "bracket operation" which is derived from convolution of measures. First, recall that if α, β are measures on the group (E, ε), then the convolution of α with β, denoted α × β, is the measure on (E, ε) defined by

$$\alpha \times \beta(\Gamma) = \int_{E\times E} 1_\Gamma(k+h)\,\alpha(dk)\,\beta(dh) \quad (\Gamma \in \varepsilon)$$

Notation: If N is a kernel on (E, ε),

- $N^\dagger$ denotes the kernel $N^\dagger(e, G) = N(-e, -\Gamma)$, $(e \in \varepsilon, \Gamma \in \varepsilon)$;
- $N_e(\cdot)$ denotes the measure $N(e, \cdot)$.

Definition: If Q and R are kernels on (E, ε), $[Q, R]$ is the kernel on (E, ε) given by

$$[Q, R](e, \Gamma) = (Q_e \times R^\dagger_e)(e - \Gamma).$$

Proposition: With notation as above, $P = [Q, R]$.

Proof.

$$P(e, \Gamma) = \int_{E\times E} 1_\Gamma(e-k+h)\,Q(e,dk)\,R(-e,dh)$$

$$= \int_{E\times E} 1_{e-\Gamma}(k-h)\, Q(e,dk)\, R(-e,dh)\,;$$

change variables so that h is replaced by $-h$:

$$= \int_{E\times E} 1_{e-\Gamma}(k+h)\, Q(e,dk)\, R(-e,dh)$$

$$= \int_{E\times E} 1_{e-\Gamma}(k+h)\, Q(e,dk)\, R^{\dagger}(-e,dh)$$

$$= (Q_e * R_e^{\dagger})\,(e-\Gamma) = [Q,R]\,(e,\Gamma)\,.$$

For the moment, let P' denote the kernel for the Markov chain of perspectives of A relative to B. On the one hand, it is geometrically evident that $P'(e, \Gamma) = P(-e, -\Gamma)$ (where, as above, P denotes the kernel for the perspectives of B relative to A). On the other hand, from Proposition we find that $P' = [Q, R]$. We conclude

Proposition: For any kernels Q, R,

$$[Q, R] = [R, Q]:$$

We close this section with several remarks. First, nothing prevents A and B from occupying the same perspective in E at a given instant. Second, where each participator channels only to the other (and not to itself) is the opposite extreme of that treated in the previous section, where a participator channels only to itself. To make the comparison appropriate, imagine two kinematical participators A and B, each channeling only to itself.

In this case we would get a Markov chain on $E \times E$; in each factor we would have a random walk, (one for A and one for B) as in the previous section. These random walks would be completely "uncoupled." In the situation treated the perspectives of A and B are completely coupled: it is very unlikely that we would get anything resembling a random walk by looking at their sequences of states separately (or jointly).

In the general setting, the question of the relative frequencies of cross-channelings and self-channelings in, say, a two participator dynamical system is described by an additional datum, called a τ-distribution, which we think of as describing the "informational conductivity" of E. Depending on the τ-distribution, the dynamical chain generated by an ensemble of participators with given action kernels will express some degree of coupling of the random walks each participator would undergo were there no cross-channelings.

TRUE PERCEPTION AMONG PAIRS

We have seen that a dynamics of perspectives arises naturally on reflexive frameworks. Intuitively, the purpose of the dynamics is to allow the participators to "perceive truly," i.e., to choose conclusion measures

$\eta(s, \cdot)$ which in fact reflect the probabilities of events on the reflexive framework. Specifically, if participator *A* channels with *B*, leading to a premise s_A, then *A* should arrive at a conclusion measure $\eta_A(s_A, \cdot)$ which correctly describes with what probability the perspective of *B*, relative to A, lies in various subsets of $\pi_a^{-1}(s_A) \cap E$.

We specify conditions in which each participator, in a system of two mutually observing participators, can perceive truly the perspective of the other.

All channelings are cross-channelings. We found in the last section that the kinematics of relative perspectives for two participators is markovian with transition probability P. The theory of Markov chains describes some interesting properties of this kinematics that are relevant to the problem of true perception.

Depending on the details of the transition probability P, one finds that the state space *E* of the markovian dynamics contains different "pockets" which act like traps, if the state of the chain happens to enter one of these pockets, then the chain will forever stay within that pocket almost surely. For this reason these pockets are called "absorbing sets." The complement in the state space of all the absorbing sets is a pool of states called the "transient states."

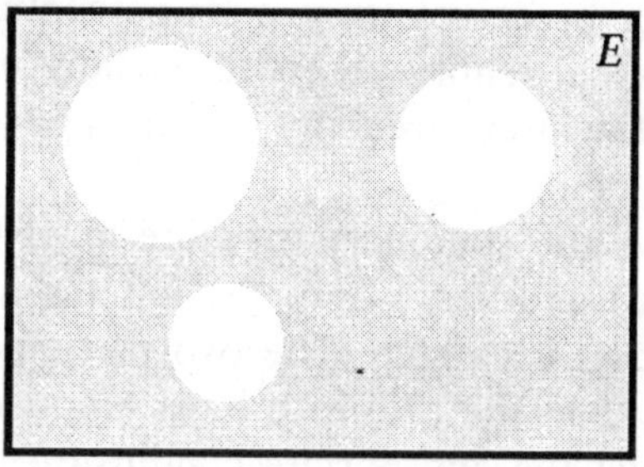

Fig. Absorbing Sets on the State Space of a Markovian Dynamics. White Disks Represent Absorbing sets. Blue Regions Represent Transient States.

and the states outside the disks are the transient states. An absorbing set may contain infinitely many states. If a chain enters an absorbing set, the chain then marches probabilistically from state to state within the absorbing set, and almost surely never enters a state outside of the absorbing set.

One finds that, for each absorbing set *C*, there is a unique probability measure supported on *C* which describes the long term behaviour of the chain, once it is trapped in *C*. This measure, say m, gives for each subset *D* of the absorbing set a probability, *m*(*D*), *m*(*D*) can be interpreted as the relative frequency that the trapped chain is found within *D* over a very long time. The measure m is called a "stationary" measure, an example of such a measure for a dynamics of two participators. Darker regions indicate higher frequency states. The little circles drawn over the stationary measure indicate the perspectives each participator happens to adopt at some instant of the dynamics.

Now if a two participator dynamical chain enters an absorbing set with stationary measure m, then each participator can reach true perceptual conclusions if its conclusions η are related appropriately to *m*.

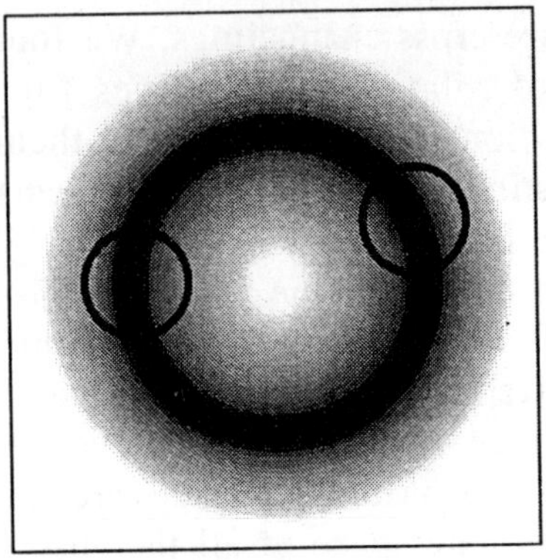

Fig. A Stationary Measure. Darker Regions Indicate Higher Probability States.

That is, a participator perceives truly if its perceptual conclusions η are matched to the dynamical reality observed, namely *m*. The way to match η to m is to make the measures η(s, ·) the appropriate conditional probability measures of m.

This figure shows the stationary measure m of the dynamics of one participator relative to another, where the latter's perspective is always taken to be the origin at each step. The two participators are channeling, leading the participator at the origin to have premises.

It can be seen that the appropriate conclusion η for this premise is the conditional probability of m when m is restricted to the line between the participators, viz., the line $\pi^{-1}(s)$. By choosing $\eta(s, \cdot)$ to be this probability measure, the participator at the origin has its perceptual conclusions matched to reality.

Thus, in the case of a two participator dynamics involving only cross-channelings, the equation that specifies when perception matches reality simply asserts that the conclusion kernel η is the rcpd with respect to π of a stationary measure *m*.

A measure m is stationary under the action of the transition probability *P* if $m = mP$, i.e., if

$$m(I - P) = 0,$$

where *I* is the identity operator. In the dynamics considered here, this equation, together with the stipulation that η is the rcpd of m, is the "perception = reality" equation. Note that there are in general many absorbing sets, each with its own stationary measure, so that the measure m is not uniquely determined even when P is fixed. Therefore, to determine if perception matches reality, we must be careful to use the appropriate stationary measure.

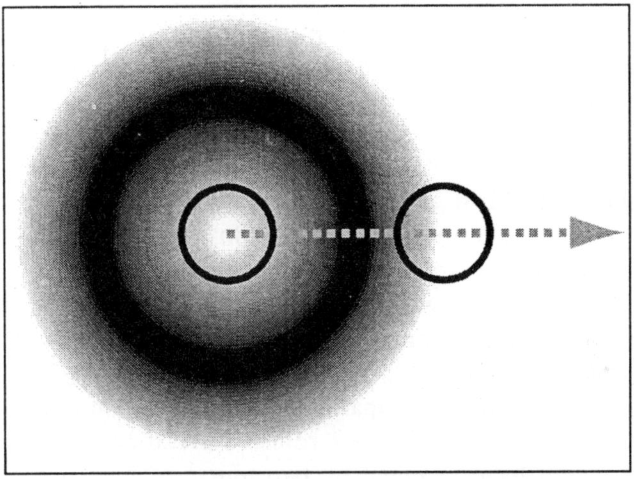

Fig. A Participator's Conclusion Measure Should be Derived as the rcpd of the Appropriate Stationary Measure.

Now if the chain never enters an absorbing set, i.e., if the dynamics is not stable, then there is no stationary probability measure to use to compute η. True perception is not possible.

There are no probability measures $\eta(s, \cdot)$ that are matched to the dynamical reality. We see that a stable dynamics of perspectives is necessary for true perception.

An example: A specific and elementary example, including a computation of its stationary measures. Consider the symmetric framework (X, Y, E, S, G, J, π), where

$$X = R, \quad E = Z \text{ (the integers)},$$
$$Y = S = \{1, 0, -1\}, \; G = \langle R, +\rangle, \; J = \langle Z, +\rangle,$$
$$\pi(x) = \operatorname{sgn}(x)$$

and where the signum function "sgn" is given by

$$\operatorname{sgn}(x) = \begin{cases} 1, & \text{if } x > 0; \\ 0, & \text{if } x = 0;. \\ -1 & \text{if } x < 0 \end{cases}$$

Suppose we have two participators labelled "1" and "2" respectively, which channel with each other at each instant of reference time. As before, we do not allow self-channeling. Both participators are assumed to have the same action kernel Q, defined as follows:

$$Q(0, \cdot) = \in_0(\cdot)$$

(where $\in_0(\cdot)$ is Dirac measure at 0), if $r \neq 0$,

$$Q(r, x) = \begin{cases} \rho, & \text{if } x = \operatorname{sgn}(r); \\ 1-\rho, & \text{if } x = \operatorname{sgn}(r); \\ 0, & \text{Otherwise} \end{cases}.$$

(Here r is the relative position before channeling and x is the participator's change in position after channeling, we assume that the quantity $\subset$ lies between 0 and 1).

In words: if a channeling came from the participator's current position, there is no movement. Otherwise, the participator moves one step in the direction from which the channeling came, with a probability of ρ, or one step away from that direction, with the complementary probability of 1 − ρ.

Imagine that the two participators are initially separated by a nonzero distance. After they channel, their relative distance will either remain unchanged, or will have changed by two units.

These are the only possibilities. (If they were initially at the same position in E, nothing will change.) This is expressed in the following derivation of the dynamical kernel P of the joint markovian dynamics (as introduced in section four above). Note that the dynamics is relativized, it is a dynamics on the group Z of the relative displacements of participator 2 with respect to participator 1.

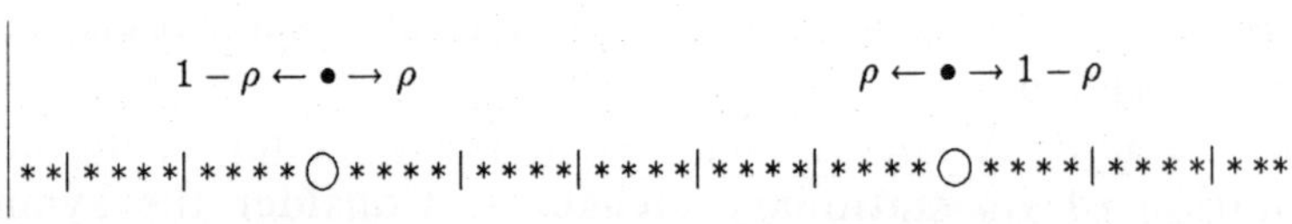

Participator 2 Participator 1

Fig. A Markovian Two-participator Dynamics with $E = Z$. The Current Relative Separation is r = −5. After a Channeling each Participator will Jump in the Indicated Directions with the Given Probabilities.

Proposition: Let r denote the current relative separation and q the relative separation after channeling. Then the kernel P of the dynamics is given by

If $r = 0$,

$$P(0, q) = \in_0(q).$$

If $q = r, r \neq 0$,

$$P(r, q) = 2\pi(1 - \pi).$$

If $q = r - 2\text{sgn}(r)$, $r \neq 0$,

$$P(r, q) = \rho^2.$$

If $q = r + 2\text{sgn}(r)$, $r \neq 0$,

$$P(r, q) = (1 - \rho)^2.$$

If $q \neq r$, and $q \neq r \pm 2$

$$P(r, q) = 0.$$

Proof: The result is a consequence of the assumption of independence between the jumps of the individual participators, as expressed in Proposition. By that Proposition we see that

$$P(r, q) = [Q, Q]\,(r, q)$$

$$= \sum_{z,w} Q(r,z)Q(-r,w)1_q(r - z + w)$$

$$= \sum_{w} Q(r,(r-q)+w)Q(-r,w).$$

The result then follows from the Definition of Q, after analyzing the possibilities into the indicated cases.

Notice that $\sum_q P(r,q) = 1$, for all $r \in Z$.

Up to an arbitrary initial probability measure ξ on the group Z, we have described the Markov chain which is the (relative) dynamics. We may now inquire into the long-term behaviour of the dynamics, as introduced in section five.

Suppose that ν is a probability measure on Z. Recall that 1 is said to be stationary for the chain with T.P. P if

$$\nu P = \nu,$$

i.e., if for all $q \in Z$,

$$\sum_r \nu(r)P(r,q) = \nu(q)$$

This is just equation 5.3 transcribed to our situation. For convenience we extract the $r = 0$ term in the sum on the left, to get

$$\sum_{r \neq 0} \nu(r)P(r,q) + \nu(0)\epsilon_0(q) = \nu(q).$$

If $\rho = 1$, the participators simply move towards each other after any channeling. Imagine that the participators are initially an even distance apart.

Then they will move towards each other until they are at the same point, thenceforth to remain there. If they were to start an odd distance apart, they would eventually find themselves one unit apart. From then on they would oscillate, with relative positions of ± 1. Thus when $\rho = 1$

there are two stationary measures: Dirac measure $\in_0(\cdot)$ at the origin and a measure μ given by $\mu(+1) = \mu(-1) = 1/2$, $\mu(q) = 0$ if $q \neq \pm 1$.

In general, the set of measures stationary with respect to a given T.P. is always a convex set.

That is, if λ and σ are stationary, so is $a\lambda + b\sigma$ whenever $0 \cdot a, b \cdot 1$ and $a + b = 1$.

In particular, in our situation when $\rho = 1$ the set of stationary measures consists of all convex combinations $a_{\in 0}(\cdot) + b\mu(\cdot)$.

Note that, regardless of the value of ρ, $\nu(\cdot) = \in_0(\cdot)$ is always a stationary measure for P. It is interesting that the only set of values of ρ for which the dynamics has a stationary measure other than $\varepsilon_0(\cdot)$ is the interval $\left(\frac{1}{2}, 1\right]$. In the rest of this section we will demonstrate this fact and explicitly determine the stationary measures.

If $\rho = 0$ it is intuitively clear from Equation that the chain wanders off to infinity, if it is not already at the origin. Thus, if $\rho = 0$, the Dirac measure at zero is in fact the only stationary measure. Henceforth we assume $\rho \neq 0$.

Now applying Proposition to equation, we identify the following cases:

- If $q = 0$,
  $\nu(0) = \rho^2 (\nu(2) + \nu(-2)) + \nu(0)$.
- If $q = \pm 1$,
  $\nu(\pm 1) = 2\rho(1 - \rho)\nu(\pm 1) + \rho^2\nu(\mp 1) + \rho^2\nu(\pm 3)$.
- If $q = \pm 2$,
  $\nu(\pm 2) = 2\rho(1 - \rho)\nu(\pm 2) + \rho^2\nu(\pm 4)$.

These cases are special, for the general case $|q| \geq 3$, we have

$\nu(q) = 2\rho(1 - \rho)\nu(q) + \rho^2 \nu(q + 2\text{sgn}(q)) + (1 - \rho)^2 \nu(q - 2\text{sgn}(q))$,

which, with a little algebra, may be re-expressed as follows:

(iv) If $|q| \geq 3$,

$$\nu(q) = c^2 \nu(q + 2\text{sgn}(q)) + s^2 \nu(q - 2\text{sgn}(q)).$$

where

$$c^2 = \frac{\rho^2}{\rho^2 + (1-\rho)^2}, \quad s^2 = \frac{(1-\rho)^2}{\rho^2 + (1-\rho)^2},$$

note that $c^2 + s^2 = 1$.

Equation is a linear difference equation with constant coefficients. Its solutions may be obtained by substituting the trial solution $\nu(q) = x^q$, $x \neq 0$.

Doing so, we get

$$x^q = c^2 x^{q+2\text{sgn}(q)} + s^2 x^{q-\Gamma 2\text{sgn}(q)}, \quad |q| \geq 3$$

Now the substitution $x \to x^{-1}$ andconverts any solution for $q \geq 3$ into one for $q \leq -3$, as may easily be checked. This allows us to concentrate on 6.18 for positive q only. So doing, and dividing out by x^{q-2}, we arrive at the characteristic equation

$$c^2x^4 - x^2 + s^2 = 0.$$

Solving this for x^2, we get $x^2 = 1$ or $(s/c)^2$. (A quick way to see this is to set $c = \cos\theta$ and $s = \sin\theta$ and to use elementary trigonometric formulas.)

If $s = c$, these two solutions are the same. This happens when $\rho = 1/2$. For the moment, assume $s \neq c$. Put

$$t = \left(\frac{s}{c}\right)^2 = \left(\frac{1-\rho}{\rho}\right)^2$$

We may immediately solve for I at the even integers. If $\rho \neq 1/2$ (i.e., $t \neq 1$), every solution to 6.16 is, at even values of q, of the form

$$\nu\ (2k) = \begin{cases} a_+ + b_+t^k, & \text{if } k \geq 2 \\ a_- + b_-t^{|k|} & \text{if } k \leq -2 \end{cases} \qquad (t \neq 1)$$

for some constants $a_\pm$, $b_\pm$.

Consider now $t = 1$. Then $s^2 = c^2 = 1/2$ and, by 6.16, $\nu(q)$ is an average of $\nu(q + 2)$ and $\nu(q - 2)$:

$$\nu(q) = \frac{1}{2}\nu(q+2) + \frac{1}{2}\nu(q-2)\,.$$

The characteristic equation of this difference equation is

$$x^4 - 2x^2 + 1 = 0,$$

so that x^2 can only be unity. In this case, we have

$$\nu(2k) = \begin{cases} a_+ + b_+k, & \text{if } k \geq 2 \\ a_- + b_-|k|, & \text{if } k \leq -2 \end{cases} \qquad (t = 1)\,,$$

for some constants $a_\pm$ and $b_\pm$.

Lemma: If ν is a stationary measure with respect to the T.P. P of Proposition, then

$$\nu\ (2k) = 0 \quad \text{for all } k \in Z,\ k \neq 0.$$

Proof: Since ν is a probability measure, $\{\nu(2k)\}_{k=0}^{\infty}$ is a summable sequence of non-negative terms. Hence $a_+ = a_- = 0$. If $\rho = 1$.

Next assume that $0 < t \neq 1$. We have $\rho^2\ (\nu\ (2) + \nu\ (-2)) = 0$. But, since $\rho \neq 0$, the non-negative quantities ν (2) and ν (−2) are both null. The same holds for ν (4) and ν (−4). When $k = 2$, 6.21 says

$$0 = \nu\ (4) = 0 + b_+t^2$$
$$0 = \nu\ (-4) = 0 + b_-t^2,$$

that is, $b_+ = b_- = 0$. Thus the result obtains if $t \neq 1$.

If $t = 1$, the same requirement of summability shows, using 6.22, that only ν (0) could possibly be nonzero.

We turn now to the computation of ν at odd integral points. Assume that $\rho \neq \frac{1}{2}$ (i.e., $t \neq 1$). We solve the formal difference equation in 6.16 for ν at the odd integers. The general solution has the form

$$\nu\,(2k + 1) = c_+ + d_+ t^{|k|}, \quad \text{if } k \geq 0;$$
$$\nu\,(2k - 1) = c_- + d_- t^{|k|}, \quad \text{if } k \leq 0,$$

for some constants c_+, d_+, c_-, and d_-. As in the even case, summability requires that $c+ = c\_ = 0$ and that $t < 1$. Thus, in terms of $q = 2k + 1$ (for $k \geq 0$) or $q = 2k - 1$ (for $k - 0$), our general solution is,

$$\text{for } \rho \neq \frac{1}{2} \qquad \nu(q) = \begin{cases} d_+ t^{|q-1|/2}, & \text{for odd } q \geq 0 \\ d_- t^{|q+1|/2}, & \text{for odd } q \leq 0 \end{cases}$$

In particular,

$$\nu\,(q) = d_+, \quad \nu\,(-1) = d_-.$$

Sinceq ν $(q) = 1$, we have that Pqodd ν $(q) \leq 1$. Thus

$$\sum_{k\leq 0} d_- t^{|k|} + \sum_{k\leq 0} d_+ t^{|k|} = \frac{d_+ + d_-}{1-t} \leq 1.$$

In terms of ρ, this says that

$$1 \geq \rho \geq \frac{1}{1+\sqrt{1-(d_+ + d_-)}}.$$

(which restricts ρ to the interval $\left(\frac{1}{2}, 1\right]\Big)$. We know that ν (2k) = 0 if $k \neq 0$. Thus,

$$\nu(0) + \frac{d_+ + d_-}{1-t} = 1$$

or

$$\nu(0) = \frac{2\rho - 1 - \rho^2 (d_+ + d_-)}{2\rho - 1}, \quad \frac{1}{2} < \rho \leq 1.$$

We are now in a position to delineate all possible stationarities of this chain. This is significant, for once we know the stationary measures it is possible to describe the ”true perception” of the dynamical situation by a given participator.

We shall not delve into such detail here, our purpose is to give a feel for how the dynamics is analyzed.

Theorem:

If $\frac{1}{2} < \rho \leq 1$, there is a one-parameter family of probability measures stationary with respect to the T.P. *P* of the dynamical chain of our example. With parameter denoted by d, this family may be described as:

$$\nu\,(q) = \begin{cases} d\left(\dfrac{1-\rho}{\rho}\right)^{|q-1|}, & \text{if } q \text{ is even and } q > 0; \\ d\left(\dfrac{1-\rho}{\rho}\right)^{|q+1|}, & \text{if } q \text{ is even and } q < 0; \\ 0, & \text{if } q \text{ is even and } q \neq 0; \\ \dfrac{2\rho-1-2\rho^2 d}{2\rho-1} & \text{if } q = 0. \end{cases}$$

The range of allowed values of the parameter d is contained in the closed interval [0, 1].

For fixed $\rho \in \left(\frac{1}{2}, 1\right]$ the range is $[0, (2\rho - 1)/2\rho^2]$.

If $0 \leq \rho \leq \frac{1}{2}$, the only stationary measure is $\in_0(\cdot)$.

Proof: Consider (i). For q odd we have equation $t^{1/2} = (1 - \rho)/\rho$, we obtain the first two formulas below.

$$\nu\,(q) = \begin{cases} d_+\left(\dfrac{1-\rho}{\rho}\right)^{|q-1|}, & \text{if } q \text{ is even and } q > 0; \\ d_-\left(\dfrac{1-\rho}{\rho}\right)^{|q+1|}, & \text{if } q \text{ is even and } q < 0; \\ 0, & \text{if } q \text{ is even and } q \neq 0; \\ \dfrac{2\rho-1-\rho^2(d_+ + d_-)}{2\rho-1} & \text{if } q = 0. \end{cases}$$

The third formula above is Lemma and the fourth is equation. Substituting the formula for odd *q* into we get

$$d_\pm = 2\rho(1-\rho)d_\pm + \rho^2 d_\mp + \rho^2 d_\pm \left(\frac{1-\rho}{\rho}\right)^{|1\pm 2|}$$

which reduces to $d_+ = d_-$. Set $d = d_+ = d_-$; the range of allowed values of the parameter d as given in the statement is computed by requiring that

$$0 \le \nu\,(0) = \frac{2\rho - 1 - 2\rho^2 d}{2\rho - 1} \le 1\,.$$

This concludes (i).

It remains to verify (ii). We have already done so for $0 < \rho < 1/2$, since shows us that the fact that ν is a probability measure requires that $\rho \ge 1/2$. Moreover, the instance $\rho = 1/2$ requires, in the same way as in equation above, that

$$\nu\,(2k+1) = \begin{cases} a_+ + b_+ k, & \text{if } k \ge 1 \\ a_- + b_- |k|, & \text{if } k \le -1 \end{cases}$$

which is only summable if it is in fact zero.

Chapter 2

Newtonian and Celestial Mechanics

Isaac Newton solved the premier scientific problem of his day, which was to explain the motion of the planets. The theory that the planets (including Earth) revolve around the sun was published by Nicolaus Copernicus in 1543.

This was a revolutionary ideaω The picture of the Universe that had been developed by astronomers before Copernicus had the Earth at rest at the centre, and the sun, moon, planets and stars revolving around the Earth. But this picture failed to explain accurately the observed planetary positions. The failure of the Earth centered theory led Copernicus to consider the sun as the centre of planetary orbits. Later observations verified the Copernican theory. The important advances in astronomical observations were made by Galileo and Kepler.

Galileo Galilei was perhaps the most remarkable individual in the history of science. His experiments and ideas changed both physics and astronomy. In physics he showed that the ancient theories of Aristotle, which were still accepted in Galileo's time, are incorrect.

In astronomy he verified the Coper-nican picture of the Universe by making the first astronomical observations with a telescope. Galileo did not invent the telescope but he made some of the earliest telescopes, and his telescopes were the best in the world at that time. Therefore he discovered many things about the the solar system and stars:

- Craters and mountains on the moon
- The moons of Jupiter
- The phases of Venus
- The motion of sunspots
- The existence of many faint stars

These discoveries provided overwhelming evidence in favour of the Copernican model of the solar system.

Johannes Kepler had extensive data on planetary positions, as functions of time, from observations collected earlier by Tycho Brahe. He analyzed the data based on the Copernican model, and deduced three empirical laws of planetary motion:

Kepler's Laws

- The planets move on elliptical orbits with the sun at one focal point.
- The radial vector sweeps out equal areas in equal times.
- The square of the period of revolution is proportional to the cube of the semimajor axis of the ellipse.

Newton started with the results of Galileo and Kepler. His goal, then, was to explain why.

Why do the planets revolve around the sun in the manner discovered by Galileo and Kepler? In particular, what is the explanation for the mathematical regularities in Kepler's laws of orbital motion? To answer this question, Newton had to develop the laws of motion and the law of universal gravitation. And, to analyse the motion he invented a new branch of mathematics, which we now call Calculus. The solution to planetary motion was published in Principia in 1687. Newton had solved the problem some years earlier, but kept it secret. He was visited in 1684 by the astronomer Edmund Halley. Halley asked what force would keep the planets in elliptical orbits.

Newton replied that the force must be an inverse-square law, which he had proven by mathematical analysis; but he could not find the paper on which he had written the calculationsω After further correspondence, Halley realized that Newton had made great advances in physics but had not published the results. With Halley's help, Newton published Principia in which he explained his theories of motion, gravity, and the solar system.

After the publication of Principia, Newton was the most reknowned scientist in the world. His achievement was fully recognized during his lifetime. Today scientists and engineers still use Newton's theory of mechanics. Classical mechanics breaks down at atomic dimensions, but for engineering systems Newton's theory is valid and extremely accurate.

How is this early history of science relevant to the study of calculus? Newton used calculus for analyzing motion, although he published the calcu- lations in Principia using older methods of geometrical analysis. Ever since that time, calculus has been necessary to the understanding of physics and its applications in engineering and science.

POSITION, VELOCITY, AND ACCELERATION

Position and Velocity

Suppose an object M moves along a straight line. We describe its motion by giving the position x as a function of time t. The variable x is the coordinate, i.e., the displacement from a fixed point 0 called the origin. Physically, the line on which M moves might be pictured as a road, or a track. Mathematically, the positions form a representation of the ideal real line. The coordinate x is positive if M is to the right of 0, or negative if to the left. The absolute value $|x|$ is the distance from 0.

The possible positions of M are in one-to-one correspondence with the set of real numbers. Hence position is a continuous function $x(t)$ of the independent variable t, time.

Example: What is the position as a function of time if M is at rest at a point 5 m to the left of the origin?

Solution: Because M is not moving, the function $x(t)$ is just a constant,

$$x(t) = -5 \text{ m}.$$

Note that the position has both a number (– 5) and a unit (m, for meter). In this case the number is negative, indicating a position to the left of 0. The number alone is not enough information. The unit is required. The unit may be changed by multiplying by a conversion factor. For example, the conversion from meters (m) to inches (in) is

$$5 \text{ m} = 5 \text{ m} \times \frac{38 \text{ in}}{1 \text{ m}} = 190 \text{ in}.$$

(There are 38 inches per meter.) So, the position could just as well be written as

$$x(t) = -190 \text{ in}.$$

This example shows why the number alone is not enough: The number depends on the unit.

A Comment on Units: In physical calculations it is important to keep track of the units|treating them as algebraic quantities. Dropping the unit will often lead to a failed calculation. Keeping the units has a bonus. It is a method of error checking. If the final unit is not correct, then there must be an error in the calculation; we can go back and figure out how to correct the calculation.

Example: A car travels on a straight road, toward the East, at a constant speed of 35 mph. Write the position as a function of time. Where is the car after 5 minutes?

Solution: The origin is not specified in the statement of the problem, so let's say that $x = 0$ at time $t = 0$. Then the position as a function of time is

$$x(t) = +\left(35 \frac{\text{mi}}{\text{hr}}\right) t,$$

where positive x is east of the origin. Equation is based on the formula

$$\text{distance} = \text{speed} \times \text{time},$$

familiar from grade school; or, taking account of the signs,

$$\text{displacement} = \text{velocity} \times \text{time}.$$

Any distance must be positive. 'Displacement' may be positive or negative. Similarly, 'speed' must be positive, but 'velocity' may be positive or negative, negative meaning that M is moving to smaller x.

After 5 minutes, the position is

$$x(5\text{ min}) = 35\frac{\text{mi}}{\text{hr}} \times 5\,\text{min}$$

$$= 35\frac{\text{mi}}{\text{hr}} \times 5\,\text{min}\frac{1\,\text{hr}}{60\text{ min}}$$

$$= 2.917\text{ mi.}$$

(We multiply by the conversion factor, 1 hr/60 min, to reduce the units.) The position could be expressed in feet, as

$$x(5\text{ min}) = 2.917\text{ mi} \times \frac{5280\,\text{ft}}{1\,\text{mi}} = 15400\text{ ft.}$$

So, after 5 minutes the car is 15400 feet east of its initial position.

It is convenient to record some general, i.e., abstract formulas. If M is at rest at x_0, then the function $x(t)$ is

$$x(t) = x_0. \qquad \text{(object at rest)}$$

If M moves with constant velocity v_0 then

$$x(t) = x_0 + v_0\,t. \qquad \text{(object with constant velocity)}$$

Figure illustrates graphs of these functions. The abscissa (horizontal axis) is the independent variable t, and the ordinate (vertical axis) is the dependent variable x. The slope in the second graph is v_0. Note that the units of v_0 must be a length unit divided by a time unit, because

$$v_0 = \text{slope} = \frac{\text{rise}}{\text{run}} = \frac{\Delta x}{\Delta t};$$

for example, the units could be m/s. But the slope in a graph is the derivative of the function! Thus, the derivative of the position $x(t)$ is the velocity $v(t)$.

So far we have considered only constant velocity. If the velocity is not

constant, then the instantaneous velocity at a time t is the slope of the curve of x versus t, i.e., the slope of the tangent line. This is nothing but the derivative of $x(t)$. Letting $v(t)$ denote the velocity,

$$v(t) = \lim_{\Delta t \to 0} \frac{\Delta x}{\Delta t} = \frac{dx}{dt}.$$

Another notation for the time-derivative, often used in mechanics, is $\dot{x}(t)$ dx/dt. Because $v(t)$ is defined by the limit $\Delta t \to 0$, there is an instantaneous velocity at every t. We summarize the analysis by a definition:

Definition: The velocity $v(t)$ is a function of time t, defined by $v(t) = dx/dt$.

Acceleration

If the velocity is changing t city,

$$a(t) = \lim_{\Delta t \to 0} \frac{\Delta v}{\Delta t} = \frac{dv}{dt}. \text{(definition of acceleration)}$$

Taking the limit $\Delta t \to 0$, the acceleration is defined at every instant. Also, because $v = dx/dt$, the acceleration a is the second derivative of $x(t)$,

$$a(t) = \frac{d^2x}{dt^2} = \ddot{x}(t).$$

Example: A car accelerates away from a stop sign, starting at rest. Assume the acceleration is a constant 5 m/s^2 for 3 seconds, and thereafter is 0.(a) What is the final velocity of the car? (b) How far does the car travel from the stop sign in 10 seconds?

Solution: Let the origin be the stop sign. During the 3 seconds while the car is accelerating, the acceleration is constant and so the velocity function must be

$$v(t) = at,$$

because the derivative of at (with respect to t) is a. Note that $v(0) = 0$; i.e., the car starts from rest. At $t = 3$ s, the velocity is the final velocity,

$$v_f = v(3s) = 5\frac{\text{m}}{\text{s}^2} \times 3s = 15\frac{\text{m}}{\text{s}}.$$

The position as a function of time is $x(t)$, and

$$\frac{dx}{dt} = v(t).$$

Equation is called a differential equation for $x(t)$. We know the derivative; what is the function? The general methods for solving differential equations involve integration. But for this simple case we can figure out the answer by guess-work.

Before we try to do the calculation, let's make sure we understand the problem. The car moves with velocity $v(t) = at$ (constant acceleration) for 3 seconds. How far does the car move during that time? Thereafter it moves with constant velocity $v_f = 15$ m/s. How far does it move from $t = 3$ s to $t = 10$ s? The combined distance is the distance traveled in 10 s.

For $t < 3$ s the velocity is $v(t)$ = at. This must be the derivative of $x(t)$. What function $x(t)$ has derivative at? The derivative is also a power of t, so $x(t)$ is also a power; recall that the derivative of t^p is pt^{p-1}. If we set

$p = 2$ then the derivative is/ t, as required. Multiplying by a constant factor does not change the power of t.

Evidently $x(t)$ should be Ct^2 for some constant C. Then $dx/dt = 2Ct$. This is supposed to be at; thus, $C = a/2$. So the distance at t, for $t < 3$ s, is $\frac{1}{2}at^2$. We could also add a constant x_0 because the derivative of a constant is 0. So, more generally, $x(t) = x_0 + \frac{1}{2}at^2$. But the stop sign is at $x = 0$, so $x(0) = 0$; that initial condition requires $x_0 = 0$. Hence the position as a function of t, for $t < 3$ s, is

$$x(t) = \frac{1}{2}at^2$$

For $t > 3$ s the velocity is constant v_f. What function $x(t)$ has derivative v_f? By similar reasoning,

$$x(t) = c + v_f t \quad (\text{for } t \geq 3 \text{ s})$$

general equations for constant acceleration

position	$x(t) = x_0 + v_0 t + \frac{1}{2}at^2$
velocity	$v(t) = dx/dt = v_0 t + at$
acceleration	$a(t) = dv/dt = a$, constant

Table: Formulae for constant acceleration. (x_0 = initial position, v_0 = initial velocity, and a = acceleration.) where c is a constant. Both equation must be correct. So, if we set $t = t_1 \_ 3$ sec in the two equations, we must have

$$x(t_1) = \frac{1}{2}at_1^2 = c + v_f t_1.$$

(Remember, $t_1 = 3$ s is the time when the car ceases to accelerate.) The only unknown in equation is c, and solving the equation gives

$$c = \frac{1}{2}at_1^2 - v_f t_1$$

$$= \frac{1}{2}(5\text{m}/s^2)(3s)^2 - (15\text{m/s})(3s) = -22.5\text{m}.$$

The position of the car at $t = 10$ s is now obtained from equation,

$$x(10 \text{ s}) = c + v_f \times 10 \text{ s}$$

$$= -22.5 \text{ m} + 15 \text{ m}15\frac{\text{m}}{\text{s}} \times 10\text{s} = 127.5 \text{ m}.$$

After 10 s the car has moved 127.5 m.

Generalization: Some useful general formulae for constant acceleration. In the table, v_0 is a constant equal to the velocity at $t = 0$. Also, x_0 is a constant equal to the position at $t = 0$. As an exercise, please verify that a = dv/dt and $v = dx/dt$. Remember that x_0 and v_0 are constants, so their derivatives are 0. The velocity and position as functions of t for constant acceleration.

Example: A stone is dropped from a diving platform 10 m high. When does it hit the water? How fast is it moving then?

Solution: We'll denote the height above the water surface by $y(t)$. The initial height is y_0 = 10 m. The initial velocity is v_0; because the stone is dropped, not thrown, its initial velocity v_0 is 0.

The acceleration of an object in Earth's gravity, neglecting the effects of air resistance, is $a = -g$ where $g = 9.8$ m/s^2. The acceleration a is negative because the direction of acceleration is downward; i.e., the stone accelerates toward smaller y. Using above Table, the equation for position y as a function of t is

$$y(t) = y_0 - \frac{1}{2}gt^2.$$

The variable y is the height above the water, so the surface is at $y = 0$. The time t_f when the stone hits the water is obtained by solving $y(t_f) = 0$,

$$y_0 - \frac{1}{2}gt_f^2 = 0.$$

The time is

$$t_f = \sqrt{\frac{2y_0}{g}} = \sqrt{\frac{2\times 10\,\text{m}}{9.8\,\text{m/s}^2}} = 1.43\text{s}.$$

Note how the final unit came out to be seconds, which is correct. The time to fall to the water surface is 1.43 s.

The equation for velocity is

$$v(t) = \frac{dy}{dt} = -gt.$$

This is consistent with the second row in Table, because $v_0 = 0$ and $a = -g$. The velocity when the stone hits the water is

$$v_f = v(t_f) = -9.8\frac{\text{m}}{\text{s}^2}\times 1.43\text{s} = -14.0\frac{\text{m}}{\text{s}}.$$

The velocity is negative because the stone is moving downward. The final speed is the absolute value of the velocity, 14.0 m/s.

Newton's Second Law

Newton's second law of motion states that the acceleration a of an object is proportional to the net force F acting on the object,

$$a = \frac{F}{m},$$

or F = ma. The constant of proportionality m is the mass of the object. Equation may be taken as the definition of the quantity m, the mass.

Vectors: For two- or three-dimensional motion, the position, velocity, and accleration are all vectors| mathematical quantities with both magnitude and direction. We will denote vectors by boldface symbols, e.g., *x* for position, *v* for velocity, and a for acceleration. In hand-written equations, vector quantities are usually indicated by drawing an arrow (→) over the symbol.

Acceleration is a kinematic quantity|determined by the motion. Equation relates acceleration and force. But some other theory must determine the force.

There are only a few basic forces in nature: gravitational, electric and magnetic, and nuclear. All observed forces (e.g., contact, friction, a spring, atomic forces, etc.) are produced in some way by those basic forces. Whatever force is acting in a system, equation states how that force in uences the motion of a mass *m*.

The mass is called the inertial mass, because it would be determined by measuring the acceleration produced by a given force. For example, if an object is pulled by a spring force of 50 N, and the resulting acceleration is measured to be 5 m/s^2, then the mass is equal to 10 kg.

The gravitational force is exactly (i.e., as precisely as we can measure it!) proportional to the inertial mass.

Therefore the acceleration due to gravity is independent of the mass of the accelerating object. Near the surface of the Earth, all falling objects have the same acceleration due to gravity, g = 9:8 m/s^2, neglecting the force of air resistance.

It took the great genius of Galileo to see that the small differences between falling objects are not produced by gravity but by air resistance. The force of gravity is proportional to the mass so accelerations by gravity are independent of the mass.

The equation F = ma tells an engineer how an object will respond to a specified force. Because the acceleration a is a derivative,

$$a = \frac{dv}{dt} = \frac{d^2x}{dt^2},$$

Newton's second law is a differential equation.

PROJECTILE MOTION

A projectile is an object M moving in Earth's gravity with no internal propulsion, and no external forces except gravity. (A real projectile is also subject to aerodynamic forces such as drag and lift. We will neglect these forces, a fairly good approximation if M moves slowly.)

The motion of a projectile must be described with two coordinates: horizontal (x) and vertical (y). Figure shows the motion of the projectile in the xy coordinate system. The curve is the trajectory of M.

Suppose M is released at $(x, y) = (x_0, y_0)$ at time $t = 0$. Figure also indicates the initial velocity vector v_0 which is tangent to the trajectory at (x_0, y_0). Let θ be the angle of elevation of the initial velocity; then the x and y components of the initial velocity vector are

$$v_{0x} = v_0 \cos \theta,$$
$$v_{0y} = v_0 \sin \theta.$$

Horizontal component of motion. The equations for the horizontal mo- tion are

$$x(t) = x_0 + v_{0x}t,$$
$$v_x(t) = v_{0x}.$$

These are the equations for constant velocity, $v_x = v_{0x}$. There is no horizon- tal acceleration (neglecting air resistance) because the gravitational force is vertical.

Vertical component of motion: The equations for the vertical motion are

$$y(t) = y_0 + v_{0y}t - \frac{1}{2}gt^2,$$
$$v_y(t) = v_{0y} - gt.$$

These are the equations for constant acceleration, $a_y = -g$. (As usual, $g = 9.8$ m/s^2.) The vertical force is $F_y = -mg$, negative indicating downward, where m is the mass of the projectile. The acceleration is $a_y = F_y/m$ by Newton's second law. Thus $a_y = -g$. The acceleration due to gravity does not depend on the mass of the projectile because the force is proportional to the mass.

Example: Verify that the derivative of the position vector is the velocity vector, and the derivative of the velocity vector is the acceleration vector, for the projectile.

Solution: Position, velocity, and acceleration are all vectors. The position vector is

$$x(t) = x(t)\hat{i} + y(t)\hat{j}.$$

Here $\hat{i}$ denotes the horizontal unit vector, and $\hat{j}$ denotes the vertical unit vector (cf. Fig. 5). For the purposes of describing the motion, these unit vectors are constants, independent of t. The time dependence of the position vector $x(t)$ is contained in the coordinates, $x(t)$ and $y(t)$.

The derivative of $x(t)$ is

$$\frac{dx}{dt} = \frac{dx}{dt}\hat{i} + \frac{dy}{dt}\hat{j}$$

$$= v_{0x}\hat{i} + (v_{0y} - gt)\hat{j}$$

$$= v_x\hat{i} + v_y\hat{j} = v$$

As required, *dx*/*dt* is *v*. The derivative of *v*(*t*) is

$$\frac{dv}{dt} = \frac{dv_x}{dt}\hat{i} + \frac{dv_y}{dt}\hat{j}$$

$$= 0 + (-g)\hat{j} = -g\hat{j}.$$

The acceleration vector has magnitude *g* and direction $-\hat{j}$, i.e., downward; so $a = -g\hat{j}$. We see that *dv*/*dt* = a, as required.

CIRCULAR MOTION

Consider an object M moving on a circle of radius R. We could describe the motion by Cartesian coordinates, *x*(*t*) and *y*(*t*), but it is simpler to use the angular position θ(*t*) because the radius R is constant. The angle θ is defined in Fig. 6. It is the angle between the radial vector and the *x* axis. The value of θ is sufficient to locate M. From Fig. 6 we see that the Cartesian coordinates are

$$x(t) = R\cos\theta(t),$$
$$y(t) = R\sin\theta(t).$$

If –(*t*) is known, then *x*(*t*) and *y*(*t*) can be calculated from these equations. In calculus we always use the radian measure for an angle θ. The radian measure is defined as follows. Consider a circular arc with arclength s on a circle of radius R. The angle subtended by the arc, in radians, is

$$\theta = \frac{s}{R}. \qquad \text{(radian measure)}$$

Angular Velocity and the Velocity Vector

The angular velocity ω(*t*) is defined by

$$\omega(t) = \frac{d\theta}{dt}. \qquad \text{(angular velocity)}$$

This function is the instantaneous angular velocity at time *t*. For example, if *M* moves with constant speed, traveling around the circle in time *T*, then the angular velocity is constant and given by

$$\omega = \frac{2\theta}{T}. \qquad \text{(constant angular velocity)}$$

To derive this result, consider the motion during a time interval Δt.

The arclength Δs traveled along the circle during Δt is $R\Delta\theta$ where $\Delta\theta$ is the change of θ during Δt, in radians. The angular velocity is then

$$\omega = \frac{\Delta\theta}{\Delta t} = \frac{\Delta s / R}{\Delta t}.$$

Because the speed is constant, $\Delta\theta/\Delta t$ is constant and independent of the time interval Δt. Let Δt be one period of revolution T. The arclength for a full revolution is the circumference $2\pi R$. Thus

$$\omega = \frac{2\pi R / R}{T} = \frac{2\pi}{T}$$

The instantaneous speed of the object is the rate of increase of distance with time,

$$v(t) = \lim_{\Delta t \to 0} \frac{\Delta s}{\Delta t} = \lim_{\Delta t \to 0} \frac{R\Delta\theta}{\Delta t}$$

$$= R\frac{d\theta}{dt} = R\omega(t).$$

But what is the instantaneous velocity? Velocity is a vector v, with both direction and magnitude. The magnitude of v is the speed, $v = R\omega$. The direction is tangent to the circle, which is the same as the unit vector $\hat{\theta}$. Thus the velocity vector is

$$v = R\omega\,\hat{\theta},$$

which points in the direction of $\hat{\theta}$ and has magnitude $R\omega$. In general, v, ω, θ and $\hat{\theta}$ are all functions of time t as the particle moves around the circle. But of course for circular motion, R is constant. We summarize our analysis as a theorem:

Theorem: The velocity vector in circular motion is

$$v(t) = R\omega(t)\;\hat{\theta}(t).$$

Acceleration in Circular Motion

Now, what is the acceleration of M as it moves on the circle? The acceleration a is a vector, so we must determine both its magnitude and direction. Unlike the velocity v, which must be tangent to the circle, the acceleration has both tangential and radial components. Recall that we have defined acceleration as the derivative of velocity in the case of one-dimensional motion. The same definition applies to the vector quantities for two- or three-dimensional motion. Using the definition of the derivative,

$$a(t) = \lim_{\Delta t \to 0} \frac{v(t+\Delta t) - v(t)}{\Delta t} = \frac{dv}{dt}.$$

The next theorem relates a for circular motion to the parameters of the motion.

Theorem: The acceleration vector in circular motion is

$$a = R\frac{d\omega}{dt}\hat{\theta} - R\omega^2\hat{r}.$$

Proof: We must calculate the derivative in for v. At time *t*, the acceleration is

$$a(t) = \frac{dv}{dt} = \frac{d}{dt}\left(R\omega\hat{\theta}\right)$$

$$= R\left(\frac{d\omega}{dt}\hat{\theta} + \omega\frac{d\hat{\theta}}{dt}\right).$$

Note that (49) follows from the Leibniz rule for the derivative of the product $\omega(t)\,\hat{\theta}(t)$. Now,

$$\frac{d\hat{\theta}}{dt} = \lim_{\Delta t \to 0}\frac{\Delta\hat{\theta}}{\Delta t}.$$

Figure shows that $\Delta\hat{\theta} \approx -\hat{r}\Delta\theta$ for small $\Delta\theta$. (The relation of differentials is $d\hat{\theta} = -\hat{r}d\theta$) The direction of $\Delta\hat{\theta}$ is radially inward. This little result has interesting consequences, as we'll seeω The derivative is then

$$\frac{d\hat{\theta}}{dt} = \lim_{\Delta t \to 0}\frac{-\hat{r}\Delta\theta}{\Delta t} = \hat{r}\frac{d\theta}{dt} = -\hat{r}\omega.$$

Substituting this result into equation we find

$$a(t) = R\frac{d\omega}{dt}\hat{\theta} - R\omega^2\hat{r},$$

which proves the theorem.

The radial component of the acceleration vector is $a_r = -R\omega^2$. This component of a is called the centripetal acceleration. The word "centripetal" means directed toward the centre. We may write a_r in another form. By Theorem, $\omega = v/R$; therefore

$$a_r = -\frac{v^2}{R}.$$

If the speed of the object is constant, then $d\omega/dt = 0$ and the acceleration a is purely centripetal. In uniform circular motion, the acceleration vector is always directed toward the centre of the circle with magnitude v^2/R. Imagine a ball attached to a string of length R, moving around a circle at constant speed with the end of the string fixed.

The trajectory must be a circle because the string length (the distance from the fixed point) is constant.

The ball constantly accelerates toward the centre of the circle ($a_r = -v^2/R$) but it never gets any closer to the centre ($r(t) = R$, constant)! This example illustrates the fact that the velocity and acceleration vectors may point in different directions. In uniform circular motion, the velocity is tangent to the circle but the acceleration is centripetal, i.e., orthogonal to the velocity.

Example: Suppose a race car travels on a circular track of radius R = 50 m. (This is quite small) At what speed is the centripetal acceleration equal to 1 g?

Solution: Using the formula $a = v^2/R$, and setting $a = g$, the speed is

$$v = \sqrt{gR} = \sqrt{9.8\,\text{m/s}^2 \times 50\,\text{m}} = 22.1\,\text{m/s}\,.$$

Converting to miles per hour, the speed is about 48 mi/hr. A pendulum suspended from the ceiling of the car would hang at an angle of 45 degrees to the vertical (in equilibrium), because the horizontal and vertical components of force exerted by the string on the bob would be equal, both equal to mg.

The pendulum would hang outward from the centre of the circle. Then the string exerts a force on the bob with an inward horizontal component, which is the centripetal force on the bob.

The equation $a_r = -v^2/r$ for the centripetal acceleration in circular motion was first published by Christiaan Huygens in 1673 in a book entitled Horologium Oscillatorium.

Huygens, a contemporary of Isaac Newton, was one of the great figure s of the scientific revolution.

He invented the earliest practical pendulum clocks. He constructed excellent telescopes, and discovered that the planet Saturn is encircled by rings. In his scientific work, Huygens was guided by great skill in mathematical analysis. Like Galileo and Newton, Huygens used mathematics to describe nature accurately.

KEPLER'S LAWS OF PLANETARY MOTION

Kepler's first law is that the planets travel on ellipses with the sun at one focal point. Newton deduced from this empirical observation that the gravitational force on the planet must be proportional to $1/r^2$ where r is the distance from the sun.

Figure shows a possible planetary orbit. The ellipse is characterized by two parameters: a = semimajor axis and e = eccentricity.

Kepler's Third Law

Kepler's third law relates the period T and the semimajor axis a of the ellipse. To the accuracy of the data available in his time, Kepler found that T^2 is proportional to a^3.

The next example derives this result from Newtonian mechanics, for the special case of a circular orbit.

A circle is an ellipse with eccentricity equal to zero; then the semimajor axis is the radius.

Example: Show that $T^2 \propto r^3$ for a planet in a circular orbit of radius r around the sun.

Solution: In analyzing the problem, we will neglect the motion of the sun. More precisely, the sun and planet both revolve around their centre of mass.

But because the sun is much more massive than the planet, the centre of mass is approximately at the position of the sun, so that the sun may be considered to be at rest.

Then the planet moves on a circle around the sun. Let m denote the mass of the planet, and M the mass of the sun. For a circular orbit the angular speed of the planet is constant, dω=dt = 0. Therefore the acceleration is $a = -r\omega^2\hat{r}$. In terms of the speed $v = r\omega$,

$$a = -\frac{v^2}{r}\hat{r}.$$

The direction is $-\hat{r}$, i.e., centripetal, toward the sun. The gravitational force exerted by the sun on the planet is

$$F = -\frac{GMm}{r^2}\hat{r},$$

which is also centripetal. Equation is Newton's theory of Universal Gravitation, in which the force is proportional to $1/r^2$.

Newton's second law of motion states that $F = ma$. Therefore,

$$\frac{mv^2}{r} = \frac{GMm}{r^2}.$$

The distance traveled in time T (one period of revolution) is 2πr (the circum-ference of the orbit), so the speed is v = 2πr/T. Substituting this expression for v into equation gives

$$\left(\frac{2\pi r}{T}\right)^2 = \frac{GM}{r}.$$

Or, rearranging the equation,

$$T^2 = \frac{4\pi^2 r^3}{GM};$$

we see that T^2 is proportional to r^3, as claimed.

In obtaining we neglected the small motion of the sun around the centre-of-mass point.

This is a very good approximation for the solar system. In this approximation, $T^{2/} r^3$ is constant, i.e., it has the same value for all nine planets.

Kepler's Second Law

Kepler's second law is that the radial vector sweeps out equal areas in equal times. In Newtonian mechanics it is a consequence of conservation of angular momentum. The next two examples show how Kepler's second law follows from Newton's theory.

Example: Conservation of angular momentum. The angular momentum L of an object of mass m moving in the xy plane is defined by

$$L = m\,(xv_y - yv_x).$$

Show that L is constant if the force on the object is central.

Solution: To show that a function is constant, we must show that its derivative is 0. The coordinates x and y, and velocity components vx and vy, are all functions of time t.

But the particular combination in L is constant, as we now show. The derivative of L is

$$\begin{aligned} \frac{dL}{dt} &= m\left[\frac{dx}{dt}v_y + x\frac{dv_x}{dt} - \frac{dy}{dt}v_x - y\frac{dv_x}{dt}\right] \\ &= mv_x v_y + xF_y - mv_y v_x - yF_x \\ &= xF_y - yF_x. \end{aligned}$$

In the first step we have used the fact that $dx/dt = v$, and dv/dt = a; also, by Newton's second law, the acceleration a is equal to F/m. The final line is called the torque on the object.

For any central force the torque is 0. What is meant by the term "central force" is that the force is in the direction of $\pm\hat{r}$, i.e., along the line to the origin.

The sign |attractive or repulsive| is unimportant for the proof of conservation of angular momentum. The components of F are

$$F_x = -F\cos\theta \text{ and } F_y = -F\sin\theta$$

where F is the strength of the force and the minus signs mean that F is toward 0. Thus the torque on the object is

$$\begin{aligned} \text{torque} &= xF_y - yF_x \\ &= r\cos\theta\, F\sin\theta + r\sin\theta\, F\cos\theta = 0. \end{aligned}$$

Since the torque is 0, equation implies that dL=dt = 0. Since the derivative is 0, the angular momentum L is constant, as claimed.

Example: Kepler's law of equal areas. Show that the radial vector from the sun to a planet sweeps out equal areas in equal times.

Solution: Figure shows the elliptical orbit. The shaded area ΔA is the area swept out by the radial vector between times t and $t + \Delta t$.

The shaded area may be approximated by a triangle, with base r and height $r\Delta\theta$, where $\Delta\theta$ is the change of the angular position between t and $t + \Delta t$. Approximating the area as a triangle is a good approximation for small Δt. Now consider the limit $\Delta t \to 0$; i.e., Δt and ΔA become the differentials dt and dA. The area of the triangle becomes

$$dA = \frac{1}{2} \times \text{base} \times \text{height} = \frac{1}{2} \times r \times r d\theta = \frac{1}{2} r^2 d\theta .$$

Thus, in the limit $\Delta t \to 0$, where we replace Δt by dt,

$$\frac{dA}{dt} = \frac{1}{2} r^2 \frac{d\theta}{dt} = \frac{1}{2} r^2 \omega .$$

We'll use this result presently.

But now we must express the angular momentum in polar coordinates. The position vector of M is $x = r\hat{r}$, and its x and y components are

$$x = r \cos\theta \quad \text{and } y = r \sin\theta.$$

The velocity vector is

$$v = \frac{dx}{dt} - \frac{dr}{dt}\hat{r} + r\frac{d\theta}{dt}\hat{\theta},$$

because $d\hat{r} = \hat{\theta} d\theta$. The x and y components of velocity are

$$v_x = \frac{dr}{dt}\cos\theta - r\frac{d\theta}{dt}\sin\theta,$$

$$v_y = \frac{dr}{dt}\sin\theta + r\frac{d\theta}{dt}\cos\theta .$$

L is defined in (60); substituting the polar expressions for x, y, v_x and v_y we find

$$L = m\,(xv_y - yv_x)$$

$$= m\left[r\frac{dr}{dt}\cos\theta\sin\theta + r^2\frac{d\theta}{dt}\cos^2\theta\right]$$

$$= -m\left[r\frac{dr}{dt}\sin\theta\cos\theta - r^2\frac{d\theta}{dt}\sin^2\theta\right]$$

$$= mr^2\frac{d\theta}{dt}(\cos^2\theta + \sin^2\theta) = mr^2\frac{d\theta}{dt}$$

The result is

$$L = mr^2\omega.$$

Comparing this we see that dA/dt is $L/2m$. But L is a constant of the motion by conservation of angular momentum. Thus dA/dt is constant. In words, the rate of change of the area is constant, i.e., independent of position on the orbit. Hence Kepler's second law is explained: The area increases at a constant rate, so equal areas are swept out in equal times.

The Inverse Square Law

Kepler's first law is that the planets travel on ellipses with the sun at one focal point. We will prove that this observation implies that the force on the planet must be an inverse square law, i.e., proportional to $1/r^2$ where r is the distance from the sun. The calculations depend on all that we have learned about derivatives and differentiation. The equation for an elliptical orbit in polar coordinates (r, θ) is

$$(r, \theta) = \frac{a(1-e^2)}{1+e\cos\theta}$$

where a = semimajor axis and e = eccentricity. Figures show graphs of the ellipse. What force is implied by the orbit equation ? The radial acceleration is

$$a_r = \frac{d^2r}{dt^2} - r\left(\frac{d\theta}{dt}\right)^2.$$

The first term involves the change of radius; the second term is the centripetal acceleration $-r\omega^2$. Now, a_r must equal F_r =m by Newton's second law. To determine the radial force F_r we must express a_r as a function of r. We know that angular momentum is constant; by,

$$mr^2\frac{d\theta}{dt} = L, \quad \text{so} \quad \frac{d\theta}{dt} = \frac{L}{mr^2}.$$

Now starting from, and applying the chain rule,

$$\frac{dr}{dt} = \frac{dr}{d\theta}\frac{d\theta}{dt} = \frac{-a(1-e^2)}{(1+e\cos\theta)^2}(-e\sin\theta)\frac{d\theta}{dt}$$

$$= \frac{a(1-e^2)e\sin\theta}{(1+e\cos\theta)^2}\frac{L(1+e\cos\theta)^2}{ma^2(1+e^2)^2} = \frac{Le\sin\theta}{ma(1-e^2)};$$

and, taking another derivative,

$$\frac{d^2r}{dt^2} = \frac{Le\cos\theta}{ma(1-e^2)}\frac{d\theta}{dt} = \frac{Le\cos\theta}{ma(1-e^2)}\frac{L}{mr^2}.$$

Combining these results in, the radial component of the acceleration is

$$a_r = \frac{L^2 e\cos\theta}{m^2 a(1-e^2)r^2} - r\left(\frac{L}{mr^2}\right)^2$$

$$= \frac{L^2}{m^2 r^2}\left\{\frac{e\cos\theta}{a(1-e^2)} - \frac{1+e\cos\theta}{a(1-e^2)}\right\} = \frac{-L^2}{m^2 a(1-e^2)r^2}.$$

By Newton's second law, then, the radial force must be

$$F_r = ma_r = -\frac{k}{r^2} \quad \text{where} \quad k = \frac{L^2}{ma(1-e^2)}$$

Our result is that the force on the planet must be an attactive inverse-square-law, $F_r = -k/r^2$. The orbit parameters are related to the force parameter k by

$$L^2 = ma(1 - e^2)k.$$

Newton's Theory of Universal Gravitation

From the fact that planetary orbits are elliptical, Newton deduced that $F_r = -k/r^2$. Also, k must be proportional to the planet's mass m because $T^2 \propto a^3$, independent of the mass (cf. Section 9.5.1).

But then k must also be proportional to the solar mass, because for every action there is an equal but opposite reaction. Therefore the force vector must be

$$F = F_r\hat{r} = -\frac{GMm}{r^2}\hat{r}$$

where G is a constant. Newton's theory of universal gravitation is that any two masses in the universe, m and M, attract each other according to the force.

Newton's gravitational constant G cannot be determined by astronomical observations, because the solar mass M is not known independently.

G must be measured in the laboratory. An accurate measurement of G is very difficult, and was not accomplished in the time of Newton. The first measurement of G was by Henry Cavendish in 1798. G is hard to measure because gravity is extremely weak,

$$G = 6.67 \times 10^{-11}\ \text{m}^3\ \text{s}^{-2}\ \text{kg}^{-1}.$$

Newton's theory of gravity is very accurate, but not exact. A more accurate theory of gravity|the theory of general relativity|was developed by Einstein.

In relativity, planetary orbits are not perfect ellipses; the orbits precess very slowly. Indeed this precession is observed in precise measurements of planetary positions, and the measurements agree with the relativistic calculation.

The examples show how calculus is used to understand profound physical observations such as the motion of the planets.

That is why calculus is so important in the education of scientists and engineers.

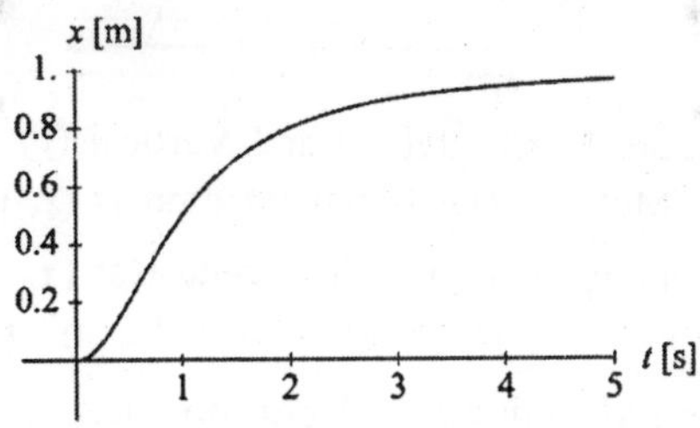

Fig. An Example of Position x as a Function of time t for an Object Moving in one Dimension

The object starts at rest at the origin at $t = 0$, begins moving to positive x, has positive acceleration for about 1 second, and then gradually slows to a stop at a distance of 1 m from the origin.

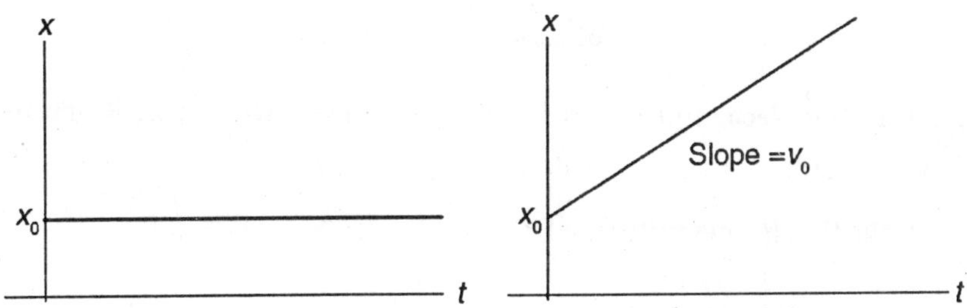

Fig. Motion of an Object: (a) Zero Velocity and (b) Constant Positive Velocity

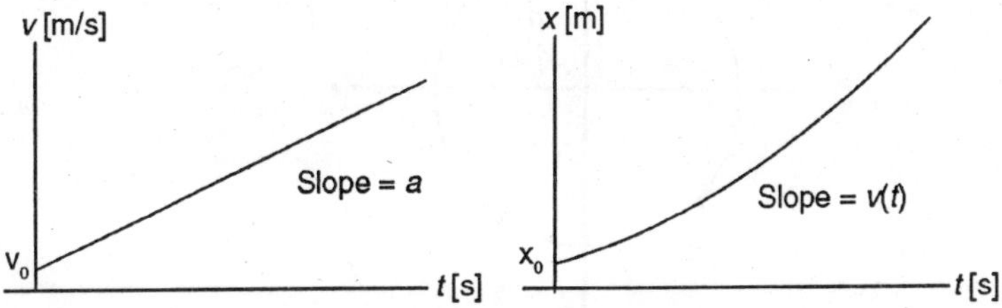

Fig. Constant acceleration. The two graphs are (a) velocity $v(t)$ and (b) position $x(t)$ as functions of time, for an object with constant acceleration a

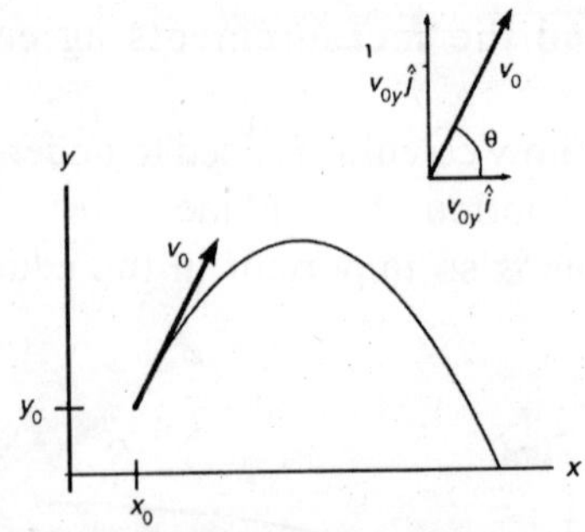

Fig. Projectile motion. Horizontal (x) and Vertical (y) Axes are set up to Analyse the Motion. The Initial Position is $(x, y) = (x_0, y_0)$

The initial velocity v_0 is shown as a vector at $(x_{0,} y_0)$. The curve is the trajectory of the projectile. The inset shows the initial velocity vector v_0 separated into horizontal and vertical components, $v_{0x}\hat{i}$. and $v_{0y}\hat{j}$; θ is the angle of elevation of v_0. $\hat{i}$ (= unit horizontal vector, $\hat{j}$ = unit vertical vector)

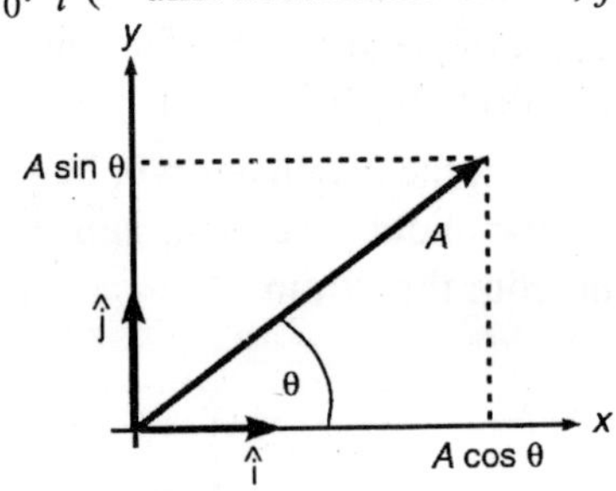

Fig. The Unit Vectors bi and bj Point in the x and y Directions, Respectively

Any vector A may be written as $A_x\hat{i} + A_y\hat{j}$ where $A_x = A \cos\theta$ and $A_y = A \sin\theta$. The magnitude of the vector is $\underline{A} = |A| = \sqrt{A_x^2 + A_y^2}$.

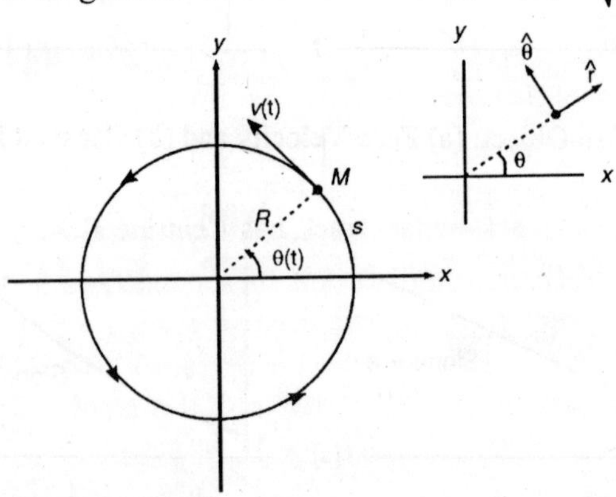

Fig. Circular motion. A mass M moves on a circle of radius R. The Angle $\theta(t)$ is used to specify the Position. In Radians, $\theta = s/R$ where s is the Arclength.

The velocity vector $v(t)$ is tangent to the circle. The inset shows the unit vectors $\hat{\theta}$ and $\hat{r}$, which point in the direction of increasing θ and r, respectively.

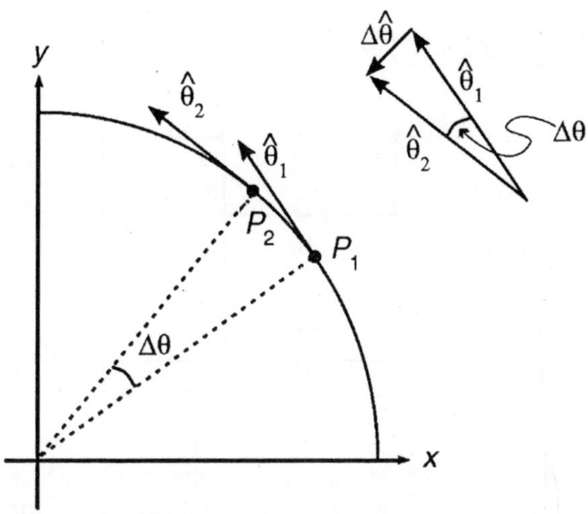

Fig. Proof that $d\hat{\theta} = -\hat{r}d\theta$. P_2 are Points on the Circle with Angle Difference Δθ.

The inset shows that $\Delta\hat{\theta}(=\hat{\theta}_2 - \hat{\theta}_1)$ is Centripetal (i.e., in the Direction of $-\hat{r}$) and has Magnitude Δθ in the Limit of Small Δθ.

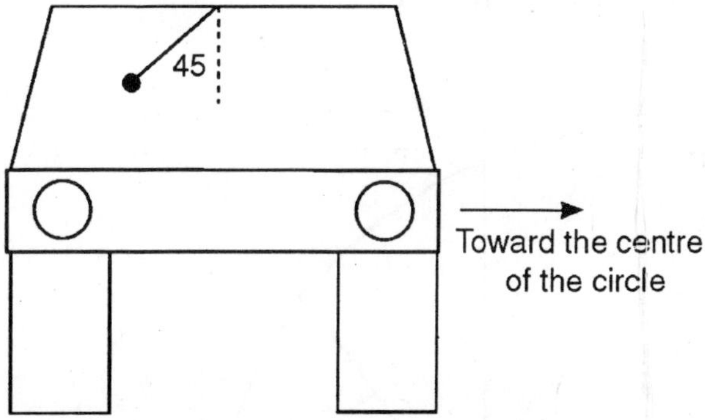

Fig. A Race car on a Circular Track has Centripetal Acceleration v^2/R.

If $v^2/R = g$, then the equilibrium of a pendulum suspended from the ceiling is at 45 degrees to the vertical.

In the frame of reference of the track, the bob accelerates centripetally because it is pulled toward the centre by the pendulum string.

In the frame of reference of the car there is a centrifugal force|an apparent (but fictitious) force directed away from the centre of the track.

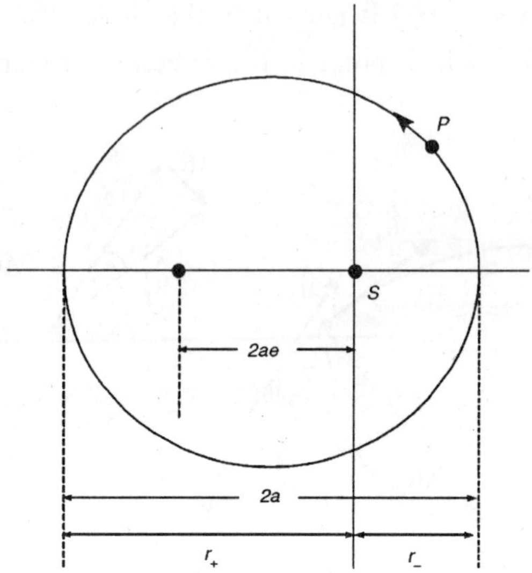

Fig. A Possible Planetary orbit. The sun S is at the Origin, which is one Focal Point of the Ellipse, and the Planet P Moves on the Ellipse

The large diameter is 2a, where a is called the semimajor axis. The distance between the foci is $2a$ where e is called the eccentricity. The perihelion distance is $r_- = a(1 - e)$ and the aphelion distance is $r_+ = a(1 + e)$. A circle is an ellipse with $e = 0$.

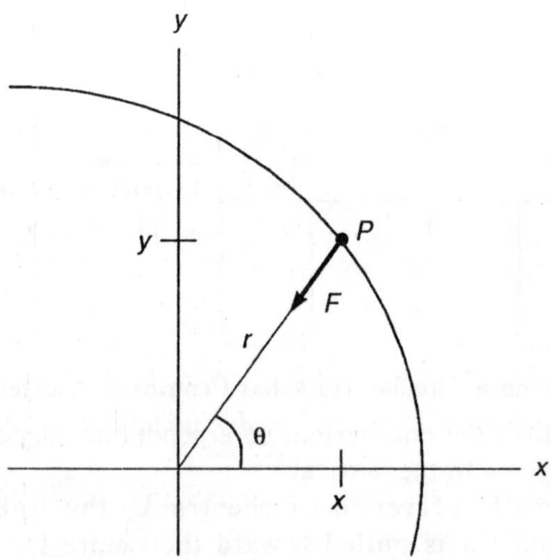

Fig. An Attractive Central Force. The Cartesian Coordinates at P are x = r cos θ and y = r sin θ. The Force Components are $F_x = -F\cos\theta$ and Fy = $-F\sin\theta$ where F is the Magnitude of the Force Vector. The Torque, $xF_y - yF_x$, is 0.

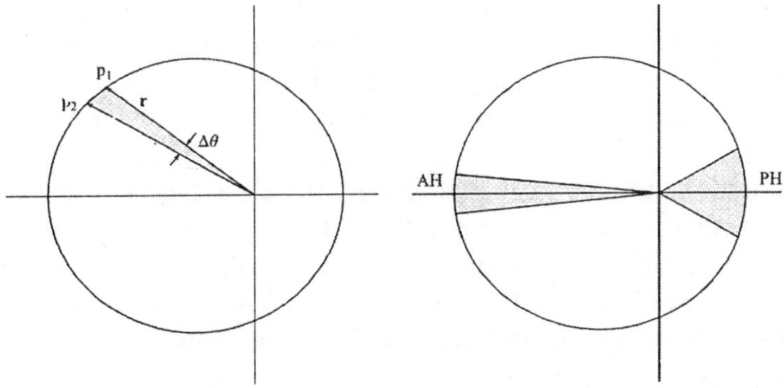

Fig. Kepler's Second Law. The Radial Vector Sweeps out Equal Areas in Equal times. (a) The Radial Vector Sweeps out the Shaded Region as the Planet Moves from P_1 to P_2. (b) The Planet moves Faster Near Perihelion (PH) and Slower Near Aphelion (AH)

CELESTIAL MECHANICS

Johannes Kepler (1571-1630) discovered that the planets moved in elliptical orbits, and his three laws permit the calculation of planetary position once the orbit is known. Although correct, the loss of ideal circular motion as a fundamental (though impossible to apply) concept was a disappointment. This was changed into beautiful triumph in 1686 by Isaac Newton (1642-1727), who showed that the observed orbital motion was a consequence of fundamental principles of universal application.

Newton's formulation of mechanics, which involved the new concepts of mass and force, was subjected to intense, but sterile, criticism by Ernst Mach (1838-1916) and others, which did not change its application to the slightest degree, and shed no light on its fundamentals. The ideas of Albert Einstein (1879-1955), on the other hand, modified the fundamentals essentially, especially in the theory of gravity and the kinematics of motion, making the theory able to explain the smallest discrepancies with the earlier predictions.

This *theory of relativity* has also been abundantly proved by observation. Except for Newton and Einstein, all the vast amount of work in mechanics has been the elaboration and application of the theory, which has proved in every respect correct. Particularly notable is the work of Joseph Louis Lagrange (1736-1813) in The Beautiful Theory, where forces are replaced by work and energy, and the laws of mechanics can be expressed as problems in maxima and minima. This branch of mechanics is usually called *analytical mechanics*, and mechanics with forces is called *vector mechanics*, from the convenient use of that formalism.

Newton's theory of motion, in its original form, is still the basis of engineering mechanics, and a fundamental part of engineering training. Indeed, it is still used in celestial mechanics. The only aspect needing relativistic corrections so far has been the time effects in the Global Positioning System. In the solar system, velocities are very comfortably less than the speed of light, 3×10^8 m/s, although distances are great. Light requires about 8 minutes to pass from the sun to the earth, and gravity travels at the same speed.

However, the sun does not care a great deal about the force exerted on it by the earth, while the earth moves in a static gravitational field, so the effects of the finite speed of gravity are not evident. These effects have, however, recently been measured and verified, so it is known that gravity is not instantaneous.

We will not need to consider any relativistic effects in the sequel, leaving these tiny effects to another article.

NEWTON'S LAWS

Newton's laws of motion are:

- A body under no forces moves uniformly in a straight line;
- The acceleration of a body is equal to the force acting on it divided by its mass; and
- The mutual forces acting between two bodies are equal and opposite.

Newton stated the laws in this form so that these *axioms* would use only common notions and not involve any results following from them, in the spirit of Euclidean geometry.

The first law gives an operational method for determining an *inertial system*, a way of describing the position of a mass point in space as a function of time.

In the absence of gravitating masses (such as the earth) it is found that a Euclidean space is such a system, in which a uniform time can be approximated by periodic events (such as the oscillations in lasers). Strictly (and uselessly) speaking, the definition is circular, but in practice it is clearly possible to establish inertial systems at least approximately, and the results obtained agree perfectly with observation.

The second law defines force as the cause of a time rate of change of velocity, by dv/dt = f/m. Again, the definition is logically faulty, but can be realized with great accuracy in practice. Whatever causes a change in velocity in an inertial system is a force. Masses can be compared dynamically, or more easily by the gravitational forces on them. A consistent array of masses and forces can be assembled, which always gives the observed motion. When the velocity is known, the position can be found from the kinematical relation dr/dt = v. Relations not involving force we call *kinematical*, those involving force, *dynamical*.

The third law guarantees that the total linear momentum and total angular momentum will be conserved; that is, that they will remain constant during the motion.

These quantities are defined later, and their conservation is the motive for the third law, which cannot be stated in a general form in other than an extremely tedious and unilluminating way. All forces obey it (magnetic forces between currents are an example), but nevertheless momenta are conserved.

The conservation of momentum is an important fundamental of mechanics; the third law is a way of introducing it in terms of forces. In celestial mechanics, all the forces are central (act along the line joining two interacting masses) and so the simple statement is sufficient.

The path traced by the tip of the velocity vector v as time elapses is called the *hodograph*, and the path traced by the tip of the position vector r is called the *trajectory*. An orbit is a special kind of trajectory, a closed (or almost closed) curve. These curves are useful in describing, visualizing and analyzing the motion.

The *momentum* is the product p = mv. The second law can be written dp/dt = f, so that force is the rate of change of momentum. If two particles m and m' interact, then f' = -f from the third law, so that (d/dt)(p + p') = f + f' = 0, or dP = 0, where P = p + p' is the total momentum. The two particles, taken as a whole, constitute an isolated system, on which no net force acts, so the rate of change of momentum is zero, and thus the momentum is constant.

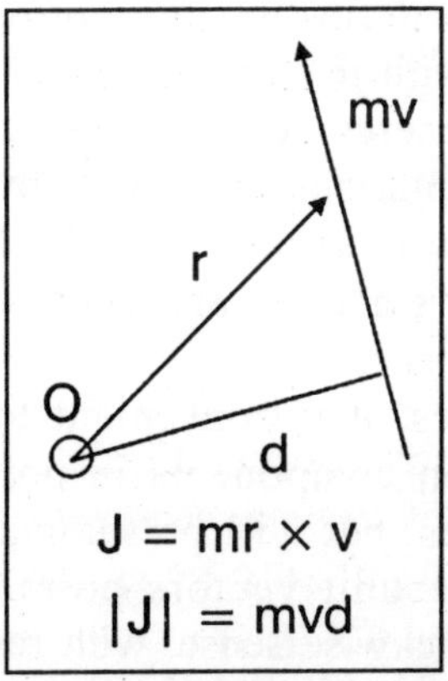

Fig. Angular Momentum

The *angular momentum* with respect to a point O is the vector $J = r \times p$, where r is the radius vector from O to any point on the line of action of $p = mv$. In the figure, J points out of the page. If we use $r \times p$ instead of p in the equations of the preceding paragraph, we can prove that the total angular momentum is conserved. The conservation of linear and angular momentum gives us six scalar constants of the motion, useful in analyzing and understanding the motion. The existence of these constants is no trivial matter, but a beautiful and fundamental result of the theory.

The second law can be formally integrated to give p = ∫ fdt = I, where I is a vector called the *impulse*, the integral of the force with respect to time. By taking moments about point O, we obtain a similar expression for the angular impulse. Of course, the limits on the integrals must be handled properly.

If the second law is scalar-multiplied by v, we find mv·(dv/dt) = f·v, or (d/dt)(mv^2/2) = f·v = -dW/dt.

This introduces two new scalar quantities, the *kinetic energy* T = mv^2/2, and the *work* -W = ∫ f·dr. The minus sign, is simply means that work is being done *on* the particle, while W is conventionally the work done *by* the particle. If W is a function of position, then in vector notation f = – grad W.

In one dimension, f = – dW/dx, which may be more familiar. What we have shown is that (d/dt)(T + W) = 0, or T + W = constant, a new conservation law. If W is a *conservative* quantity (that is, its value is independent of path and depends only on position), it is called the *potential energy* V.

Then, the *total energy* E = T + V is conserved in the motion. Energy is a widely used and misused quantity that could be discussed in great detail, but only its bare definition is required here, and the reader is assumed to be familiar with it.

The introduction of energy permits the use of *generalized coordinates*, and the derivation of equations of motion by the Lagrange procedure, which uses the Lagrangian function L = T– V. This facilitates the solutions of very many problems, since we are liberated from carrying around a basket of vector components.

For vector mechanics, it is convenient to have expressions for the velocity and acceleration components in polar coordinates. These are derived in the References, but will be summarized here. v = (dr/dt)r' + (rdθ/dt)θ'. r' and θ' are unit vectors pointing radially outward, and tangentially in the anticlockwise sense, with respect to the position vector r. The rectangular unit vectors i and j are also shown. Note that dr'/dθ = θ' and dθ'/dθ = –r', while the derivatives with respect to r are zero. dv/dt = [d^2r/dt^2 – r$(d\theta/dt)^2$]r' + (1/r)(d/dt)(r^2dθ/dt)θ'.

The reader should carefully note the distinction between, for example, d^2r/dt^2 and d^2r/dt^2.

The first is the second derivative of a scalar quantity, a simple thing.

The second is a vector, and since the direction of r can change, the complete expression is somewhat elaborate, as we have just seen. It is essential to distinguish carefully between scalars and vectors, which are arrays of scalars.

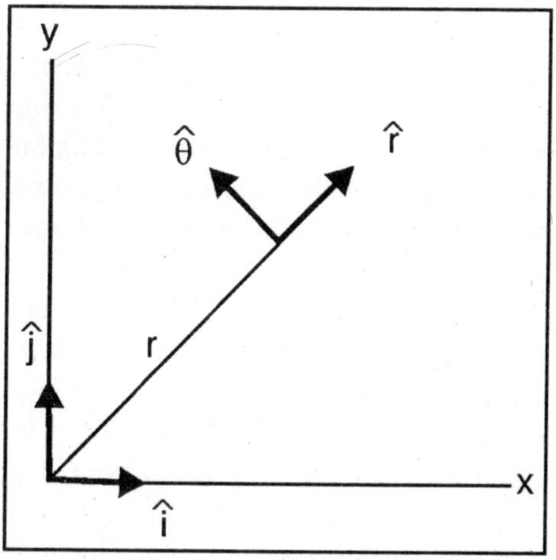

Fig. Unit Vectors

These relations are very useful. For example, suppose that the force is central, so that the tangential component of the acceleration is zero. This means that $(d/dt)(r^2 d\theta/dt) = 0$, or $r^2 d\theta/dt$ = constant = h. Now, half of r times r $d\theta$ is the area dA swept out by the radius vector in dt, so h is 2dA/dt. Also, $mrd\theta/dt$ is the component of the momentum perpendicular to the radius vector, so $mr^2 d\theta/dt = mh$ is the angular momentum.

We then have that the angular momentum L = mh is a constant. Also, this means that the motion takes place in a plane, and L is perpendicular to this plane. We have found out a lot here with very little effort. The constancy of areal velocity is one of the most useful properties of orbital motion, not just a curiosity. It relates the rate of change of the angle θ to the radius r.

KEPLERIAN ORBITS

Kepler's three laws of planetary motion, deduced by prolonged and tedious consideration of the observed position of Mars, are:

- The planets move in ellipses with the sun at one focus;
- The areas swept out by the radius vector in equal time intervals are equal; and
- The cubes of the mean distances (half the major axis of the orbit) are proportional to the squares of the periodic times. These laws are sufficient to determine the position of a planet at any later

time if its position is known at one time, and the dimensions and orientation of the orbit are known.

For a deeper understanding, and the power to attack an arbitrary problem in orbital motion, such as the movement of earth satellites, we should consider the dynamics, on the basis of Newton's theory.

The force acting between two spherical, radially symmetric bodies of masses M and m a distance r apart is, by Newtonian gravitation, GMm/r^2 in magnitude, and is directed along the line joining the centres of the two bodies. G is the Newtonian gravitational constant, 6.67259×10^{-11} $m^3/kg\text{-}s$. Since the force is central, the angular momentum is conserved, and the bodies revolve about one another in a plane with the centre of mass (the barycentre) fixed.

The motion can be analyzed as the rotation of a *reduced mass* $Mm/(M + m)$ about the centre of gravity. The potential energy V is $-GMm/r$, and is negative because the particles attract. For the total energy to be zero, $T = mv^2/2 = GMm/r$, or $v = \sqrt{(2GM/r)}$.

This is the *escape velocity* from the particle of mass M at the distance r. At the surface of the earth, the escape velocity is 11,200 m/s or 25,200 mph. This is also the velocity of a meteor that comes in from infinity. The escape velocity from the *sun* at the distance of the earth is 42,000 m/s or about 95,000 mph. It is not easy to get away from the earth; it is even harder to escape from the sun.

If we consider a circular orbit of radius a, then $GMm/a^2 = [Mm/(M + m)]a(d\theta/dt)^2$, since $v = a(d\theta/dt)$ and the centripetal acceleration is v^2/a. This is the equation of motion for the relative motion, with r the distance between M and m.

The centre of gravity of M and m moves with constant velocity, not affected by the relative motion. Thus, $G(M + m) = a^3(d\theta/dt)^2 = 4\pi^2(a^3/P^2)$, where P is the period of the motion.

This is a derivation of Kepler's third law for a special case, but it is a general result when a is the mean distance of any orbit. The quantity $G(M + m)$ is usually called μ, so that $r^3/P^2 = \mu/4\pi^2$. If r is in astronomical units (radius of the earth's orbit, about 150×10^6 km), and the period P is in years, then $G(M + m) = 4\pi^2 = \mu$.

The quantity GM is called the Gaussian gravitational constant k^2 if time is in days instead of years, so we get approximately $k^2 = 4\pi^2/(365.25)^2$, and $k = 0.017202$.

A more precise value is 0.01720209895. Untangling G from GM means knowing the mass of the sun accurately, which is impossible, so celestial mechanics uses k instead of G. The value should be corrected for the finite mass of the earth, but since the ratio of the sun's to the earth's mass is 332,946 the correction is small.

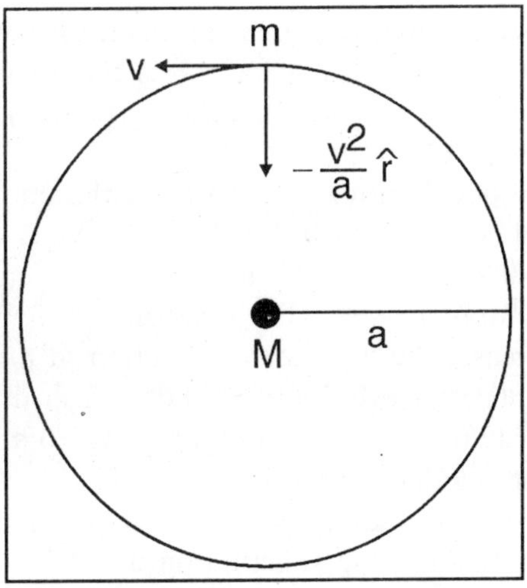

Fig. Circular Orbit

The polar equation of an ellipse, with the origin in a focus, is $r = p/(1 + e \cos \theta)$. The *eccentricity* of the ellipse is e, and p is a length called the semi-latus-rectum. In connection with orbits, the angle θ is called the *true anomaly*. In an elliptical orbit, 0 d" e < 1. When e = 0, we have a circular orbit r = p. The same formula gives a parabola for e = 1, and a hyperbola for e > 1. In an ellipse, the maximum and minimum radii are p/(1 – e) and p/(1 + e), respectively, and the corresponding points on the orbit are called *aphelion* and *perihelion*, respectively. The distance from perihelion to aphelion is the major axis of the ellipse, twice the mean distance a.

Therefore, $2a = p/(1 - e) + p/(1 + e) = 2p/(1 - e^2)$, or $p = a(1 - e^2)$. The perihelion distance is a(1 – e), the aphelion distance a(1 + e). The distance from the centre to a focus is a – a(1 – e) = ea = c. This defines the eccentricity as e = c/a. Since the ellipse is the locus of points the sum of whose distances to the two foci is a constant, 2a, the length of the radius from a focus to the end of the minor axis is a. Since this is the hypotenuse of a right triangle, $a^2 = c^2 + b^2$, where b is the semiminor axis. Hence $b = a\sqrt{(1 - e^2)}$. The reader can show that the average value of r over one revolution is a. This is about all we need to know about the geometry of the ellipse for present purposes.

If we know the orbit is an ellipse, and the areal velocity is constant, we should be able to prove that the force is central and varies as the inverse square of the distance. On the other hand, if we know that the force is central and inverse-square, then we should be able to prove that the orbit is an ellipse and the areal velocity is constant. Either of these things can be done fairly easily.

In the first case, differentiate the expression for r with respect to time, obtaining $dr/dt = (e/p) \sin\theta\ 2A$, where A is the areal velocity (this was called dA/dt above; now A represents the derivative). Then differentiate again to find $d^2r/dt^2 = (4A^2e/pr^2) \cos\theta$.

The tangential acceleration is zero from the constancy of the areal velocity. The radial acceleration is $a_r = (4A^2e/pr^2) \cos\theta - 4A^2/r^3 = -4A^2/pr^2$, which shows that the force is inverse square, and we have already shown that it is central. In fact, $GM = 4A^2/p$.

In the second case, the equations of motion in the plane of the orbit are $d^2r/dt^2 - r(d\theta/dt)^2 = -GM/r^2$ and $rd^2\theta/dt + 2(dr/dt)(d\theta/dt) = 0$.

Now, $(d/dt)(r^2d\theta/dt) = 2r(dr/dt)(d\theta/dt) + r^2d^2\theta/dt^2 = 0$, so the areal velocity $A = (1/2)r^2(d\theta/dt) = \text{constant}$.

Using this result, the radial equation becomes $d^2r/d\theta^2 - (2/r)(dr/d\theta)^2 - r = -r^2GM/4A^2$. Making the substitution $u = 1/r$, this equation can be thrown into the form $d^2u/d\theta^2 + u = GM/4A^2$. This second-order linear equation with constant coefficients is easy to solve. We find $u = 1/r = GM/4A^2 + C \cos\theta$.

This is the equation of a conic section, as we have seen above. If $p = 4A^2/GM$ and $e = pC$, we have $r = p/(1 + e \cos\theta)$, which is what we wanted to prove.

Now we can find the components of the velocity. The transverse component is $rd\theta/dt = 2A/r = 2A(1 + e \cos\theta)/p$, where A is the areal velocity. The radial component is $dr/dt = 2A(e/p)\sin\theta$.

The square of the velocity is the sum of the squares, or $v^2 = 4A^2[(1 + e^2 + 2e \cos\theta]/p^2] = 4A^2[(1 - e^2)/p^2 + 2/pr] = -4A^2/pa + 4A^2/pr$, from which the kinetic energy per unit mass (i.e., setting m = 1 for simplicity) is: $v^2/2 = -GM/2a + GM/r$, since $4A^2/p = GM$.

Rearranging, $v^2/2 - GM/r = -GM/2a$, which is to say, $T + V = E$, with $T = v^2/2$, $V = -GM/r$ and $E = -GM/2a$.

The total energy is negative (i.e., the orbiting particle is "trapped") and a function of the mean distance a only. The eccentricity is determined by the angular momentum h through $e^2 = 1 - h^2/aGM$. When h = "aGM, the orbit is circular.

The quantity a, called the "mean distance," is not the average value of the radius vector r. It is the average of the perihelion and aphelion distances, however. The time average value of 1/r turns out to be 1/a, which is its real definition.

This means that the time average potential energy is -GM/a, and the time average kinetic energy is -GM/2a + GM/a = +GM/2a. For a circular orbit, the kinetic and potential energies are constant.

The time average value of r is $a(1 + e^2/2)$, and the average of r over the true anomaly is b, the minor axis of the orbit. The average of 1/r over the true anomaly is 1/p. We obtained the relation between the mean distance a and the orbital period P for the special case of a circular orbit. A more general proof is as follows.

The areal velocity $dA/dt = h/2 = $ constant, so integrating from $t = 0$ to $t = P$ we find $A = hP/2 = \pi ab$. We also know that $p = a(1 - e^2) = h^2/GM$. Therefore, $P^2 = 4\pi^2a^2b^2/h^2 = [4\pi^2/GM]a^3$, since $b^2/p = a$. This is the desired general result. If we know the period and mean distance of any orbit, we can calculate the gravitational constant GM, which in the general case is G(M + m).

COORDINATES AND ORBITS

The angular coordinates that are used to describe directions in space are shown in the figure at the left. The equator and ecliptic planes are shown. The dihedral angle between these two planes is $\varepsilon = 23° 26' 21''.412$, or thereabouts.

The equatorial coordinates are the right ascension α and the declination δ, while the analogous ecliptic coordinates are the longitude λ and the latitude β. The relations between the coordinates are best found by considering the rectangular components of the vector OP, considered as of unit length, and then performing the rotation about the x-axis that takes one system into the other. We will not require these relations at the present time.

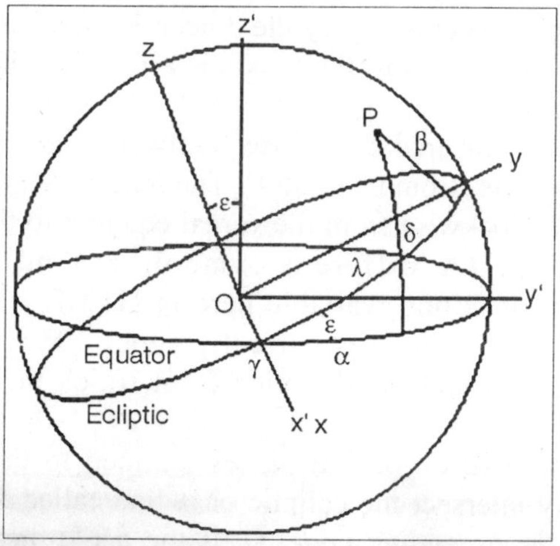

Fig. Equatorial and Ecliptic Coordinates

The position of a body in the solar system can be specified in terms of a gigantic rectangular coordinate system centered on the sun. The x-axis is always taken as pointing to the vernal equinox, a direction specified by the line of intersection of the orbital plane of the earth (the ecliptic plane) with a plane parallel to the equator of the earth, the equatorial plane. Both contain the centre of the sun. This is made interesting by the precession of the normal to the equatorial plane around the pole of the ecliptic. In the diagram, this is the movement of z' (the pole of the equator) around z (the pole of the ecliptic). Fortunately, this is not too rapid, taking about 26,500 years for one revolution.

The equatorial plane has to be considered because our earth-based measurements use it as a reference, when directions are given in terms of right ascension and declination.

If we know a vector whose components are referred to this plane, we can find the right ascension and declination of its direction. When the earth is located in its orbital plane on the line of intersection of the ecliptic and equatorial planes, the sun is seen in front of the vernal equinox about March 21, climbing from below the equator to above it. At this time, when we look at the sun (careful!) we are looking in the direction of the x-axis of the heliocentric coordinate system.

The y-axis of the ecliptic coordinates is in the plane of the ecliptic, at right angles to the x-axis, and in the direction in which the planets move, in which a screw would advance in the direction of the north pole of the earth. The ecliptic z-axis is then perpendicular to the x- and y-axes, and makes a right-handed coordinate system with them.

The equatorial rectangular coordinates are defined analogously. If we represent these coordinates by primes, x and x' are the same, while x',y' is rotated with respect to x,y by the dihedral angle between the planes. The rotation from the z' axis to the z axis is a right-handed rotation about the positive x,x' axis.

Longitudes are measured clockwise (eastwards) on the ecliptic plane from the vernal equinox, from 0° to 360°. The *longitude of perihelion*, ψ, is the angle measured clockwise from the vernal equinox to the radius passing through perihelion. (A different symbol, a kind of ω, is used conventionally, but it is not available here in HTML.) The longitude of perihelion of the earth's orbit is currently about 103°. This locates the major axis of the orbit and the direction from which the true anomaly is measured.

Other planets move on orbital planes inclined to the ecliptic by the *inclination*, i. They intersect the ecliptic on a line called the *line of nodes*. The *longitude of the ascending node*, Ω, is the angle measured clockwise from the vernal equinox to the line of nodes, on the end where the orbit is climbing above the ecliptic plane, called the *ascending node*.

The node 180° away is, not surprisingly, the *descending node*. Note in the figure that the traditional symbol for the ascending node is used, which we represent in text by Ω, which looks similar.

The symbol for the descending node is inverted, with the two little loops on the top. The longitude of perihelion in this case is the longitude of the ascending node, *plus* the angle ω in the orbital plane from the node to the perihelion. This may be a little odd, but it does locate the perihelion. That is, $\psi = \Omega + \omega$.

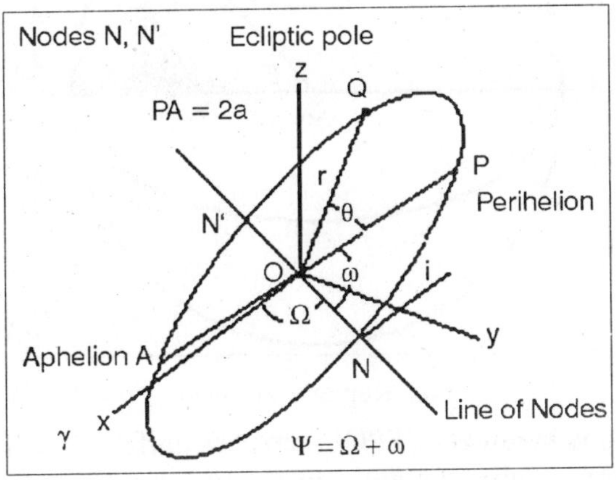

Fig. Orbital Elements

Now that the orbital plane and the direction of the perihelion have been specified by the three orbital elements i, Ω and ψ, we proceed to specify the shape of the orbit by its mean distance a and its eccentricity e. The position of the body in its orbit at a specified time is given by a sixth element, the mean longitude at the epoch.

The "epoch" is just the reference time assumed, usually by its Julian date. The mean distance a determines the orbital period P through Kepler's third law. Therefore, only one of a or P need be specified. The *mean daily motion*, 360° divided by the orbital period, is often given in place of P.

The constancy of the areal velocity is used to locate the body in its orbit at a specified time. This is called Kepler's Problem, which Kepler also solved.

A reference circle of radius a is drawn around the orbital ellipse. Point Q is projected onto the auxilary circle by a perpendicular to the major axis to point Q'.

The line CQ' then defines the angle E at the centre of the circle, called the *eccentric anomaly*. We can show that it is related to the radius vector r and the true anomaly θ by $r = a(1 - e \cos E)$ and $\tan (\theta/2) = \sqrt{[(1 + e)/(1 - e)]} \tan E/2$.

The shaded sector PFQ is the area swept out by the radius vector, and the area is proportional to the time. The area FPQ' is proportional to the area FPQ, since it is merely magnified vertically in the ratio $a/b = 1/\sqrt{(1 - e^2)}$. Indeed, the total area of the orbit is πab, while the total area of the circle is πa^2.

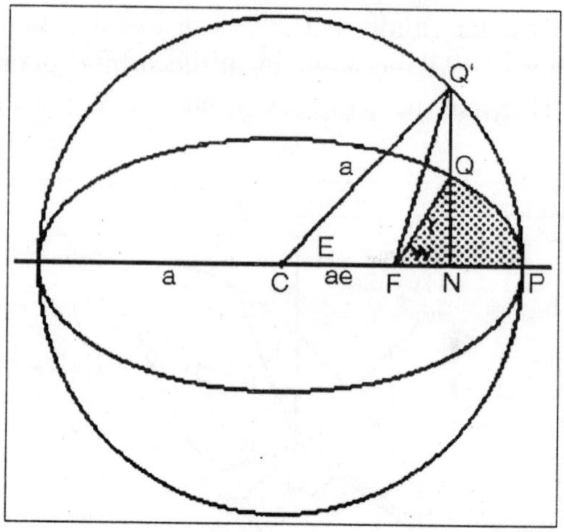

Fig. Kepler's Problem

Therefore, this larger area FPQ' increases uniformly with time as well. Since the total area is πa^2, the area swept out is given by $(\pi a^2/P)t$, where t is the time since perihelion passage. This area can also be found in terms of E, since it is the area of the whole sector PCQ' less the area of the triangle CQ'F.

The area of the sector is $\pi a^2(E/2\pi) = (a^2/2)E$. The area of the triangle is $(1/2)(a \sin E)(ae) = (a^2/2)e \sin E$. Therefore, we have $E - e \sin E = 2\pi t/P = M$, called Kepler's Equation. M is called the *mean anomaly* and increases proportionally to time after perihelion passage. This is really a beautiful result, allowing us to find the position in orbit in terms of a uniformly increasing quantity.

Kepler's Equation can be easily solved by iteration. If it is written $E = M + e \sin E$, we find the iteration equation $E_{n+1} = M + e \sin E_n$ (e is the eccentricity here, not the base of natural logarithms). Starting with $E_0 = M$, a few interations are enough to get good precision, especially when e is small.

The iterations are very easy, and can be performed on a pocket calculator. For e = 0.1, six iterations give 8 digits. The Newton-Raphson method can also be used. This method finds the roots of the equation f(x) = 0 by improving an initial guess x_0 by the formula $x_{n+1} = x_0 - f(x_n)/f'(x_n)$. A sketch will reveal the reasoning behind this formula. In this case, $f(E) = M - E + e \sin E = 0$ is the function, and $f'(E) = e \cos E - 1$. Fewer iterations are required, but each requires more work, than in the case of simple iteration. In practice, an expansion in powers of e is often used for small e, but the iteration methods are no more trouble, especially for a computer.

Although the determination of position in orbit may sound complicated, it is really quite straightforward, and after you have done it a few times, it will be easy. Let's find the position of the earth on 2003 April 2. The mean longitude of the earth was 100.2440°, and the longitude of perihelion was 103.0°. April 2 is 91 days later, so M = 100.2440 – 103.0 + 91 × 0.9856 = 86.93° = 1.5172 radians. A first approximation to E is M + e sin M, and e = 0.0167, so the correction is 0.01668 radians, and E = 1.5339 radians or 87.89°. There is no need to iterate again.

The mean distance can be taken as 1.00000, so r = 0.9994 and θ = 88.84°. The earth's longitude is 103.0 + 88.84 = 191.84°. On p. C8, the solar ephemeris gives longitude 11.8506° and r = 0.9994. The solar longitude is 180° less than the earth's, or 11.84° according to our calculation. In both cases, we are using the mean ecliptic of date, and came quite close using Keplerian orbital elements instead of the more elaborate calculations used to calculate the solar ephemeris. Precise orbits require the consideration of planetary perturbations, and this is a very difficult subject indeed.

The orbit of the moon is an example, since it is subject to perturbations by the earth's spheroidal shape and the sun that make its orbit vary in a very complicated way.

In the case of the earth's orbit, we really calculate the position of the earth-moon barycentre. This point is 4671 km from the centre of the earth, about two-thirds of the way to the surface.

The mean distance of the moon is 3.844 × 10^5 km, and its sidereal period is 27.321661 days, from which G(M + m) can be calculated and used to relate mean distance and period for other earth satellites after correcting for the mass of the moon.

The moon's radius is 1738 km, and its mass 7.3483 × 10^{22} kg. m/M = 0.0123, which is not quite negligible. The eccentricity of the moon's orbit is about 0.0549, and its inclination 5.145396°. The plane of the orbit and the position of perigee change rather rapidly. The small eccentricity of the earth's orbit suggests that the earth-moon system did not suffer any external disturbance during its formation, and in particular any collision which created the moon. This was a one-time event, so speculations on what happened can never be proved, since no similar state is known, or is ever likely to become known.

Except for Mercury and Pluto, orbital inclinations are less than 4°, and eccentricities less than 0.0936, the value for Mars. Kepler was fortunate to have picked Mars for study because of its relatively large eccentricity, about twice those of Jupiter and Saturn. Venus's orbit is practically circular, with e = 0.0067.

Mercury not only has a large inclination, 7.005°, but an eccentric orbit, with e = 0.2056. Only Pluto has a greater inclination and eccentricity, but it is probably a special case. Observations of Mercury and Pluto were not available to Kepler, who had to work with Tycho's naked-eye observations.

The angle made by r with the x-axis in our problem was 191.84°, so the coordinates will be x = 0.9994 cos 191.84° = -0.97814, and y = 0.9994 x sin 191.84° = -0.20506, in astronomical units. In the more general case of a planet in an inclined orbital plane, the radius is first projected on the line of nodes [$a = r\cos(\omega + \theta)$] and on a perpendicular (dip) line [$b = r\sin(\omega + \theta)$]. Then $z = b \sin i$, $y = b\cos i \cos\Omega + a\sin\Omega$, $x = a\cos\Omega - b\cos i\sin\Omega$. It is just a matter of projecting r on the coordinate axes.

The necessary formulas are given on page E4 of the almanac. There is an expansion for the true anomaly directly in terms of the mean anomaly, without going through the eccentric anomaly.

To find the equatorial rectangular components, rotate about the x-axis through an angle of i. This gives $y' = y\cos i - z\sin i$ and $z' = y\sin i + z\cos i$. If x', y' and z' are the components of a vector in equatorial rectangular components, then the right ascension α is given by $\cos\alpha = x'/$"$(x'^2 + y'^2)$, and the declination δ by $\sin\delta = z'/$"$(x'^2 + y'^2 + z'^2)$. The proper quadrants have to be determined. In this way, the direction in which a planet is seen from the earth can be found.

COMETARY ORBITS

Comets can be divided into two classes, the *short-period* comets, like Halley's Comet, with a period of 76 years, and *long-period* comets, like Hale-Bopp (1995 O1), which do not return for thousands of years, if at all. A period of 200 years is the conventional dividing-line. Short-period comets have elliptical orbits like the planets, except that the eccentricity is larger, the inclination can take any value, and the comets can move in a *direct* or *retrograde* direction. Retrograde motion is usually expressed by an inclination greater than 90°.

Comets were the first real test for Newton's theory, which finally showed that they were normal members of the solar system, not mysterious atmospheric happenings, as had always been supposed. Positions of short-period comets are calculated in the same way as planetary positions, and the orbital elements are presented the same way.

Long-period comets come from the periphery of the solar system, where they wander in the hypothetical Oort Cloud of cometary debris, normally water and carbon dioxide ice and dust. Their total energy is about zero, so the eccentricity of their orbits when they make an excursion towards the sun is about 1.

That is, their orbits are parabolic. Short-period comets are those that have suffered an energy-losing collison with a planet (usually Jupiter) and have dropped into an elliptical orbit. There must also have been comets that have gained energy, entered hyperbolic orbits, and were ejected from the solar system. Few, if any, comets have eccentricities significantly greater than 1, which would indicate that they were encountered by the solar system in its path through space, and are not part of the family.

The parabola is a considerably simpler orbit than the ellipse. Its polar formula is $r = p/(1 + \cos\theta) = (p/2)\sec^2(\theta/2)$. The areal velocity $A = \sqrt{(GMp/4)}$. Setting $x = \tan(\theta/2)$, direct integration gives Kepler's Equation as $x + x^3/3 = \sqrt{(GM/2)}(2/p)^{3/2}(t - T)$.

The one real root of this equation gives the position in orbit. The time to move from perihelion to the end of the latus rectum is $t_1 = (1/2)$”(p^3/GM). $2t_1$ is a good measure of the time that the comet will spend near the sun.

The orbital elements include the inclination i, longitude of the ascending node Ω, and argument of perihelion ω. The perihelion distance p/2 is given, and the JD of perihelion passage. For Hale-Bopp, the perihelion distance was 0.91399384 and the time of perihelion passage was JD 245 0539.60742. The inclination was 89.42064850°, longitude of ascending node 282.47215310°, and argument of perihelion 130.59561740. From these elements, it was possible to find the position of the comet as seen from the earth.

Cometary orbital elements are available on the internet soon after the discovery of a comet. The Jet Propulsion Laboratory supplied the elements for Hale-Bopp shortly after it was discovered.

Halley's comet was the first periodic comet to be recognized. Edmund Halley (1656-1742), friend of Newton's and later Astronomer Royal, suspected that the comets of 1531 and 1607 were the same as the comet of 1682, since they had similar orbits.

He predicted the return of the comet in 1758, which duly occurred. Halley had published Newton's *Principia* at his own expense in 1678, and made an extensive study of comets using the new theory. Halley's comet is the brightest of the short-period comets; most are quite dim, and can even have orbits of small eccentricity like asteroids.

To show how to use the orbital elements given by JPL, let's look at them for Halley's comet. The time of perihelion passage, T_p is shown as 19860209.45895, which we interpret to mean 1986 February 9, at 0.45895 part of the day, or 11h 0m 53s UT. The "epoch" is the date to which the elements apply, which is given as 46480. This is a modified JD. The JD can be obtained by adding 24400000.5, or JD 2446480.5, which is 1986 February 18.

The size and shape of the orbit is specified by the perihelion distance q = 0.58710374, in AU, and e = 0.96727724. The eccentricity is nearly, but not quite, unity. Near perihelion, the orbit will be indistinguishable from a parabola, so the parabolic formulas can be used for predicting its position while it is near the sun and visible. The orientation of the orbital plane is specified by the longitude of the ascending node, 58.86004°, and the inclination i = 162.24220°.

When the numbers are located, look at the symbols used in the website, which may not be the usual ones because of font limitations. Since i > 90°, the motion of the comet in its orbit appears to be retrograde, opposite to the direction of motion of the planets, if the inclination is taken to be 180° – 162.24220° = 17.75780°.

Finally, the orientation of the orbit in its plane is given by the argument of perihelion, 111.8656°, measured from the ascending node in the direction of movement. It is not easy to comprehend the orbital position from the bare numbers; a drawing, made with some care, will show exactly what is going on.

As the perihelion distance is q = a(1 – e), it is a simple matter to determine a. In fact, a = q/(1 – e) = 0.58710374/0.03272276 = 17.941755 AU. From this, the orbital period P can be found. Since $P^2/a^3 = 1$ if P is in years (earth = 1) and a is in AU (earth = 1), P = 75.99716 years, very close to 76.0 years. The aphelion distance is 2(17.941755) – 0.58710374 = 35.296 AU.

This is outside the orbit of Neptune (a = 30) but not as far as Pluto (a = 39.5). In accordance with the law of areas, Halley's comet spends most of its time drifting in this dark region, periodically darting in to the sun to have a little more of itself boiled off into space.

Although orbital elements are given to high precision, the accuracy is not necessarily as high. Especially for comets, they can be changed by planetary perturbations, sometimes by large amounts. The ejection of gases from comets also leads to reaction forces that can affect the orbit. Also, the reference coordinates may change with time, because of precession of the equinoxes and other effects, and this must be taken into account in accurate calculations. These changes are not changes in the orbit, of course.

DETERMINATION OF ORBITS

Finding the orbital elements from observations, and predicting the changes in orbital elements due to perturbations, are two of the most important problems in celestial mechanics, and have received close attention from Newton's time onwards. We cannot give any reasonable account of this work here, but we can show how orbital elements come from observed motions by a graphical analysis that is very instructive, though of little practical use where high precision is required. The reader with drawing supplies is encouraged to follow along.

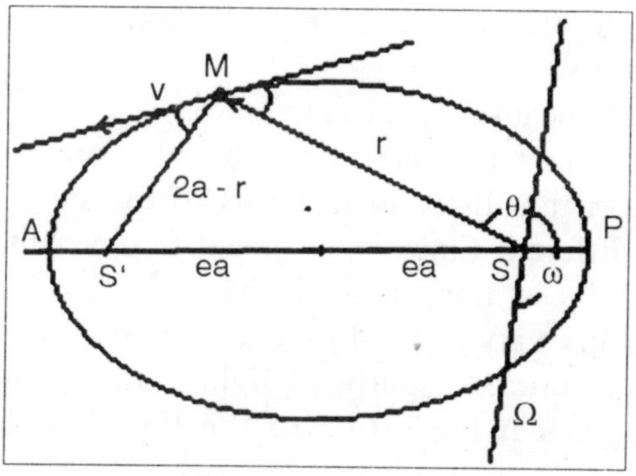

Fig. Determination of Orbital Elements

We suppose known the position r of a body M from the sun and its velocity v relative to the sun. These are six parameters that will serve to determine the six orbial elements. The orbital plane is defined by the plane of r and v, and can be represented in two views by means of orthographic projection. The line of intersection of the orbital plane with the ecliptic plane can then be found, and the dihedral angle between the two planes. Since the line of intersection will be the line of nodes, we have now found two of the orbital elements, Ω and i, which determine the plane of the orbit.

The next step is to calculate the mean distance a, using the energy relation $v^2 = GM(1/r - 1/2a)$. From a we can find the orbital period by $P^2/a^3 = 4\pi^2/GM$. The radius vector SM and the line from M to the other focus of the ellipse make equal angles with the tangent to the ellipse at M, which is in the direction of v.

In a view showing the orbital plane in true size, Lay off SM and MQ, where MQ is an arbitrary line in the direction of the empty focus. The empty focus S' is located by laying off a distance 2a – r along MQ', since the sum of the distances from the foci to M is 2a.

Now we can draw the line SS', find its centre at O, and then the locations of perihelion and aphelion. The eccentricity of the orbit is e = OS/a = OS'/a. The orbit can now be drawn, and the angle from the line of nodes to the perihelion, the argument of perihelion, ω, can be measured, as well as the true anomaly θ, which is the angle PSM. We now have the five elements Ω, i, ω, e, and a.

The remaining element is the time of perihelion passage, which can be found from the true anomaly. First, we find the eccentric anomaly by $\tan(E/2) = [(1-e)/(1+e)]^{1/2} \tan(\theta/2)$, and from it the mean anomaly $M = E - e \sin E$. The mean anomaly is $M = 2\pi(t - T)/P$, where T is the time of perihelion passage, and t is the time of observation. That is, the body passed through perihelion at a time $t - T = MP/2\pi$ earlier. We have now determined all six orbital elements from the six components of the initial position and velocity.

This was easy because the distance r was one of the given quantities. In general, orbits must be determined by observations of the angular position from earth, not directly in terms of distance and velocity. The position at three different times is sufficient to determine the orbit in this case, but the analysis is more difficult than what we have done.

Orbital elements vary only slowly in the solar system, except when there are near encounters, usually of light bodies with massive planets or satellites, that affect the orbits of the light bodies, but not of the heavier ones.

If we know the orbit at one instant, then we can predict the velocity and position at a later time from the orbit. We can also integrate the actual changes in position and velocity at the later time, taking into account forces exerted by bodies other than the sun.

The difference will be reflected in the orbital elements, which will slowly change. This is a useful way to take perturbations into account, and is widely used.

The maximum errors in using the Keplerian orbits are on the order of 5" to 30" for the inner planets, somewhat more for Jupiter and Saturn. Neptune was discovered in 1846 as a result of its perturbations of the motion of Uranus. This was a remarkable demonstration of the accuracy of Newtonian mechanics.

The earth's mean distance is decreasing by 5×10^{-8} AU per century, about 75 m per year. The eccentricity is decreasing by 3.804×10^{-5} per century. The inclination is decreasing by 46.94" per century.

These are average rates, and are simply the current values of changes that may be periodic. Nevertheless, they show how gradual any changes are, and allow the calculation of the earth's position with reasonable accuracy for thousands of years either side of the present. The longitude of the node, however, moves at -3.038" per *year* because of the precession of the equinoxes, which moves the reference point.

The argument of perihelion is increasing at 11.9828" per year. It is typical for an orbit to show a relatively rapid change in longitude of the node and motion of the perihelion; the moon is an excellent example, and also earth satellites, which are perturbed by the equatorial bulge of the earth and tidal forces exerted by the sun and moon, as well as by atmospheric drag if they are in low orbits.

MONGE'S PROCEDURE

Gaspard Monge, comte de Pıluse (1746-1818) was a mathematician in the golden age of French mathematics, and a founder of the Ecole

Polytechnique, the first modern engineering college. He is known for the science called Descriptive Geometry, which is a method of representing objects in a two-dimensional drawing in such a way that all of their geometric properties can be investigated and described. These methods were taught in classic form in every engineering course of study until computer graphics changed the approach.

The fundamentals are the basis of computer graphics and design, however. The classic methods were intended as much as training for the mind as means of calculation. In addition, they are inherently fascinating and enjoyable.

The first attempt to represent a three-dimensional object in a two-dimensional drawing is by means of some pictorial view, such as perspective, isometric, or oblique views. In any such attempt, some information is necessarily lost, and the object is only partially or ambiguously represented.

Antiquity used graphical methods almost exclusively in engineering calculations, a fact that was almost unrecognized, because of the perishability of such representations, and that the work was for practical, not literary, ends.

Here and there, survivals have given a glimpse into the extensive development of these methods, in architecture and shipbuilding. Among these methods was the use of two coordinated drawings to present three-dimensional information, as mentioned by Vitruvius.

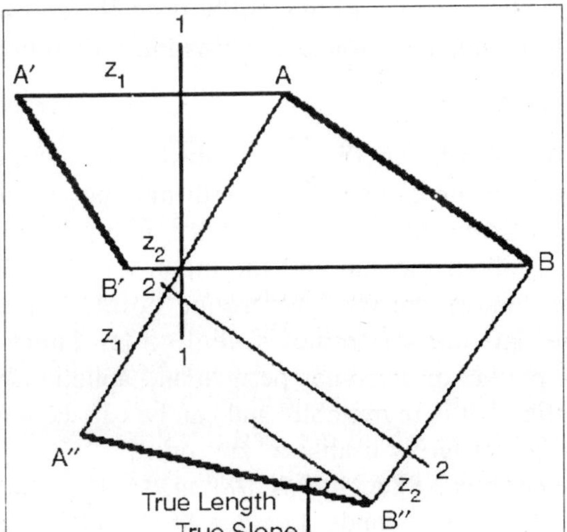

Fig. Monge's Procedure

Monge revived this science, extending it to the solution of many types of problems, and showing how it gives a complete solution of the problem.

We will show a simple application here that will give an idea of how it is carried out. AB is a *plan view* of, say, a mine shaft. The x and y coordinates of points A and B are represented in this view. Line 1-1 is called a *folding line* that separates the view showing the same points as A' and B'.

This is a *side view* showing the coordinates y and z of the same two points. Note that the horizontal *projection lines* connecting A and A', and B and B', are perpendicular to the folding lines and show that the y-coordinates are the same. These two views give all three coordinates of any point in an object, and describe it completely. Of course, any two such views will suffice; we have made an arbitrary choice here. Suppose we want to know the length and slope of the tunnel between A and B. To find them graphically, we draw a folding line 2-2 parallel to the plan vies of the line.

The direction perpendicular to this line must represent the z-coordinate. Now we draw the two projection lines AA" and BB", and measure out from folding line 2-2 exactly the same distances from folding line 1-1 to A' and B'. We have now located the projections of A and B in this view. The shaft lies in a plane parallel to this plane, so the true length and the true slope of the shaft, and can be scaled off.

We have shown the essential parts of Monge's Procedure, which are the selection of folding lines, and the measurement of equal distances in alternate views, by which any questions of size, shape, or orientation can be solved. This is the basis of orthogonal projection for engineering drawings, where three views are generally used to represent an object to make use of the redundancy for added vividness of presentation.

EARTH SATELLITES

The mass of an earth satellite is infinitesimally small compared to the mass of the earth, so the centre of the earth may be considered as a fixed point. A reference frame fixed to the earth's centre is not an inertial frame, due to revolution about the sun and the moon's motion. The earth-moon barycentre is a "more inertial" reference point, but in any case the difference from an inertial frame is negligible. There is a change in terminology, as perihelion becomes perigee and aphelion becomes apogee. This useless distinction is regrettable, and can be carried to excess. It would be good if there were terms usable in any orbit.

GM for the earth is 3.98600440 X 10^{14} m^3/s^2. Therefore, $a^3/P^2 = 10097$ if a is in km and P is in seconds. For P = 86164.0989 s (sidereal rotation period of the earth), a = 42,164,172 m, or an altitude above the surface of 22,232 miles.

A satellite with this mean distance will return to the same point in the sky each day. If the inclination and the eccentricity are zero, the satellite will appear to hang motionless in the sky. Such an orbit is called *geostationary*, obviously of great value as a communications relay location. An area of very choice space real estate is thereby created, for which there is considerable demand. Satellites occupy fixed stations around the equator.

They remain in a constant direction from an observing point on the earth's surface, so an antenna can be permanently pointed at them without having to be continually redirected, which would not be easy nor cheap.

Their positions may be adjusted slightly to keep the satellites on station by on-board thrusters during their effective life, and the thrusters are used to remove them from orbit when their fuel runs out, since space is at a premium in the geostationary belt.

From a geostationary orbit, the earth subtends an angle of 17.4°, the same as a 12"-diameter globe at a viewing distance of 40", so it is quite reasonable to take photos of the whole visible hemisphere at once. Examples are available at the Dundee University website in the References. The poles are not visible, since the view extends only to latitude 81.3°.

The first artificial earth satellite was Sputnik I, launched on 4 October 1957 and weighing 83.6 kg. Its orbital period was 96.15 min, so its mean distance was 6953 km. The eccentricity was 0.0517, giving a perigee altitude of 228 km and an apogee altitude of 947 km. The inclination of the orbit was 64.26°, so it could easily be seen in every part of the earth when the sun shone on it.

Sputnik remained in orbit until January 1958, making 1350 revolutions. Because of the low perigee, it was considerably affected by atmospheric drag, its final period being near 90 min. The period of a satellite that just grazes the surface of the earth would be 84.48 min, and its velocity 7906 m/s.

The Global Positioning System (GPS) uses satellites that continually radiate very accurate timing information, corrected for relativistic time dilatation due to the satellite's speed, so that their distances from an observer can be found to within a few centimetres from the time differences.

The orbits must be known to a similar accuracy, so the effectiveness of this system is evidence of the correctness of the dynamics. Orbit information is broadcast along with timing information. The carrier frequencies are 1.57542 and 1.22760 GHz. The complete GPS system was deployed in 1993, consisting of 21 operational satellites and three spares, on circular orbits inclined at 55° and with a period of one-half a sidereal day (a sidereal day is about 23h 56m), so the satellites are seen to rise and set about 4 minutes earlier each day, and appear twice a day. Four satellites in good postions are intended to be visible from any point at any time, giving a good intersection, and eliminating the necessity for accurate calibration of the receiver clock (the offset of the receiver clock is one of the four unknowns that can be determined).

The altitudes, about 20,200 km (a = 26,560 km), are high enough to make atmospheric drag negligible.

There are six orbital planes, with four satellites spaced equally in each. The GPS is by no means the only satellite navigation system that has been developed, but it has become the most used, and is replacing the others.

Earth satellites are affected by many small forces in addition to the main inverse-square force directed toward the centre of the earth, and these *perturbations* cause the orbital elements to change slowly with time. In the case of earth satellites, the mean distance a, the eccentricity e and the inclination i do not change on the average (they may fluctuate slightly over the short term).

The longitude of the line of nodes, Ω, the argument of perigee, ω, and the rate of change of mean anomaly change steadily with time. The drag of the atmosphere is negligible for satellites that do not come lower than about 1000 km at perigee, which includes most practical satellites. For lower satellites, atmospheric drag causes a loss of energy at perigee that pulls in the apogee position, making the orbit less eccentric and decreasing the orbital period.

Paradoxically, the drag speeds up the satellite! The pulls of the moon and sun, *tidal effects* cause orbital changes. The direct tidal effect is due to the force exerted directly on the satellite, while the indirect tidal effect is due to the changes in mass distribution of the earth caused by tidal motions. Pressure of solar radiation and solar wind is another disturbing effect. The most important perturbing force, however, is that exerted by the equatorial bulge of the earth.

The gravitational potential of the earth can be expressed approximately as $V = GM/r\ [1 - (a_E/r)^2 J_2 P_2(\sin\varphi)]$, where a_E is the equatorial radius of the earth, 6 378 137 m, $J_2 = 0.001082630$, and $P_2(x)$ is the Legendre polynomial of second degree, $(1/2)(3x^2 - 1)$. The argument of this polynomial is usually seen as $\cos\theta$, but here we use the latitude $\varphi = 90° - \theta$.

This is the first part of an expansion in spherical harmonics, which can even allow for a lumpy earth that is not symmetric about the polar axis. The correction to the 1/r field shown here is a zonal harmonic of order 2, related to the flattening $f = a/(a - b) = 1/298$ of the earth. However, the gravitational potential depends on the distribution of mass in the earth, so it cannot be expressed simply in terms of f. The radial component of the force is $-\partial V/\partial r$, and the tangential part is $(1/r)\partial V/\partial\theta$. The motion of earth satellites has led to a much better knowledge of the earth's gravitational potential.

If a satellite has a mean motion $n = 2\pi/P$, a mean distance a and an inclination i, the movement of the line of nodes is $d\Omega/dt = -(3n/2)(a_E/a)^2[\cos i/(1 - e^2)^2]J_2$, and the change of the argument of perigee is $d\omega/dt = +(3n/4)(a_E/a)^2[(5\cos^2 i - 1)/(1 - e^2)^2]J_2$.

The quantities are in MKS units, and the results are in radians per second. For the moon, the line of nodes rotates once in 18.6 years. For a typical GPS satellite, the line of nodes moves about -0.03° per year, the perigee about 0.01° per year.

All the other perturbations are much smaller in amount, but must be taken into consideration in accurate work. Relativistic perturbing accelerations are inversely proportional to the fourth power of the distance r. Their magnitudes for GPS satellites are about 3×10^{-10} m/s^2, less than other small perturbations, so they have no practical effect, although they have been considered.

The International Space Station (ISS), shown at the right, is a famous earth satellite. The image is from the NASA website. The first two modules, Zarya and Unity, were assembled in orbit in 1998, and now there are 14. It is 73 m across the solar arrays, 44.5 m long and 27.5 m high, with 425 m^3 of habitable space. Power comes mainly from the 892 m^2 of solar arrays. There are thrusters to adjust the orbit when necessary. The total mass is about 179 metric tons. Although this is called "space travel∂ by NASA, it is not even outside the earth's atmosphere!

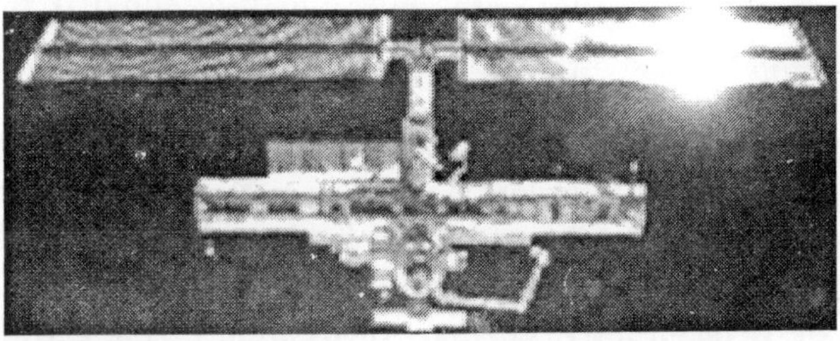

Approximate elements of the ISS orbit at 13.49Z, are i = 51.6°, Ω 19.99°, e = 0.00083, ω = 52.91°, M = 307.29°, and mean motion 15.59579861 rev/day, or 0.0649825°/s. ("Z∂ is another designation for UT.) Exact orbits, predicted about 10 days in advance,

The period is P = 5539.95 s (1.54 hr), and the mean distance is 6767009 m. This is a low, almost circular orbit at an altitude of about 400 km. Because the orbit is low, the ISS is seldom illuminated by the sun at night, so it is not frequently seen, though a frequent, large and prominent object.

The longitude (RA) of the ascending node changes rapidly, by about -5.00° per day, as does the argument of perihelion, by -3.64° per day. Perturbation by the oblateness of the earth is the reason for most of this change. The inclination, eccentricity and mean distance do not change rapidly. The orbit is just above the maximum ionization in the F layer of the ionosphere, and so is affected by atmospheric drag. The density of the atmosphere at this height is about $\rho = 9 \times 10^{-12}$ kg/m^3.

The drag is given by $F = C\rho V^2 A/2$, where C is the drag coefficient, V the satellite velocity, and A the effective projected area. NASA gives A = 344 m^2 and C = 2.36. $V = 2\pi a/P$ = 6541 m/s (assuming a circular orbit), so F = 0.16 N, which will produce an acceleration of 8.9×10^{-7} m/s^2. This drag causes the mean distance and period to decrease.

NASA notes that the "decay∂ is 4.11 X 10^{-4} rev/day^2. In 100 days, the mean motion will increase by 0.0411 rev/day, or P = 5527 s, a decrease of 13 s, or 0.23%. If uncorreoted, the satellite would spiral inward at an increasing rate, eventually burning up catastrophically.

It may be interesting to find out where you should look for a satellite in a known orbit.

The idea is to find the rectangular coordinates of the satellite in an approximately inertial system with its origin at the centre of the earth, and then to find the rectangular coordinates of the point of observation in the same system.

The differences of the coordinates then give a vector from the point of observation to the satellite. The motion of a satellite as seen from a fixed location on earth may be very complex, because of the interaction of the two motions involved, the revolution of the satellite and the rotation of the earth. Only when these are approximately equal is the situation more or less simple.

Let's suppose the satellite is in a circular orbit with a = 26 500 000 m and i = 55°, like a GPS satellite. Let the longitude of the ascending node be 40°. Since the orbit is circular, there is no perigee point, so we measure the true anomaly from the line of nodes in the orbital plane. Let the true anomaly at the time we are considering be 70°.

The z-axis is taken along the axis of rotation of the earth, and the x-axis in the direction of the vernal equinox. The y-axis then makes a right-handed coordinate system. This is just like the case of planetary motion, except that the role of the ecliptic plane is played by the equatorial plane. The orientation of these axes remains fixed in space as the earth revolves about the sun.

First, resolve the radius vector along and perpendicular to the line of nodes. The components are $a \cos \theta$ = 9 084 055.0 m and $a \sin \theta$ = 24 958 236.0. Then, $z = a \cos \theta \sin i$ = 20 444 590.0 m, $x = a \cos \theta \cos \Omega - a \sin \theta \sin \Omega$ = -9 084' 055.0 m, and $y = a \cos \theta \sin \Omega + a \sin \theta \cos \Omega$ = 24 958 236.0 m. This is the instantaneous position of the satellite in the inertial system, and it could obviously be found for any time equally easily.

Now for the point of observation. Let's use my own location, which is longitude λ = -104.92583°, latitude φ = 39.72694 and height above the ellipsoid (or mean sea level) h = 1633.7 m. Your own coordinates, if you do not know them already, can be found from a USGS topographic map. A meridional section of the earth is shown at the right, where φ is the geographic or spheroidal latitude.

The dimensions of the earth are a = 6378136 m and b = 6356752 m. There is a number of reference spheroids, any one of which will work satisfactorily. The eccentricity of the elliptical cross-section of the spheroid can then be found to be e = 0.006694167. The meridional radius of curvature at latitude φ is $N = a[\cos^2\varphi + (1 - e^2)\sin^2\varphi]^{-1/2}$, or 6 378 194.38 m at my location. The length of N below the equatorial plane is e^2N = 285.8 m.

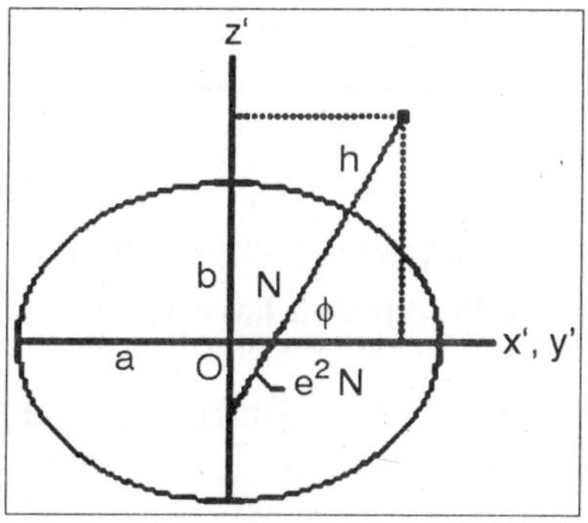

Fig. Geographic to Rectangular

We now choose an earth-fixed rectangular coordinate system with the z'-axis along the rotational axis of the earth, the x'-axis in the meridian of Greenwich, and the y'-axis making a right-handed system. The rectangular coordinates of my location are then x' = (N + h)cos φ cos λ = -1 263 816.21 m, y' = (N + h)cos φ sin λ = -4 741 167.76 m, z' = $(N + h - e^2N)\sin\varphi$ = 4 077 353.73 m. These coordinates do not change as the earth rotates. The distance from the centre is 6 379 711.32 m.

The relation between the inertial and earth-fixed coordinate systems is shown at the left. The earth-fixed system rotates steadily in the positive direction about the common z and z' axes, and the angle of rotation Θ is the local apparent sidereal time at Greenwich, which is easy to find from the Astronomical Almanac.

The rotational period of the earth is 23h 56m 04.09890369732s. When the time is stated to such precision, there are many small corrections to be considered, but we shall ignore them here so the principle is not lost.

We shall assume that the inertial and earth-fixed coordinates are related simply by the rotation through an angle Θ, not correcting for polar motion, precession of the equinoxes and nutation. These are simply rotation matrices applied in addition to the one for the main rotation. For accurate work, however, they cannot be ignored.

Since sideral time goes from zero to 24 hours in one rotation, we must multiply an ordinary time interval by 24/23.9344719 or 1.00273781195 to find the equivalent sidereal time interval. A sidereal hour is simply equivalent to 15° of angle; it is not an "hour∂ in the usual sense of time.

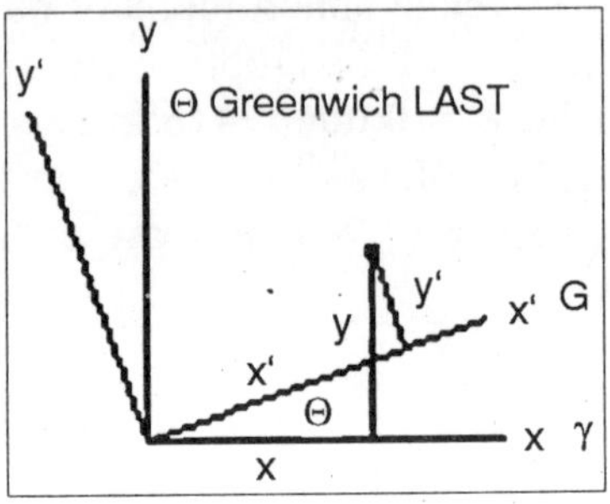

Fig. Inertial to Earth-Fixed Coordiantes

The transformation equations are the familiar x = x' cos Θ – y' sin Θ and y = x' sin Θ + y' cos Θ. All you have to do is check that the signs of the sines are correct.

If we put in coordinates in the earth-fixed system (x',y',z') we will get out coordinates in the inertial system (x,y,z) which we can compare with the satellite coordinates. Let's pick 10.00 am MST on 2003 January 1. Since my time zone is +7, the UT is 17.00h January 1. From the Astronomical Almanac, page B8, the sidereal time at 0h UT was 6.61651428 hours.

The sidereal time elapsed is then 1.00273781195 × 17.00 = 17.04654280 hours. Adding the two, the Greenwich LAST at my 10.00 am will be 23.66305708 hours, or 354.945856°. If it seems easier, this can also be expressed as –5.0541438°.

The results are x = –1 676 585.46 m, y = –4 611 395.05, z = 4 077 353.73 m. The square root of the sum of the squares is 6 379 711.32, so the arithmetic checks.

Now we can subtract the coordinates of the observation point from the coordinates of the satellite to find the relative vector (X,Y,Z). We find X = –7 407 469.54 m, Y = 29 569 631.05 m, Z = 16 367 236.27 m. The direction of this vector can now be expressed in terms of right ascension and declination, since it is a vector in the inertial system referred to the vernal equinox.

The distance from the observer to the satellite is 34 599 423.53 m. The projection on the equatorial plane is 30 483 334.55 m, so that the declination δ = 28.2324° and the right ascension is 75.9363° or 5h 3m 45s.

At the date and time specified, this direction is beneath the earth. In fact, the antipodeal point of right ascension 18h and declination -29° is in the southern sky at an altitude of about 20° above the horizon.

Although these calculations are tedious and subject to error, even when done with an electronic calculator, a computer programme can do them with ease, speed and correctness, and even the more precise calculations are not much bother. It would not be difficult to write a programme to determine the visibility of all of the 24 GPS satellites, and the directions in which they are to be seen.

OTHER ORBIT LORE

The total energy E of a body of unit mass moving under the influence of a potential energy per unit mass V(r) is the kinetic energy $T = v^2/2$ plus V(r). In polar coordinates, the components of velocity are dr/dt and r dθ/dt, so the total energy becomes $E = (dr/dt)^2/2 + r^2(d\theta/dt)^2/2 + V(r)$. The constancy of angular momentum for a central force gives $r(d\theta/dt)^2 = h =$ constant.

Substituting for dθ/dt, we find $E = v^2/2 + h^2/2r^2 + V(r)$. This is the energy equation for a particle of unit mass moving in one dimension with the *effective* potential $h^2/2r^2 + V(r)$.

The additional term is called the *centrifugal potential*. In the case of orbital motion under gravity, $V(r) = -GM/r$. The form of the resulting potential is shown in the figure. For E = 0, a particle coming in from infinity accelerates steadily until $r = h^2/GM$, then is rapidly decelerated, coming to rest at $r = h^2/2GM$, and afterwards retracing its path in reverse. Of course, the particle swings round the centre of force in this motion. For $E = -(GM)^2/2h^2$, r has the constant value that gives the minimum of the curve, and the orbit is circular. For intermediate energies, r oscillates between a minimum and a maximum value, corresponding to perhelion and aphelion.

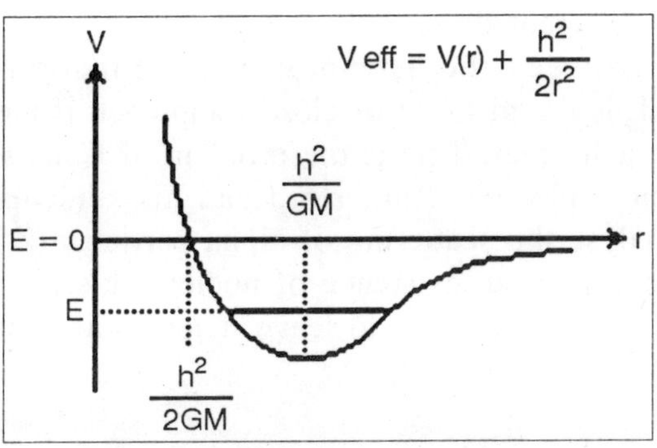

Fig. Effective Radial Potential

Consider the vector B = v X h + (GM/r)r. The angular momentum per unit mass, h, is here taken to be a vector perpendicular to the orbital plane. The cross product then lies in the orbital plane, and so then does B. Taking its time derivative, we have (dv/dt) X h – (GM/r^3)r(v·r) + (GM/r)v.

Now dv/dt = -(GM/r^3)r, and h = r X v, so the first term becomes -(GM/r^3)[r(r·v) – r^2v]. All the terms cancel, so dB/dt = 0, and B is a *constant of the motion*. This is extraordinary, since we already have one constant of the motion, the total energy E, and this is usually the only one that can be found.

For an inverse-square force, we have found a second one. It is known that this happens when we have an additional symmetry in the problem—here it happens to be symmetry under rotation in a four-dimensional space, but there is no room to explain this here..

Since B is a constant, it can be evaluated at any point of the motion, and it is easiest to choose perihelion.

At this point, v X h = $v_p^2 r_p$, so B is a vector parallel to the line between the focus and perihelion. We will find the same vector if we evaluate B at any point of the motion. Its constancy is a good proof of an exact inverse-square force.

We have seen that with an attractive inverse-square force, closed orbits exist for negative energies. For zero or positive energies, the orbits are open. We have considered briefly the zero-energy parabolic orbits that divide the two classes, but not the hyperbolic orbits, since they are not of much use in the solar system.

For a repulsive inverse-square force, there are no closed orbits of negative energy at all, only the hyperbolic orbits of positive energy. On these orbits, the body comes in from infinity along a path asymptotic to a line of a certain direction.

It then interacts with the other body, in what is called a "collision,∂ though the bodies do not come into contact, and leaves to go to infinity asymptotic to a different direction.

The angle between the asymptotes is a function of the *impact parameter*, which is the distance of closest approach if the bodies did not exert forces on each other. This is the problem of *Coulomb scattering* in atomic and nuclear physics. Rutherford and his students used classical mechanics to analyse the scattering of alpha particles (Z = 2, A = 4) by atomic nuclei to verify the existence of nuclei. This is an area different enough from celestial mechanics to leave it for another article.

Chapter 3

Gravitation and Hodograph

GRAVITATION

ELASTIC COLLISIONS IN THE SOLAR SYSTEM

The article by Van Allen quoted in the Reference recalled my attention to the subject of encounters between orbiting bodies and the interesting effects on their orbits.

Apparently, this is a subject of some mystery to physics students, though it is really quite simple in principle. The reader is strongly encouraged to read Dr Van Allen's excellent article, in which the encounter of Pioneer 10 with Jupiter in December 1973 is discussed as an example. The article, despite its clarity, still has the aura of mystery, or at least complicated physics, here and there.

One source says that the first use of gravitational assist was with the Mariner 10 probe of 1973-1975.

An encounter with Venus on 10 March 1974 was used to decelerate the vehicle and send it towards Mercury. Mariner 10 made three passes around Mercury, giving us most of the information on the surface of that planet that we have to date. This date is later than the December, 1973 date for the Pioneer 10 encounter with Jupiter, so perhaps Pioneer 10 was actually the first.

Jupiter is about 5 AU (the AU is the radius of the earth's orbit) from the sun, and revolves at an orbital speed of about 13.1 km/s on the average, completing its orbit in 11.857 years. Pioneer 10 approached Jupiter on an elliptical orbit at a speed of 9.8 km/s, not far from aphelion. After its closest approach of 2.027×10^5 km, it swung away, approximately tangent to Jupiter's orbit, at a speed of 22.4 km/s, which was sufficient to put it into hyperbolic orbit about the Sun. Since then, Pioneer 10 has left the solar system, and just recently its final radio transmission was received.

This is an example of *gravitational assist*, which has been frequently used in planetary probe trajectories, performing feats that would be impossible, or very costly, if done directly.

Although we do deal with three bodies here, the sun, Jupiter and the spacecraft, none of the complicated results of the celebrated "problem of three bodies∂ is required for an understanding. There is a massive body, Jupiter, and a body of negligible mass, the spacecraft, in independent orbits about the even more massive Sun.

A close encounter between Jupiter and the spacecraft, which is an elastic collision in which the bodies interact gravitationally, takes place in a short time and limited space to change the instantaneous velocities of the participants, sending them into new orbits.

The new orbit of Jupiter is indistinguishable from the old one; though slightly different, the difference is extremely small. The new orbit of the spacecraft, however, can be very different, and its kinetic energy can change by a considerable fraction. This energy comes from Jupiter, of course, which does not notice the theft.

In the elastic collision, the total kinetic energy and total momentum of the two bodies is conserved.

However, it may be distributed differently before and after the collision, as is very well known. Let us reduce the problem to its simplest state, that of a one-dimensional collision, as on an air track. An actual collision in three dimensions may approach this more or less closely when the impact parameter (distance of closest approach if the bodies did not interact) is small.

First, let's assume that the large mass M is moving with velocity w to the right, and the small mass m is moving with velocity v to the left. The magnitudes of the vectors are shown above the arrows giving the directions. The relative velocity is v + w, and in the elastic collision is reversed in the centre of mass system in which the centre of gravity is fixed. This system moves with velocity w to the right, very closely, since M >> m.

Therefore, after the collision, the mass m is moving to the right with a velocity v + 2w. In a head-on collision, the light body can pick up twice the velocity of the heavy body.

In the figure, momentum and energy are apparently not conserved, but this is only because the very small decrease in w is not shown. An accurate figure would differ imperceptibly.

Now suppose the large mass is still moving with velocity w, but the small mass comes up behind it with velocity v > w, approaching with relative velocity v – w. In the centre of mass system this velocity is reversed after the collision, and the final velocity of the small mass is v – 2w. In a rear-end collision, the small mass can lose twice the velocity of the heavy body.

We have now found the limits of the velocity of the light body after collision, and how these limits depend on the velocity of the heavy body.

In a three-dimensional collision, the velocity change will be smaller, and can easily be calculated explicitly for any case. In the case of Jupiter, the speed of Pioneer 10 after the encounter could be as high as 9.8 + 26.2 = 36 km/s, or as low as zero.

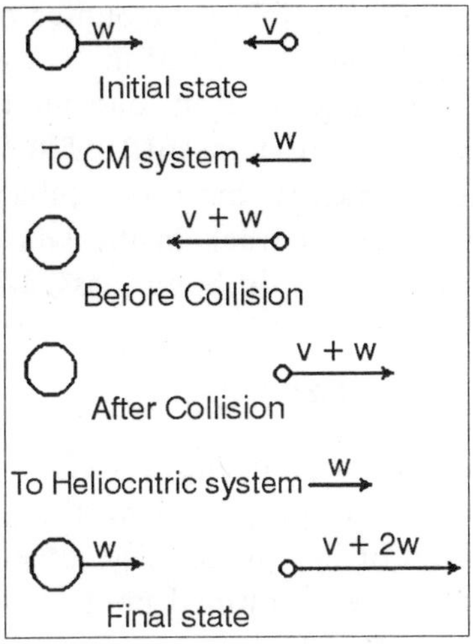

Fig. Head-On Collision

The actual result of 22.4 km/s is well within the possible range. An unlucky encounter could send the spacecraft plummeting rapidly into the Sun. Comets have long been known to interact with Jupiter in this way. If they come in behind Jupiter, they lose energy and are put into short-period elliptical orbits, many of which do indeed come perilously close to the sun. If they come in ahead, they are flung well out of the solar system, like Pioneer 10. Saturn exerts a very similar effect.

Saturn's orbital velocity is 9.67 km/s, so the changes it can make are not as large as those of Jupiter. If Pioneer 10 got a kick in the apse from Saturn, it could only with difficulty attain the required escape velocity. Uranus and Neptune move even more slowly, but still could perturb a parabolic comet into an elliptical one. These massive planets can create a tight orbit in an encounter that leads to orbit modification.

Mars, with 24.1 km/s; Earth, with 29.8 km/s; and Venus, with 35.0 km/s orbital velocities could transfer large velocities in an encounter, but the smaller masses do not reach as far. Any planet is sufficiently massive relative to a comet, spacecraft or meteoroid to modify its orbit with no change to its own. It should be noted that a planet cannot *capture* a body in this way, in a two-body encounter.

It is the velocity relative to the Sun that is changed. An exception is when the collision is not elastic, and the body encounters some atmosphere or even the solid surface of the planet. This is, of course, of frequent occurrence on Earth, and has occurred elsewhere as well, as the pock-marked surface of the Moon attests. We also note that encounters in which kinetic energy is lost generally give rather elliptical orbits; circular orbits are not to be expected.

The gain or loss of energy in an encounter of a light body with a heavy one is only a natural consequence of the conservation of momentum and energy, and the transformation from one inertial system to another, with no mystery. The infinite possiblities introduced with three-dimensional orbits adds complexity, but no essential difference. The effect could easily be seen with an air track in the laboratory, or indeed with pendulums of elastic balls of very different mass, which is the basis of the argument in this article.

PROJECTILES

The first internal combustion engine was the gun, its piston the projectile that left its muzzle, propelled by the gases at high pressure produced by the explosion of its charge. The intention of the gunner was to send the projectile to a predetermined point, to execute its destructive or evil purpose.

The importance of this process in the affairs of mankind has led to the study of the paths of projectiles, or *ballistics*. At school, ballistics is a problem in uniformly accelerated motion, and the trajectory is a parabola, determined by the initial velocity and the angle of projection. In practice, this description falls far short of reality because of the influence of the atmosphere on the motion of the projectile. In studying the consequences of air resistance, we will come across many peculiar facts, and expert guidance in the matter of streamlining.

Suppose that we measure the velocity of a projectile in level flight at two points separated by a distance x, finding values V and v. The average retarding force at velocity $(V + v)/2$ is then $F = m(V^2 - v^2)/2x$, where m is the mass of the projectile.

This force is called *drag*. If the drag as a function of the velocity is known, then the trajectory of the projectile can be calculated, at least in principle. We require, therefore a means of measuring the velocity of a projectile.

A means was devised by Benjamin Robins (1707-1751), the *ballistic pendulum*, around 1842. The projectile is fired into a heavy pendulum, to which it transfers its momentum. From the swing and weight of the pendulum, the velocity of the projectile can be found. This is still an experiment performed in the Physics laboratory. Robins measured the velocity of the projectile at various distances from the gun, using the same charge each time.

He showed that the drag was roughly proportional to the square of the velocity, the first time this was done, but accurate results were impossible due to the lack of reproducibility of the muzzle velocity even when the same charge was used.

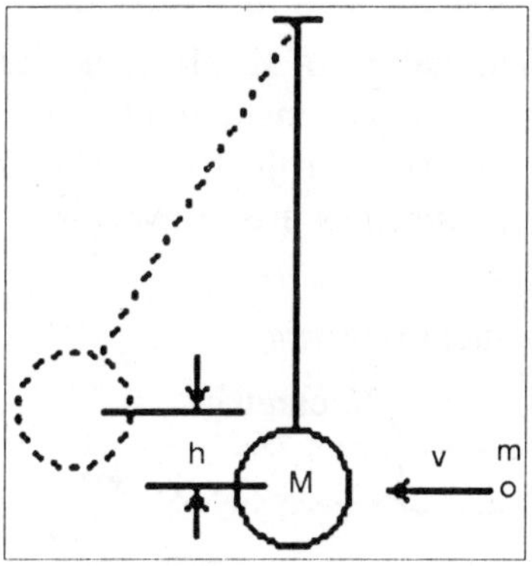

$$v = \left(1 + \frac{M}{M}\right)\sqrt{2gh}$$

Fig. The Ballistic Pendulum

Hutton, at Woolwich Arsenal around 1812, suspended the gun as a pendulum itself. From its recoil, the muzzle velocity could be determined, so that accurate results for the drag as a function of velocity could be obtained.

Bashforth, also at Woolwich, devised an electrical timer that gave the time at which the projectile passed through each in a sequence of screens. From these times, the speed as a function of distance can be determined.

Around the time of the First World War, wind tunnel measurements began to measure the drag directly. So important were these measurements regarded that they were initially military secrets.

The drag on a projectile can be expressed as $F = \rho d^2 v^2 f(v/c, vd/\nu)$. Here, ρ is the density of the air, d is the diameter of the projectile, v its velocity, and the dimensionless quantity f(M,R) is called the *drag coefficient*, depending on the dimensionless parameters M = v/c, the Mach number, and vd/ν, the Reynolds number, where ν is the kinematic viscosity of the air, about 0.144 cm^2/s under normal conditions. When comparing drag coefficients, carefully note the defining equations, which may differ in the constants used.

For values of R below some critical value, on the order of 100 or 1000, the drag coefficient $f = C/R = C\nu/vd$, so $F = C\rho\nu dv$, where C is some constant. The drag is proportional to the velocity, and to the linear dimension of the projectile. This *viscous* regime is rather unimportant for projectiles, since even for a projectile as small as 5 mm diameter, the critical speed is below 2.8 m/s.

Above the critical value of R, f is roughly constant, say f = C', and F = $C\rho d^2v^2$, proportional to the square of the velocity and the cross-sectional area of the projectile. The nature of the flow is shown in the Figure. It is the same whether the projectile is considered to move in still air, or whether the air is moving past the fixed projectile at the same speed. Streamline flow occurs where the streamlines are converging.

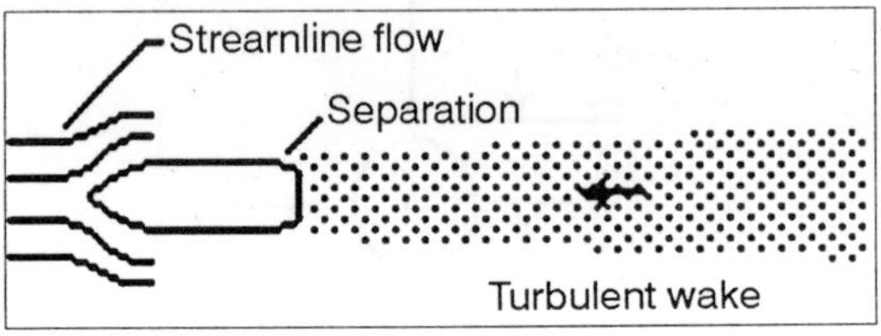

Fig. Subsonic Projectile

There is a stagnation point exactly at the nose of the projectile, whatever its shape, and otherwise the air flows smoothly around the obstruction. If there were no viscosity, the streamlines would close in again behind, and the air would exert no net force on the projectile: there would be no drag! Instead, flow in the diverging region is unstable, and a separation surface between streamline flow and the turbulent wake appears.

The air in the wake has been accelerated in the direction of the motion, representing a loss of kinetic energy of the projectile, and this movement is dissipated in turbulent eddies.

The pressure is low at the tail of the projectile, and this is the major contribution to the drag.

A body that is not spherical tends to rotate so that its wake is as large in cross-section as possible, and the drag a maximum. This can be seen by dropping light objects, such as a leaf.

At around the speed of sound, Mach 1 or about 350 m/s, the drag increases abruptly, doubling or more, as we enter the transonic region, and then usually decreases slightly as the velocity is increased further. The reason for this change is a radical change in the flow around the projectile as M = 1 is passed.

The new feature is the appearance of shock-wave cones at the nose and tail, making an angle Ω with the trajectory. The Mach number M = $1/\sin\Omega$. These are sudden jumps in pressure, carrying considerable energy away, which explains the increased drag.

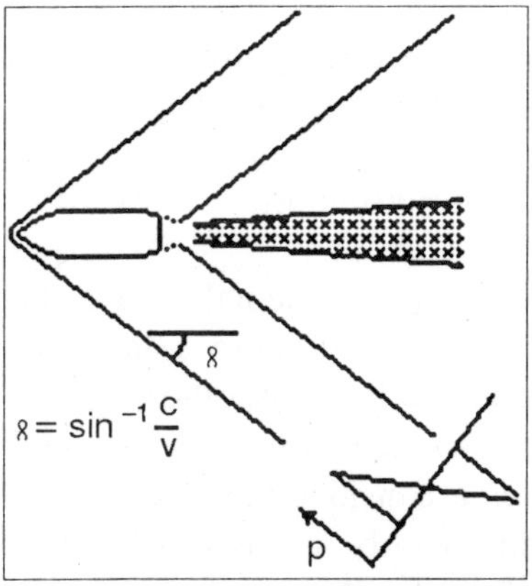

Fig. Supersonic Projectile

The tail drag from the turbulent wake still exists, but it is dominated by the shock-wave dissipation. The pressure signature of the shock waves is a kind of N, which is heard as a distinct crack that appears to originate some distance behind the projectile.

A strong shock wave travels at faster than the speed of sound, because of nonlinear propagation.

The curve of the shock wave around the nose of the projectile is accompanied by a high pressure, which appears as drag. The idea is to create as weak a shock wave as possible to reduce drag, and this is best done with a finely pointed nose.

Although aircraft must travel at faster than 350 m/s to enter the supersonic regime, wave speeds on the surface of water are small enough that ships are strongly affected by wave drag. The bow wave is analogous to the shock wave of a projectile, and carries away energy that must be replaced by the propulsion system.

There is also a turbulent wake, that sucks at the stern of the ship. The bows of the ship are pointed to reduce wave resistance. A canoe, which moves at less than the wave velocity is slender at each end to encourage streamline flow.

Although there are obvious analogies, there are essential differences, such as the two-dimensional nature of waves on water and their dispersive propagation. The kinematic viscosity of water is only 0.01 cm^2/s, a factor of 10 less than that of air.

The streamlining of subsonic vehicles is a different matter, which should concentrate on encouraging streamline flow as far as possible, and reducing the wake to as small a cross-section as possible. The shape of the front is of little consequence, so long as it does not generate turbulence. Almost any form will create an invisible region of dead air ahead of it that acts as smooth fairing. The rounded fairing seen on the fronts of road trailers is almost useless in reducing drag; it would have a greater effect on the rear, where it might discourage separation.

Any projections into the air stream create turbulence and consequently the drag. At the speed of automobiles, streamlining is only an esthetic matter.

Indeed many of the 'fast' cars actually have shapes that increase drag, and appurtenances that are applied in a largely useless fashion. Any smooth outline will do as well as any other.

On railways, the resistance to motion from other causes is so small that air resistance makes a considerable, even a major contribution, above 50 mph, and streamlining can be economical, a fact often overlooked. As in many other cases, additional power is more cheaply applied than streamlining. If you want a fast car, a big engine will do the job better than excellent streamlining.

When $M > 1$, the undisturbed air is separated from air influenced by the projectile by a strong shock wave. There is one shock cone at the front, and another at the rear, both compressive. The turbulent wake exists as before. If the shock wave curves around the nose, high pressures result that are chiefly responsible for the increase in drag. By making the nose sharp and pointed, this force is reduced.

Making the tail slender has no effect, the turbulent wake being much as before. In contrast with subsonic streamlining, the body should now be slender in front, and blunt at the tail. This shows up in the shapes of subsonic and supersonic aircraft, where it is a critical matter to reduce drag as far as possible in either case.

The drag coefficient depends strongly on the angle between the axis and the direction of motion, called the angle of *yaw*. The drag coefficient also depends on the size of the projectile. It is greater per unit area for small cross sections than for large.

A projectile has an axis of symmetry, and the gun generally projects it so that this axis is parallel to the velocity as the projectile leaves the muzzle. This situation does not long remain, however, as a nonrotating projectile tumbles head over heels, greatly increasing the drag, and randomly deflecting the projectile to one side or another. Therefore, all projectiles were made spherical, the one shape that cannot tumble. There is, therefore, a very good reason why smoothbore guns fired only spherical bullets.

An alternative is to provide fins or feathers at the back of the projectile to force alignment by aerodynamic forces.

It was known as early as 1500 that imparting a spin along the axis stabilized a projectile, and prevented it from tumbling. This was done by cutting spiral grooves in the barrel, or *rifling*. These grooves became fouled so rapidly that the advantages of rifling were doubtful. In addition, the bullet had to fit the barrel tightly, and this was also a drawback, especially when the gun was loaded from the muzzle.

Although rifled hand weapons were successfully used by 1800, their general adoption had to wait for the development of non-fouling smokeless powder.

The adoption of rifling meant that projectiles could become lengthened, achieving a larger projectile for the same diameter of barrel or *calibre*. The increase to supersonic muzzle velocities favoured the sharp-nosed shape with only a slight contraction at the tail, the *boat-shaped* projectile.

The nose of a projectile is usually *ogival*, that is, formed by rotating a circular arc tangent to the side of the projectile about the axis. The radius is specified in calibres (multiples of d).

The nose may be rounded off by a spherical surface, also specified in calibres. When attempts were made to streamline the tail, it was found that such bullets tumbled and were useless. The shape of the bullet reflects its aerodynamics, not artistic design.

When the gun is considerably elevated, to achieve great range, or for the projectile to drop from above, the curved trajectory means that the axis of a spinning projectile cannot remain tangent to the trajectory, and complicated gyroscopic and drag effects enter. One result is *drift*, the departure of the trajectory from a plane, so that the projectile is deviated to right or left.

Golfers, and others, are well aware of this phenomenon that accompanies a spinning ball. Mortar rounds, which are always used in high trajectories, have fins and do not rotate.

For the accurate computation of a trajectory, the forces on the projectile as a function of yaw angle must be known, and the projectile treated as a rotating rigid body.

Then, with the drag known, the trajectory can be found by numerical integration, beginning with the initial velocity and spin. Not only is the problem a difficult and involved one, but the calculations are exceedingly tedious. In fact, they were so tedious that they stimulated the creation of digital computers in the 1940's, leading to the general-purpose computer, which has now revolutionized technology and life.

As an aside, the Reference mentions a "service time-fuze,∂ which consists of a detonator pellet exploded by the shock of firing, and a train of gunpowder leading to the magazine of the fuze, of a length determined by the setting of the fuze. The problem of reliable time-delay fuzes is a difficult one.

Note the spelling fuze, which is peculiar to ballistics. Such fuzes were used by scientists to determine the pressures acting on the nose of the projectile by the pressure effect on the rate of burning of the gunpowder.

DYNES AND SLUGS

Scientific and engineering work these days is mostly done in *Systeme Internationale* (SI) units, otherwise known as the MKS system, based on the metre, kilogram and second as the fundamental units, and enforced by the Unit Police that tell us what to do through many regulations and deprecations.

The reader may encounter the good old units based on the centimetre, gram and second (cgs), or on the foot, pound and second (fps), and wonder what they are about.

The cgs system was once standard in Physics, the fps system in Engineering in Britain and the United States, and to use them strikes a blow for freedom and tradition.

They are, of course, just as good as SI, and not enforced by tedious pedants.

An *absolute* system of units lets the units or "dimensions"of all quantities be determined by the defining equations, without any arbitrary choices. Arbitrary choices were once the fashion, in order to have convenient units for particular activities.

For example, we have leagues, miles, chains, rods, perches, fathoms, yards, feet, inches and barleycorns, all expressing distance in English traditional units. A similar variety is available for most other quantities in daily use.

The benefits of having units suited to the task at hand should not be depreciated, but when we are doing physics it is extremely tedious to put in conversion factors for each equation, and often obscures what we are studying.

For this purpose, we require an absolute system of units. After our calculations are done, we can convert the results to any convenient unit. The millimeter is an excellent unit for things like optics and machining, but to say "give him and inch, and he'll take 1,609,300 millimeters (approximately)" does not appeal. "How far is it to Boulder?" you may be asked. If you answer 5.15 X 10^7 mm, your friend will still be effectively in the dark if she is wondering how long it will take to drive there. If you ask her how old she is, an answer of 9.5 X 10^8 seconds will not give an immediate impression.

We use kilometers and years for good reasons. Granted, SI does use decimal multiples that are often not inconvenient (like the millimeter), and conversions depending only on moving the decimal point are especially convenient. We now have calculators, so this benefit is somewhat blunted.

The SI is more French than it is scientific. The Celsius temperature scale is included, but it is no more metric or decimal than the good old English Fahrenheit scale or the old German Rιaumur scale. There are exactly 100° between 0° and 100° in each system, the cynical may note. All the scales are arbitrary.

It is easier to remember the freezing and boiling temperatures of water at a standard atmospheric pressure in Celsius, of course, but this is not of pressing significance to most people. The metre was designed when the French still thought the earth was perfectly spherical, and made it 1/10000 of a quadrant. This was a perfectly silly definition, impossible to measure accurately.

When sphericity of the earth proved erroneous, the metre became the length of a bar in Paris, just another arbitrary unit. The seconds pendulum is about 1 m long, and this was probably what really suggested the metre. The definition of the kilogramme as the mass of a 10-cm cube of water was another poor definition, impossible to measure accurately. Water and cubes expand and contract thermally.

The kilogramme became the mass of a lump of metal in Paris. The second was still 1/86400 of a mean solar day, somewhat better but still not easy to determine with precision.

The French originally wanted it to be something like 1/100000 of a mean solar day, but redefining people's time (with 10 months in a year, and other absurdities) was not on. So metric units, which became the SI, were just another arbitrary choice.

The French public, incidentally, had to be forced by law to use the metric system when they did not embrace it, sticking to *toise*, *pouce* and *livre* until threated by violence. Things have not changed a lot. Note that we use the French spelling metre, kilogramme, since this is required by the SI standards. Nobody outside France seems to use "seconde∂ though. Enough is enough.

THE CGS SYSTEM

The cgs system is closely related to SI, differing only by powers of 10, since the standard units are centimeter (L), gram (M) and second (T), where we represent the "dimensions" by the letters in parentheses. The SI unit of force is the newton (N), and its dimensions are MLT^{-2}. To find the equivalent in cgs, multiply by 1 = (1000g/1kg)(100cm/1m) and find 1 N = 10^5 cgs units of force. The cgs unit of force is the *dyne* (from the Greek for "force"), so we have 1 N = 10^5 dyne. Sometimes "dy" is used as an abbreviation for "dyne."

The reader is probably quite familiar with this easy way of converting with unit ratios, which always works. The SI unit of work or energy is the joule (J), which is defined as N-m (W = fd). Its dimensions are ML^2T^{-2}. Then 1 J = 10^5 dyne X 100 cm = 1 X 10^7 erg, where "erg" (from the Greek for work) is the cgs unit's name. Pressure is force per unit area, dimensions $M^{-2}LT^{-2}$.

The SI unit is the pascal (Pa), 1 N/m^2. Then, 1 Pa = $10^5(100)^{-2}$ dy/cm^2 = 1000 dy/cm^2. The cgs unit of pressure does not have a special name. It happens that the normal atmospheric pressure at sea level is about 10^7 Pa, which is called a *bar*.

That pressure is really about 1.01325 bar. The *millibar*, often seen in weather reports, is then 10000 dy/cm^2. In traditionally metric countries, the kg/cm^2 is a common unit of pressure, which is also about 1 atm.

Newton's Second Law states that the acceleration of a body is equal to the ratio of the force acting on the body to its mass, f/m = dv/dt. This elegant statement introduces two new quantities, force and mass, neither of which is independently defined, and both of which have been the subjects of extended philosophical discussion. If mass is a fundamental unit, then [f] = MLT^{-2} (the square brackets signify "dimensions of"). If force is taken as a fundamental unit, then [m] = $FL^{-1}T^2$. Mass seems to be a fundamental quality of a body, while force appears to be accidental, so mass is a reasonable choice as a fundamental unit. On the other hand, mass cannot be experienced directly, while force is constantly encountered in practice, so a case can also be made for force as a fundamental unit.

GRAVITY, DENSITY, FORCE AND MASS

Newton's Law of Gravitation does give a certain force directly, the gravitational force f = Gmm'/r^2 between two masses. If m and m' are in gram, r in cm, then G = 6.672×10^{-8} dy-cm^2/g^2. The universal gravitation constant G is not known to great precision because it is equivalent to weighing the earth—not a convenient thing. It, too, gives only a ratio of force to mass, since f/m = Gm'/r^2.

If r is the radius of the earth, 6378 km, and m' its mass, 5.98×10^{27} g, then f/m at its surface is g = 981 cm/s^2, called the acceleration due to gravity. This is a convenient standard acceleration for us to use to relate force and mass, where now the force is called *weight*, to imply the standard acceleration. Actually, we measure g and use it to determine the mass of the earth, of course.

A problem with this is that g varies with position. At any one point, it is constant enough, but has different values at different points, which gives the same mass different weights at different places.

The g we observe is the joint result of the gravitational attraction of the earth and the centrifugal acceleration due to its rotation. At the equator, g is about 978.039 gal ("gal" stands for Galileo, and is used in gravitational studies for cm/s^2, together with the smaller milligal = 10^{-3} gal). Here we are furthest from the centre of the earth, and the centrifugal force is greatest. At the pole, g = 983.217 gal. The difference is about 0.5%.

We are closest to the centre of the earth, and the centrifugal force is zero. At any latitude φ, $g = 978.049(1 + 0.0052884 \sin^2 \varphi + 5.9 \times 10^{-6} \sin^2 2\varphi)$ gal. Here, Heiskanen's 1928 value for the equatorial gravity is used. If we are above sea level, the *free-air* correction 0.3086h mgal must be subtracted, where h is the altitude in metres.

There are other corrections depending on the masses above and around the points, and when all this has been considered, there are still "anomalies" of up to about 100 mgal that cannot be explained by anything obvious. All this is to show that the actual acceleration of gravity is not a uniform value. The value g = 981 cm/s^2 is a good value for everyday use.

Since we know the mass and the radius of the earth, we may itch to know its average *density*, or ratio of mass to volume. Its density $d = m/(4\pi r^2/3) = 5.5$ g/cm^3. The ratio of the mass of a body to the mass of an equal volume of water is called its *specific gravity*, sp.gr., and in the cgs system is numerically equal to the density, but is dimensionless (and so the same in any system of units).

The specific gravity of granite is 2.64-2.76, of dolomite 2.84, of limestone 2.68-2.76, of sandstone 2.14-2.36, of feldspar 2.55-2.75 and of mica 2.6-3.2. A good characteristic value for surface rocks is about 2.7. The earth can't be made up of surface rocks! Hematite, an iron ore, has a sp.gr. of 4.9-5.3.

Rock salt, halite, is 2.18, glass 2.4-2.8, bituminous coal 1.2-1.5, crude oil 0.8, oak 0.7 and pine 0.5. The sp.gr. of aluminum is 2.8, zinc 7.1, cast iron 7.2, steel 7.9, brass 8.5, copper 8.9 and lead 11. The sp.gr. of ice is 0.917, so it floats.

The sp.gr. of water is 1.00 at 3.98°C, 0.99823 at 20°C, and 0.95838 at 100°C. Mercury, used in barometers and such, has 13.5953 at 0°C and 13.5462 at 20°C. Dry air has a density of 1.2929 g/liter at 0°C, and its density at any other temperature and pressure can be estimated from the gas laws. These are representative values that show the range for common materials, and can be used in approximate calculations.

If we must relate mass and force, however, some such value must be assumed as a constant conversion factor.

It is no longer the actual acceleration of gravity, but an arbitrary constant. Nevertheless, it should be close to an average value of g. The value 980.665 gal has been chosen to fill this role. It would have been just as easy to have chosen 981 gal, but no, we have to make it look like something precise. When we use this value, try to eliminate any suggestion in your mind that it involves some kind of physical measurement. It is just an arbitrary conversion constant.

THE FPS SYSTEM

The absolute foot-pound-second system of units (fps) retains the second as the unit of time. The unit of length is the foot, 30.48 cm (exactly). Whoever decided that an inch should be exactly 2.54 cm has my thanks. Another definition, 1 metre = 39.37 in, which gives 1 inch = 2.54000508 cm, was used to define the *US survey foot*, used in geodesy. There is no difference practically, only an annoyance in measurements of high precision.

The third fundamental unit is the pound, the unit of *force*. The Handbook of Chemistry and Physics tells that 1 pound = 0.45359237 kg. Now kg is mass, so this really means that 1 pound force = 1 lbf = (0.45359237)(980.665) dyne = 4.4482216 × 10^5 dynes = 4.4482216 N, or, that a pound force is the weight of 453.59237 g.

Someone went to a lot of trouble to compare the standard kilogram in Paris with the standard pound in London, and came up with all those digits for us to admire.

A choice of 1 lb = 454 g would have done just as well to define the fps system, with only a negligible amount of restandardization. The standard acceleration of gravity in the fps system is 980.665/30.48 = 32.17405 ft/s^2. In everyday problems, the value 32.2 ft/s^2 is normally used. The foot-pound is 1.35582 J.

It is usual to refer to the weight of a body in the fps system, not its mass. One cc of water has a mass of 1 g, and so a weight of 1/453.59237 lb in fps. A cubic foot of water is 28316.8 cc, so it weighs 62.428 lb at 3.98°C or 62.317 lb at 20°C.

Using the specific gravity, we can now find the weight of a cubic foot of anything. Since w/m = g, m = w/g in the fps system, so we have f = (w/g)dv/dt for Newton's Law. The appearance of g caused the name "gravitational system" to be applied to this kind of fps system, but it is a misnomer. There is nothing gravitational about it, and it does not depend in any way on the acceleration of gravity.

The "g" is only a conversion constant that happens to be approximately equal to the acceleration of gravity. It is *never* anything other than 980.665. The unit of mass has been named the *slug*. The weight of a slug is about 32.17 lb. It is equal to 14.593 kg. Mass densities should be specified as slug/ft^3.

Civil engineers adopted the slug, mechanical engineers didn't appreciate the problem, and chemical engineers did something else which shall mention in a moment.

The symbol γ is used for weight density, $\gamma = w/V$ in, for example, pounds per cubic foot, pcf. Mass density is still ρ or d.

There are two general ways to determine the weight or mass of a body. One way is to use a *balance*, where the mass is compared with a known mass using levers or the equivalent. This includes the once-common laboratory beam balance.

The result is unaffected by variations in gravity. The other way is with a device that measures the force directly, such as a spring scale or strain gauge. Such a *spring scale* must be calibrated, and the calibration will depend on the local value of gravity.

Household spring scales are not sensitive enough to detect variations in gravity. Most small "balances" are now electronic spring scales.

There are other ways of defining fps systems. One way lets the unit of mass be the pound, or lbm. Then f = mdv/dt defines the unit of force, called a *poundal*, pdl. The force acting on 1 lbm in standard gravity is, then, 32.2 pdl, which must be its weight. 1 pdl = 13,810 dynes, we can easily discover. This system was used in school physics for many years, but never really caught on except with physics teachers. Another way keeps both the pound mass (lbm) and the pound force (lbf), which means that Newton's law has to be written

$$f = (m/g)dv/dt,$$

causing all kinds of mischief. Note that this is not just

$$f = (w/g)dv/dt,$$

which is harmless, since w/g is just mass. By contrast, m/g isn't anything holy. Here, g is not just a conversion factor, but is an essential part of the equation, contributing units as well as magnitude. Chemical engineers used this system.

Since they don't use Newton's Second Law much, it doesn't do a lot of harm, and allows the carefree confusion of two different units with the same name.

In using the fps system, it is best to use feet and pounds consistently, and to divide them decimally, just like metres and kilograms. If it is not otherwise obvious, whether factors of g should be introduced can be determined from the units. For convenience, things like inches, miles, tons and acres can be brought in later, but not used in theoretical calculations.

For example, standard atmospheric pressure is 2117 psf, and water weighs 62.3 pcf, air 0.075 pcf. A speed of 30 mph is 44 fps. Quantities can be expressed in consistent units by factors such as these before performing a calculation.

It is a subsidiary matter that the cgs system generally uses the Celsius thermometer, and the Kelvin for absolute temperature,

$$K = °C + 273.15.$$

The fps system generally uses the Fahrenheit thermometer and degrees Rankine for absolute temperature,

$$°R = °F + 491.69.$$

Of course, $5K = 9°R.$

Either temperature scale can be used with either system of units. Specific heats are numerically the same in cal/K and Btu/°R. 778 ft-lb = 1 Btu, and 4.186 J = 1 cal.

The gas constant is 8.31441×10^7 erg/K-gmol, or 1545.33 ft-lb/°R-lbmol.

INTEGRALS OF THE EQUATIONS OF MOTION AND CONSERVATION LAWS

Let's now look at integrals of the equations of motion. If we multiply by dt and integrate, we find

$$\begin{aligned}\int fdt &= \int m(dv/dt)dt \\ &= m\int dv \\ &= mv_2 - mv_2 \\ &= \Delta mv.\end{aligned}$$

$\int fdt$ is called *impulse*, and $\underline{m}v$ is the *momentum*. Neither of these quantities has any special name, but the units are ML/T: gm-cm/s, dyne-s or lb-s. Multiplying by dx instead,

$$\begin{aligned}\int fdx &= \int (dv/dt)dx \\ &= \int (dv/dx)(dx/dt) \\ &= \int vdv \\ &= mv_2^2 - mv_1^2 \\ &= \Delta(mv^2/2).\end{aligned}$$

These quantities are much used: $\int fdx$ is *work*, and $mv^2/2$ is *kinetic energy* T. In cgs, the unit of mechanical energy is the erg (from the Greek for work), the dyne-cm, and in fps, it is the ft-lb. The MKS unit is joule = N-m, so $1 \text{ J} = 10^5 \times 10^2 = 10^7$ erg. The time rate of change of kinetic energy is $dT/dt = mv(dv/dt) = fv$, called *power*. It can be expressed as erg/s or ft-lb/s, and 10^7 erg/s = 1 W. An erg/s is 0.1 μW. In fps, we may use the colorful horsepower, 1 hp = 550 ft-lb/s. Since 1 ft-lb = 1.356×10^7 erg, 1 hp = 745.7 W.

If f is a function of position, $f = f(x,y,z)$, and $\int fdx$ around any closed curve is zero (as it is in the gravitational field of the earth), then it a function $V(x,y,z)$ can be found such that $f_x = -\partial V/\partial x$, or, in vectors, f = -grad V. This procedure is very familiar from electrostatics. The function V is the *potential energy*.

Then,

$$\begin{aligned}\int fdx &= -\int (dV/dx)dx \\ &= V_1 - V_2 = -\Delta V.\end{aligned}$$

Combining this with our earlier expression, we have $\Delta V + \Delta T = \Delta(T + V) = 0$, or $T + V = E =$ constant. This is the principle of *conservation of mechanical energy*, which proves very useful.

Energy leads the way into advanced dynamics, where it replaces the concept of force. For example, the potential energy function $V = kx^2/2$ yields the force $f = -kx$, which we know gives us the harmonic oscillator. We use Newton's Law to give $-kx = md^2x/dt^2$, or $x'' + (k/m)x = 0$, where the prime indicates differentiation with respect to t.

The solution of this equation is $x = A \sin \omega t + B \cos \omega t$, with A and B constants that are determined by the initial conditions, and $\omega = \sqrt{(k/m)}$. In advanced dynamics, we form the Lagrangian $L = T - V$ and use Lagrange's equations to get the equations of motion. This is only added complication in this simple case, but the method is capable of great generalization. Another important conservation law comes from Newton's Third Law, which states that if one body exerts a force on another, then the second body exerts an equal and opposite force on the other. This says that if

$$f = dp_2/dt,$$

then $-f = dp_1/dt$, from which it follows that

$$d(p_1 + p_2)/dt = 0,$$

so that $p_1 + p_2 =$ constant.

This is the principle of the *conservation of momentum*. Each component of the momentum is separately conserved. The conservation of energy and momentum are powerful tools for the study of collisions. An *elastic* collision is one in which the energy is conserved. Even in an inelastic collision, the momentum is still conserved.

If the force is constant, then we can integrate the equation of motion twice and find the position as a function of time. With $(f/m) = a$, we find $x = at^2/2 + v_0t + x_0$ with two constants, the initial velocity and position. This solution is studied in every elementary physics course, and is applicable to the everyday problem of projectile motion under gravity. There are many useful consequences of this solution, such as

$$v^2 - v_0^2 = 2a(x - x_0),$$

where the time has been eliminated between it and its first derivative. In the form $v = \sqrt{2as}$ it is very useful, even in things like hydrodynamics.

DYNAMICS IN POLAR COORDINATES

So far we have not emphasized the vector nature of Newton's Second Law, since in rectangular coordinates we can write one equation for each dimension and work entirely with components. In fact, many practical problems are one-dimensional, and the subject does not come up.

When we consider motion in a plane described by polar coordinates, however, we must look more deeply.

If a point is given by P(r,θ), then the unit vector r' is r/r, and the unit vector θ' is at right angles to it toward the direction of increasing θ. The primes will have to do here for identifying the unit vectors. Now, since

$$r = rr',$$
$$v = dr/dt$$
$$= (dr/dt)r' + r(dr'/dt)$$
$$= (dr/dt)r' + r\omega\theta', \text{ since } dr'/dt = \omega\theta',$$

where ω is dθ/dt.

Note also that dθ' = –ωr', because we need this for the acceleration: $a = [d^2r/dt^2 - r\omega^2]\, r' + [2(dr/dt)\omega + r(d^2\theta/dt^2)]\theta'$. This expression has many applications.

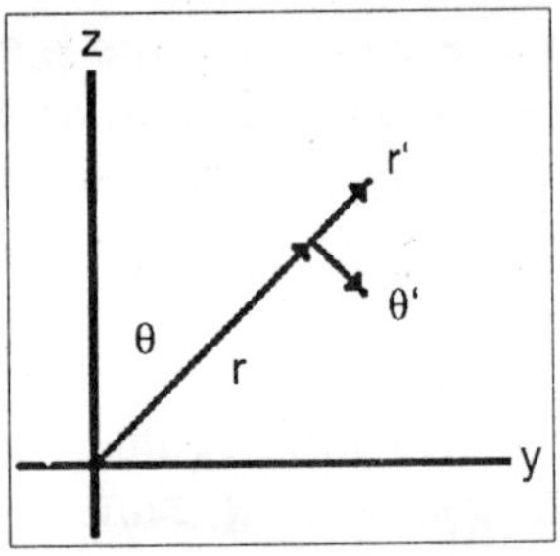

Fig. Polar Unit Vectors

Suppose first that the force acting on a particle moving in a plane is *central*, that is, directed toward the origin. Then the θ component of the acceleration must vanish, or 2(dr/dt)ω + r(dω/dt) = 0. If this is multiplied by r, we find that $d(r^2\omega)/dt = 0$, or $r^2\omega = h$, a constant.

Then, $mr^2\omega = mv_pr = H$, a constant. Here, v_p is the component of velocity perpendicular to the radius. In vector notation, H = mr × v is called the *angular momentum*, the moment of the linear momentum about the origin. With a central force, then, the angular momentum is a constant. One consequence of this is Kepler's Area Law.

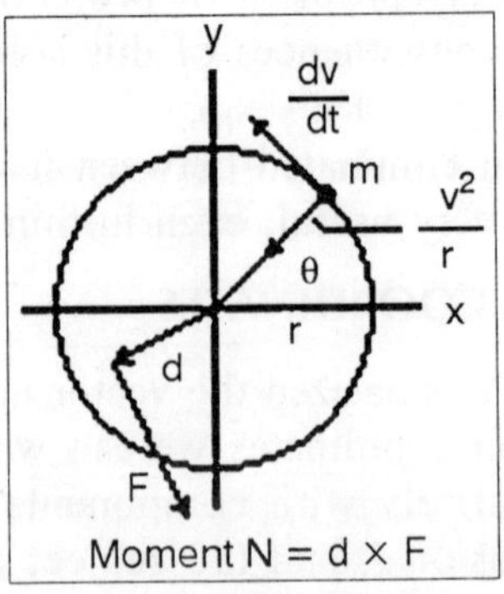

Fig. Circular Motion

Now suppose the particle is moving in a plane with constant r. This is the case for all the points in a rigid body rotating about fixed axis normal to the plane.

Then a = $-r\omega^2 r' + r d\omega/dt \theta'$. The first term is an acceleration directed toward the origin of amount v^2/r, the *centripetal acceleration* that must be supported by a force directed toward the orgin. The second term is zero unless there is an *angular acceleration*, which requires a force in the direction of dv/dt.

This force is f = mr $d\omega/dt$, and its moment about the origin is N = $mr^2 \, d^2\theta/dt^2$, which is an angular form of Newton's Second Law. If we are dealing with a rigid body, then I = Σmr^2 is called the *moment of inertia of mass*.

We can then write N = I $d^2\theta/dt^2$, with N analogous to force. In the diagram, the applied moment N balances the accelerating moment. The forces do not have to have the same direction or magnitude, just the same moment.

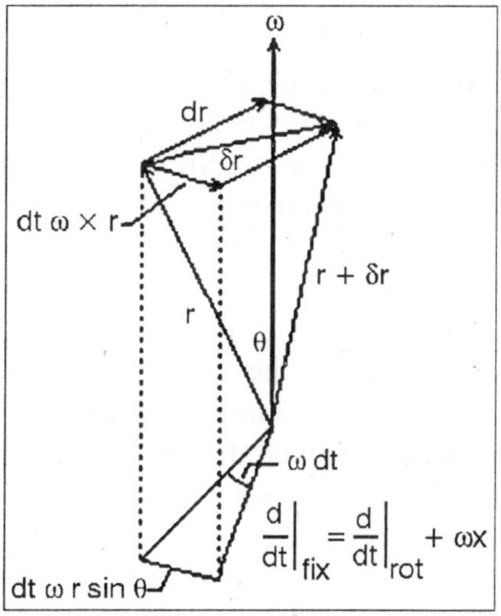

Fig. Time Derivative in a Rotating System

In vectors, if f = dp/dt, then N = r X f = r X (dp/dt) = (d/dt)(r X p) = dH/dt, since v × p = 0 because v and p = mv are in the same direction. This shows that N = dH/dt in general.

INERTIAL SYSTEMS, ROTATING SYSTEMS AND THE CORIOLIS FORCE

So far we have only considered one coordinate system, in which Newton's Second Law holds. For this reason, it is called an *inertial system*. Newton's First Law defines what is meant by an inertial system in which the Second Law can be applied. Consider a second system, moving with uniform velocity u with respect to this system. If primes refer to this system, then $r' = r + ut$.

Then, the velocity and acceleration in the moving system are $v' = v + u$ and $a' = a$. Since the acceleration is the same, Newton's Law holds as well in the moving system.

Systems moving with constant relative velocities to an inertial system are also inertial. This is the *principle of Galilean relativity*, which is valid so long as the relative velocities are much smaller than the speed of light. At greater velocities, we must use Einstein's principle of relativity instead.

Now let's consider a system rotating with a uniform angular velocity ω with respect to an inertial system.

The diagram shows that in addition to the change in a vector in the rotating system, dr, the change $\omega \times r$ must be added to get the full change δr in the inertial system. This is a general result for the time derivatives of a vector, that can be applied to the velocity as well. We have $v' = v - \omega \times v$, and then $a' = a - \omega \times (\omega \times r) - 2\omega \times v$. Obviously, the rotating system is not inertial.

However, it will seem like an inertial system if we add two *fictitious* forces to the acceleration, as defined by the second and third terms. The first is the *centrifugal force*, the second the *Coriolis* force.

The centrifugal force is independent of the velocity and proportional to the square of the angular velocity, and is in a plane normal to the axis of rotation. The Coriolis force is proportional to the angular velocity and the linear velocity, and is normal to the plane defined by them.

A coordinate system fixed to the surface of the earth is such a rotating system. The angular velocity $\omega = 2\pi/86\ 164.1\ s = 7.29211 \times 10^{-5}$ radian/s. The sidereal (with respect to a fixed direction in space) period is used, not the period relative to the sun that we use for time. This is a small value, but has noticeable effects.

The centrifugal force is included in the effective gravitational force mg, so the only fictitious force to be considered is the Coriolis force. The surface of the earth is normal to the direction of the effective gravity.

Centrifugal force is the reason the earth is a spheroid. The effective component of the angular velocity is normal to a horizontal plane at the point of observation, with magnitude $\omega \sin \varphi$, where φ is the latitude. Its positive direction is upward in the Northern hemisphere, downward in the Southern hemisphere.

The direction of the Coriolis force is normal to a horizontal velocity, so that it would turn the velocity clockwise in the Northern hemisphere, and anticlockwise in the Southern.

The Coriolis force has little effect on the water draining out of a bathtub, but a great effect on large-scale winds, where it causes winds to blow along isobars, instead of directly into a low-pressure area or out of a high-pressure area.

The result is that winds circle a low-pressure area anticlockwise and a high-pressure area clockwise in the Northern hemisphere and that these areas persist as they move about. This is certainly the most visible effect of the Coriolis force, and additional proof that the earth rotates relative to an inertial system.

The Coriolis force also causes the plane of a pendulum to rotate clockwise in the Northern hemisphere, as can be seen by considering the velocity of the bob.

The rate of this rotation is proportional to sin φ, so it vanishes at the equator and is a maximum at the poles. At the poles, the angular velocity is equal to that of the earth. At my latitude, 40°N, the pendulum makes one rotation in about 37 hours and 20 minutes. Foucault first observed this rotation, in 1851, in a famous experiment, and it is displayed in every good Science Museum. It is very difficult to create an unbiased support for the pendulum, and to release it properly, so it is a difficult home experiment. A ball dropped from a height in the Northern hemisphere winds up to the East of the point from which it was dropped (for a 200 ft drop, the distance is 0.31 inches).

FRICTION AND VISCOSITY

Coulomb not only verified the inverse-square force between electric charges and measured them absolutely, but also established the laws of friction. Friction is an extremely variable phenomenon, but one of considerable practical interest to engineers. Coulomb's main result was that the force preventing two solids with plane faces in contact from sliding was independent of the area of contact, but proportional to the normal force pressing the two solids together. The ratio of the frictional force to the normal force is called the *coefficient of friction*, $\mu = F/N$. Moreover, when the two solids were sliding, the coefficient was independent of the velocity of sliding. However, the *static* coefficient was somewhat greater than the *kinetic* coefficient. Frictional forces that obey these rules are called Coulomb friction. The coefficient of friction is dimensionless, the same in every system of units.

The coefficient of friction depends greatly on the preparation of the surfaces in contact. Clean, smooth surfaces will sometimes seize together in a vacuum so that they cannot be easily separated. It is never a good idea to have the same metal in contact at a bearing. For these reason, measurements of the coefficient of friction are not of great use, except to show average values. Nevertheless, we should have some idea of the range of values. Dry wood on wood shows $\mu = 0.25 - 0.50$, wood on metal $0.2 - 0.6$. Steel on steel is 0.58 dry, but falls to 0.16 when lubricated with light oil. Steel on brass, lubricated, gives 0.19.

The viscosity of a liquid or gas expresses the force necessary to move surfaces with respect to one another with the liquid or gas between them. It is assumed that the fluid adheres to the metal on either side, establishing a velocity gradient. The viscosity is the ratio of the force per unit area in the direction of motion, divided by the velocity gradient,

or $$\eta = (F/A)/(v/d) = Fd/Av.$$

In cgs, the units are dyne-s/cm^2, given the name of poise, after Poiseuille. In fps, they are lb-s/ft^2. Using the known unit ratios dyne/lb and ft/cm, it is easy to find that 1 lb-s/ft^2 = 478.8 poise. In fps, the lb-s/in^2 or Reyn is often used, which is 3.325 poise. The viscosity of water at 20°C is 0.01 poise, or 1 centipoise, easy to remember.

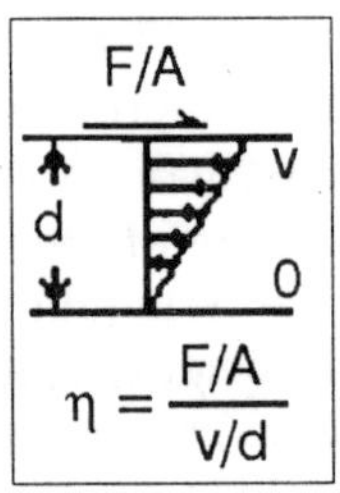

Fig. Coefficient of Viscosity

The viscosity of liquids is a strong function of temperature. In water, it varies from 1.787 cp at freezing to 0.2818 cp at boiling. Glycerine has 1490 cp at 20°C, hexane (0.6603) 0.326 cp and methanol (0.7914) 0.597 cp. Light machine oil has 34.2 cp at 100°F (37.8°C).

These figures give an idea of the viscosity of common liquids at normal temperatures. The viscosity divided by the density is found in enough formulas that it has a name, *kinematic viscosity*, and a unit, *stokes*, of its own.

Since the density of water is 1, its kinematic viscosity in stokes equals its viscosity in poise. The densities given above allow you to find the kinematic viscosity of a few liquids. Air, at 18°C, has a viscosity of 182.7 μpoise, which shows a typical value for a gas viscosity. Since its density is much lower than that of a liquid, its kinematic viscosity is not much different from that of a liquid (about 18 cp).

Hagen and Poiseuille studied the flow of water and blood, respectively, in capillary tubes around 1839-40. The flow through a tube of radius r and length s with a pressure difference of p dyne/cm^2 between its ends is $q = \pi p r^4/8s\eta$ cm^3/s, where η is the viscosity in poise. The *Reynolds number* R for this flow is $R = DV\rho/\eta$, where D is the diameter of the pipe, and V the velocity of flow. Poiseuille's formula is valid when R is less than about 2000. Above this value, the flow becomes turbulent and different rules apply. Another famous case of viscous flow is the movement of a sphere of radius a through a viscous fluid.

In this case, the force resisting the motion is $F = 6\pi\eta aV$ dynes, where η is in poise, a in cm, and V in cm/s. A glass marble 13 mm in diameter, falling in glycerine, will have a terminal velocity of $V = (4\pi/3)(0.65)^2 (981)(1.24)/6\pi(14.90) = 7.67$ cm/s. This is found by equating the viscous force to the weight of the marble (sp.gr. = 2.5) less the buoyant force of the glycerine. The Reynolds number in this flow is $R = DV\rho/\eta$ as with Poiseuille flow, and in this case is only 0.84, so viscous flow is guaranteed. Millikan used the same theory in his oil drop experiment to determine the charge on the electron. He measured the weight of a drop by the rate at which it fell in air.

Examples

Let's solve the pendulum with a small amplitude from what we have found above. This is motion in a plane with constant r, and the force on the bob of mass m is downward, and of amount mg. The moment of this force about the origin is $N = mgl \sin\theta$, tending to decrease θ, so the equation of motion is $mgl \sin\theta = -ml^2\, d^2\theta/dt^2$. Note that the tension in the cord has no moment about the origin, and so does not enter into the equation of motion.

For small oscillations, sin θ H” θ, so we have $\theta'' + (g/l)\theta = 0$, where the primes indicate differentiation with respect to time. The solutions of this differential equation are well known, and θ oscillates sinusoidally with angular frequency $\sqrt{(g/l)}$. A seconds pendulum, then, has a length $l = g/\pi^2 = 99.4$ cm. The period of a seconds pendulum is 2 seconds, one second for each swing that releases the escapement one notch, so the clock ticks once a second. The pendulum was once used to measure g, but now the fall of a ball is measured electronically, to an accuracy of a milligal or better.

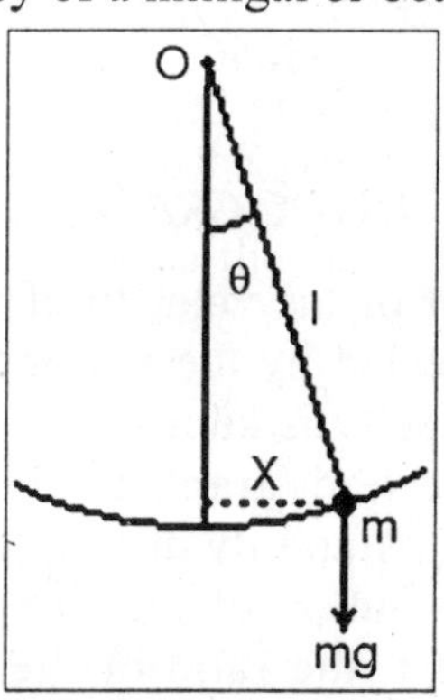

Fig. Pendulum

A ball leaves a cannon with a speed v_0 of 1800 fps and an elevation of $\theta = 15°$. How far will it go on a level field? The initial upward velocity is 466 fps, so the ball will reach the summit of its trajectory at a height h in 466/32.17 = 14.48 s, and fall back to earth in 28.97 s.

In this time, it will travel horizontally d = 1739 fps x 28.97 s = 50,362 ft, or 9.54 miles. Of course, it will fall well short of this, because air resistance has a large effect, especially on supersonic cannonballs. If the muzzle velocity is only 180 fps, then it will travel for only 2.897 s, and plow up the ground only 504 ft away. This would be closer to correct, but it would be a really puny charge. Note that the range is not proportional to the muzzle velocity, but to its square.

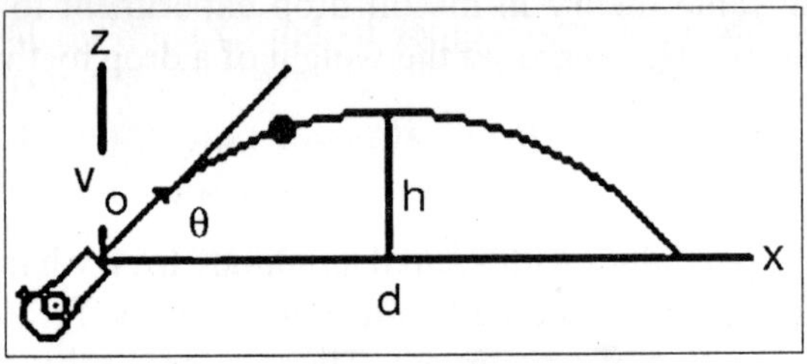

Fig. Trajectory

Such simple problems as these could be multiplied indefinitely. They are not only excellent exercises, but also build confidence in your ability to understand at least this much of the universe.

Physical Constants In CGS

Speed of light $c = 2.9979 \times 10^{10}$ cm/s
Electron charge $e = 4.80325 \times 10^{-10}$ esu
Electron mass $m = 0.9109534 \times 10^{-27}$ g
Planck's constant $h/2\pi = 1.05459 \times 10^{-27}$ erg-s
Bohr magneton $\mu_B = 0.9274078 \times 10^{-20}$ erg/gauss
Boltzmann's constant $k = 1.38066 \times 10^{-16}$ erg/K
Universal gas constant $R = 8.31441 \times 10^{7}$ erg/K-mol

HODOGRAPH

The *hodograph* is a plot of the velocity of a particle as a function of time. The velocity is represented by the vector from the origin to a point on the hodograph. If the velocity is known as a function of the time, then it is easy to pass from the hodograph to the *trajectory* by a simple integration, since $dx/dt = v_x(t)$ and $dy/dt = v_y(t)$.

The hodograph is the solution of the first-order equation $dv/dt = f$, which is Newton's Law. If f, the ratio of the force to the mass of the particle, is given as a function of t, then v is easily found. In most cases, f is a function of the position of the particle, which makes the problem more difficult. If the trajectory is known, then it is easy to find the hodograph by differentiation with respect to time. In most Physics courses, particle mechanics is based on the trajectory, while the hodograph is ignored, which is a shame.

Hodograph is from Greek "hodos∂ and "grapho,∂ meaning path-writing. Hodos is the general word for road or path, beginning with an "o∂ with rough breathing (if you want to look it up in a lexicon). In form, it looks masculine, but is of feminine gender.

It is not a particularly well-chosen term ("tachygraph" might have been better) but it is the traditional one. The hodograph was invented and named by William Rowan Hamilton, the brilliant Irish mathematical physicist, inventor of quaternions and Hamiltonian mechanics, in 1847, and was used by James Clerk-Maxwell to analyse orbital motion and to show in an easy way that Kepler's Laws imply an inverse-square law of force. The hodograph is drawn through the ends of vectors drawn from a point to represent the velocity of the particle at successive instants. There is not much danger of learning anything from this definition. Therefore, this article may well supply a need.

MOTION WITH CONSTANT ACCELERATION

The usual case of motion with constant acceleration is motion under the influence of gravity, The equation of the hodograph is then dv/dt = g, which integrates to the straight line gt, whose length is proportional to the time of the motion. If the initial velocity v_0 is drawn from the origin to a point P, then the hodograph is a vertical line lengthening downwards.

The velocity at any time is easily read off, and if the motion ends at a point Q, the corresponding velocity is $v_f = v_0 + gt$, making a triangle OPQ with side PQ vertical. The trajectory, of course, is a parabolic arc that is more difficult to work with than the simple triangle.

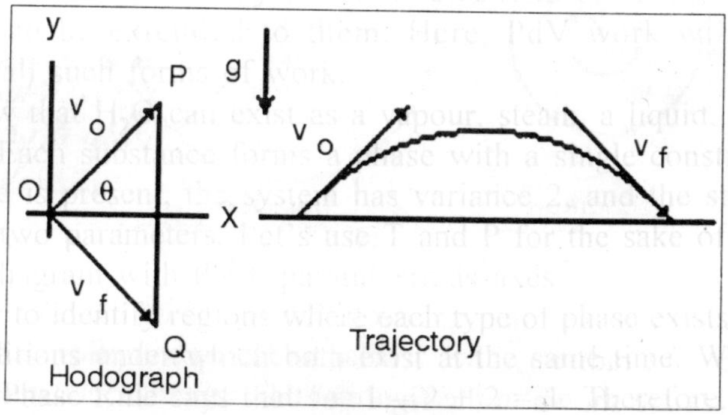

Fig. Projectile Motion

The area of the triangle is half the base, gt, times the altitude $v_o \cos \theta$, where θ is the angle between the initial velocity and the horizontal, the angle of projection.

Also, gt/2 is $v_o \sin \theta$. Eliminating v_o, we find the altitude to be gt/(2 tan θ), and its area $(gt/2)^2/\tan \theta$. This area is proportional to the horizontal distance covered by the trajectory, since the altitude is the (constant) horizontal component of the initial velocity, and the base is proportional to the time. The proportionality constant of the area of the triangle to the horizontal distance covered is g/2, of course.

To find the condition for the maximum range as a function of the angle of projection, differentiate the area with respect to θ and set the derivative equal to zero.

This gives $\cos 2\theta = 0$, or $2\theta = 90°$, or $\theta = 45°$, the well-known result. Therefore, the initial and final velocities are perpendicular, so that the hodograph is an isosceles triangle.

For a particle projected directly upwards, the hodograph will be a vertical line of length 2v with its centre at the origin. The time of flight t is given by 2v = gt, or t = 2v/g. At time t/2, we are at the origin, and the velocity is zero.

The area of the triangle is zero, meaning that the particle goes nowhere horizontally, as we know. The hodograph of a particle moving with constant velocity is a point, and the corresponding trajectory is a straight line. These simple relations are easy to visualize, and help to illustrate the meaning of the hodograph.

The hodograph of a uniform circular motion with angular velocity ω and radius r is a circle of radius v = rω, the peripheral velocity, with its centre at the origin.

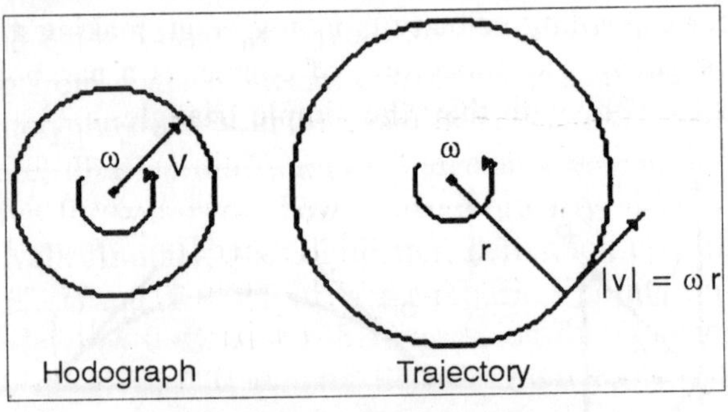

Fig. Uniform Circular Motion

The hodograph is a quarter-revolution ahead of the trajectory, which is also a circle, of radius r. The circular hodograph implies steady rotation. These relations are shown in the figure at the right. The constant magnitude of the acceleration is $v^2/r = \omega^2 r$, directed toward the centre of the trajectory circle and perpendicular to v. Note that this *centripetal* acceleration is in the same direction in the hodograph and the trajectory. The radii of the circles differ by a factor of ω.

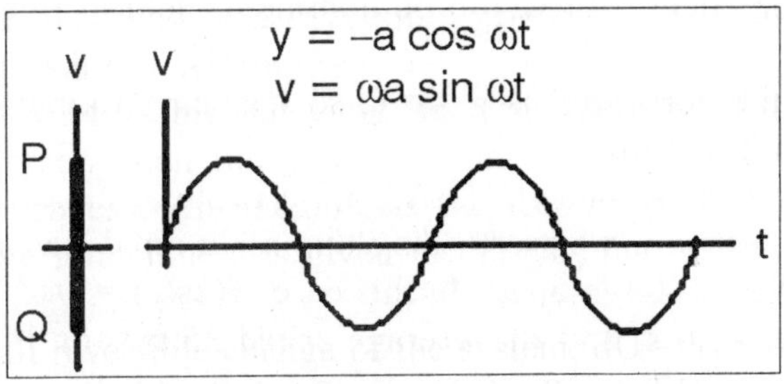

Fig. Hodograph: Simple Harmonic Motion

The hodograph of a simple harmonic motion is shown at the left, with the time dependence displayed. The amplitude of the motion is a and its angular frequency is ω = 2π/T = 2πf, where T is the period and f is the frequency.

The velocity also varies sinusoidally, with an amplitude ωa. Points P and Q correspond to the passage of the particle through y = 0, and the centre of PQ to the extreme positions.

THE BRACHISTOCHRONE

This is a famous problem of constrained motion under gravity. The problem was published by Johann Bernoulli (1667-1748) in 1696, and was solved by him, Isaac Newton (1642-1727) and G. W. von Leibniz (1646-1716) shortly afterwards.

It was the stimulus for the development of the calculus of variations by Euler and Lagrange, in which an unknown function is determined that minimizes a definite integral. The name is well-chosen from Greek, from "brachistos,∂ shortest, and "chronos,∂ time.

The particle is imagined to move from point A to a second point B at the same level, but some distance away, sliding on a certain curve without friction. It is clear that the particle will move away from A if the path slants downward in the direction of B, speeding up until it reaches the lowest point, then with decreasing velocity until it just reaches B. If the path is the straight horizontal line AB, the particle will never start moving.

The particle could be dropped from A, and when it attained a certain velocity by dropping a distance h, be deflected horizontally in the direction of B.

When it was directly under B, it would again be deflected upwards, finally stopping when it just reaches B. In this case, the minimum time required is T = 2"(2s/g), if s is the horizontal distance from A to B. T is infinite for h = 0 or for h = ∞, so there must be a minimum in between. The value h = s/4 gives the minimum T, which can be found by minimizing the expression for T with respect to h.

A more practical path would be a straight inclined ramp for half the distance on each end. The effective acceleration on a ramp making an angle θ with the horizontal is g sin θ, so it is easy to calculate the total time in this case as well.

The answer is the same, 2"(2s/g), for a depth at the centre of h = s/2 (a 45° ramp). Suppose we were using this for a rapid transit system, and two stations were 1000 m apart. In this case, √(2s/g) = 14.3 s, so the total time would be 28.6 s, and the average speed 35 m/s or 142 km/h or 89 mph.

This would be rapid transit indeed, with the maximum speed no less than 223 mph (the speed attained by falling 500 m). Friction, however, stands in the way of this energy-efficient transport system. Proposals have been made to use gravity for this purpose, but no bold ones have been realized.

If the path is given by the function y = y(x), then the speed at any point x is v = √(2gy), and the path length ds = √(1 + y'²) dx, where y' = dy/dx, so the time required to go from 0 to x can be expressed as the definite integral T = ∫(0,x) [√(1 + y'²)/∂(2gy) dx. We search for the path y(x) that gives a minimum value for the integral.

What sort of path would give us the absolute minimum time? We can anticipate that it would be a smooth path, replacing the linear ramps and abrupt falls and rises, and that the minimum depth would be between s/4 and s/2.

Bernoulli, Newton and Leibniz found the surprising result that the path is a cycloid, and the maximum depth is s/π. The time required is √π∂(2s/g), so √π = 1.7725 replaces 2. The time for our rapid transit is then 25.3 s instead of 28.6 s.

This path is the *brachistochrone*, which gives a shorter time than any neighboring path.

It is clear that since the time is a minimum for the brachistochrone, the time will not be greatly different for any approximate path. Even the ramps give only a 13% difference.

A parametric equation for the cycloidal path is x = (s/2π) (θ – sin θ), y = –(s/2π)(1 – cos θ). The angle θ is the angle of rotation of a circle of radius s/2π rolling on the underside of the x-axis, and goes from 0 to 2π. This curve is shown at the right, with the generating circle indicated. It is also the path of an electron starting from rest at A in an electric field in the direction of -x, and a magnetic field out of the plane, in the z-direction. It is not easy to find this result with the usual methods of the calculus of variations because of algebraic complexity. However, the hodograph offers an easier solution.

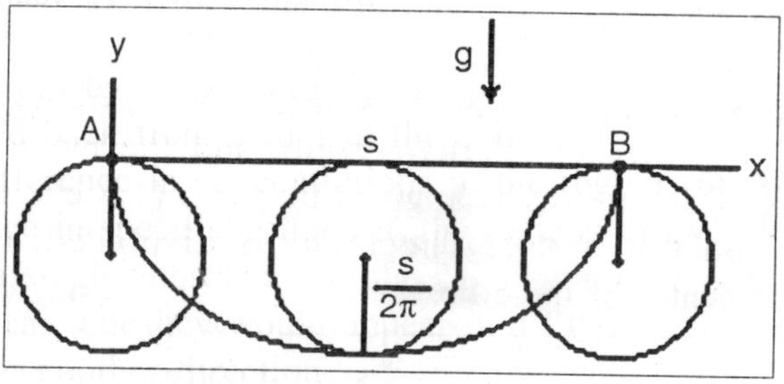

Fig. The Brachistochrone

The hodograph will begin at the origin, sloping downward as the particle picks up speed. Note that the acceleration is parallel to the path, not directly downward as in unconstrained motion. The hodograph will then loop around, reaching the horizontal axis at the minimum of the curve, where the velocity will be greatest, then looping around above as velocity decreases, and finally approaching the origin with a downward slope. Integrals can be written for the area of the hodograph (time of the motion), as well as for the horizontal distance covered, in terms of a hodograph $u(\theta)$. [This is not the parameter θ of the preceding paragraph, but the inclination of the velocity vector.] The area of the hodograph will be proportional to the time of the motion, so we look for a curve that will minimize the area inside the hodograph, while covering the necessary horizontal distances.

The first integral is to be minimized, with the second as a condition, for which the method of the Lagrange multiplier is convenient. Apostulatos shows that the hodograph will be a circle tangent to the vertical axis at the origin. It is, in fact, $v(\theta) = \sqrt{(2gs/\pi)} \cos \theta$. The coefficient of the cosine is the diameter of the circular hodograph.

If we add a constant velocity to the left so that the hodograph becomes a circle centred at the origin, that is, a velocity $\sqrt{(gs/2\pi)}$, we get the hodograph of a uniform circular motion. Therefore, the actual motion is a circular motion translating to the right at a speed equal to the peripheral speed of the circular motion. This is, of course, a cycloid. The total time is then the distance s covered by the rolling circle divided by its peripheral velocity, $\sqrt{(gs/2\pi)}$, which gives the expression quoted above.

PLANETARY ORBITS

The first two of Kepler's Laws of Planetary Motion are: (1) the planet moves in an ellipse with the sun at one focus; and (2) the area swept out by the radius vector is the same in equal time intervals.

An elliptical orbit is shown at the right, in which the planet is located by its radius vector r and the true anomaly θ. The velocity is greatest at perihelion and least at aphelion, and in both cases is perpendicular to the radius, which attains a minimum and maximum value, respectively. In between, there is a component of the velocity dr/dt parallel to the radius. The quantity p is the semi-latus-rectum, and the mean distance, or semimajor axis of the ellipse, is a, given by $p = a(1 - e^2)$, where e is the eccentricity, and $0 \le e < 1$. The focus is displaced by a distance c = ae from the centre of the ellipse.

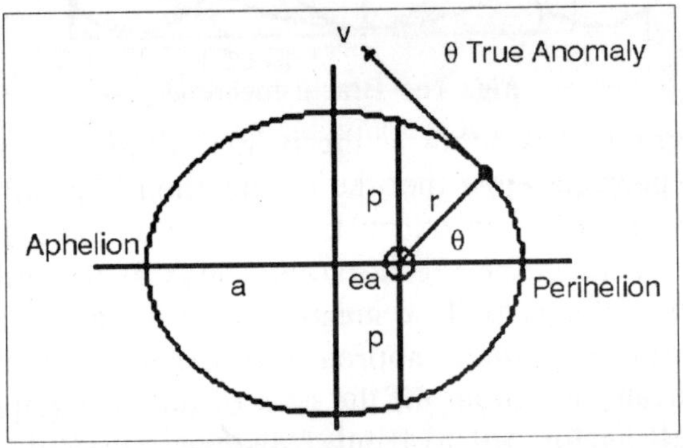

Fig. Trajectory

The polar equation of the orbit is shown at the left, together with the mathematical expression of the Law of Areas. The Law of Areas is equivalent to the conservation of angular momentum, which is proportional to the constant h.

These two expressions are derived directly from Kepler's Laws. If we differentiate r with respect to time, we can find the radial and transverse components of the velocity. The parameter θ is the true anomaly, and the radial and transverse components are relative to this direction. Plotting these components will give us the hodograph of the motion.

$$r = \frac{p}{1 + e\cos\theta} \qquad p = a(1 - e^2)$$

$$r^2\dot{\theta} = h \text{ Law of Areas}$$

$$v_r = \dot{r} = (he/p)\sin\theta$$

$$v_\theta r\dot{\theta} = (h/p)(1 + e\cos\theta)$$

The hodograph is shown at the right. It is, remarkably, a circle with centre at O, which does not coincide with the origin unless e = 0 and the orbit is a circle.

To draw the hodograph, all that is necessary are the velocities at perihelion and aphelion. These are laid off above and below the velocity origin, and form the diameter of the hodograph. The velocities marked 90° and 270° are the velocities at the ends of the latus rectum. Note that they fall at the intersections of a horizontal line through O with the hodograph.

If we add a downward velocity of he/p, the centre O is brought to the origin, and the hodograph of this motion is a uniform circular motion. This fact can be used to solve the problem of the position in orbit as a function of time (Kepler's Problem), and the angle of rotation is the mean anomaly, which increases uniformly with time. The details will not be mentioned here, but the relation of the method to the hodograph is clear from what has been said.

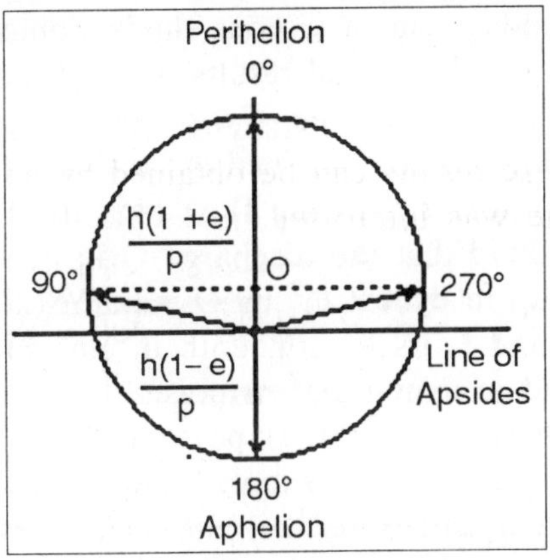

Fig. Hodograph of an Orbit

Bringing in Kepler's Third Law, that the cubes of the mean distances are proportional to the squares of the orbital periods, allows the determination of the law of force, which turns out to be the familar inverse square.

To obtain this result, note that the acceleration in the circular motion is the total acceleration, since the velocity he/p is a constant, and it is easy to find the force for a uniform circular motion. The reader is encouraged to solve the problem of a circular orbit under an inverse-square force, and from it to show that Kepler's Third Law results. Then, it is not difficult to see how to prove the inverse.

Chapter 4

Stokes's and Poiseuille's Laws

Let's begin by stating the formulas that are the goal of this analysis. They deal with the movement of viscous fluids, commonly air and water, and are some of the rare analytical results of fluid mechanics of practical value. We will study the dynamics of viscous fluids in this article, and show how these results can be obtained by analysis.

Poiseuille was interested in the flow of a fluid, actually blood, in narrow tubes, and found that the discharge Q in cc/s was related to the pressure difference Δp in dynes/cm^2 by $Q = (\pi a^4/8L\mu)\,\Delta p$. Here, a is the radius of the tube, and L its length, both in cm, and μ is the absolute viscosity of the liquid in dyne-s/cm^2 or poise.

The viscosity of water is 0.0101 poise at 20°C. Blood, with which Poiseuille was concerned, is not a simple Newtonian fluid, and its flow is a complicated problem, but water, oil and such obey the equation very well. The catch is that the flow must be rather slow, or the flow will become turbulent, obeying different rules and the resistance to flow will increase.

The flow is guaranteed to be laminar (not turbulent) provided that the Reynolds number $R = DV\rho/\mu$ is below 2000, as found from experiment. Here, D is the diameter of the tube, 2a (cm), V is the average velocity of flow $Q/\pi a^2$ (cm/s), μ the absolute viscosity, and ρ the density in gm/cc. This number is dimensionless, and is the same in any consistent system of units.

For a 1/2" pipe, R is 2000 for an average velocity of 15.7 cm/s of water, or Q = 20 cc/s, which is not a very rapid flow. We should also be aware that the laminar flow is not established immediately at the entrance to the tube, but takes a little length for this to happen, in which the liquid at the centre speeds up, and that at the wall of the pipe is retarded.

Stokes solved the problem of the steady motion of a sphere with velocity V through a still viscous fluid (or the flow of the fluid past a stationary sphere, which is the same thing).

He found that the force exerted on the sphere by the fluid was $F = 6\pi\mu aV$, where F is in dyne, a in cm, V in cm/s, and μ is the absolute viscosity. Note that the force is proportional to the velocity, the rule we often assume for a frictional force. Again, the motion must not be too fast, and a Reynolds number $R = DV\mu/\rho$ can be used to decide what is too fast.

This has the same form as in Poiseuille's equation, but the flow is different, and the critical value may be different. If V is too great, again the flow becomes turbulent, with the formation of a turbulent wake behind the sphere, and now the force is proportional to the square of the velocity. We usually write $F = C\ A(\rho V^2/2)$, where C is the dimensionless *drag coefficient*. Where Stokes's formula applies, $C = 24/R$.

The critcal Reynolds number is about 1 for air, and the drag coefficient for high R is about 0.4, dropping to about 0.2 when the boundary layer changes from laminar to turbulent. This is a complex and interesting subject that you can read more about in the references.

The streamlines in Stokes flow dip in toward the sphere, so to avoid disturbing the flow pattern, there must be no nearby boundaries. A ball dropping in a tube about twice its diameter would be significantly affected (the drag would increase).

Its area of application is really to rather small spheres in air, such as cloud particles, or to slow motion in a viscous fluid, such as glycerin or oil. It was famously used by R. A. Millikan in the determination of the electronic charge, to measure the weight of the drops by how fast they fell. The drops were so small that a correction for the mean free path of air molecules was necessary for an accurate result.

The problem can also be solved neglecting viscosity, and the results are entirely different. Some are applicable to reality, but others are not. When a sphere moves in an ideal fluid, it suffers no drag whatsoever, but once started in motion, would continue forever. The effective mass of the sphere is increased by half the mass of the fluid displaced, a result that is quite valid and observable.

It can be applied to the motion of bubbles in beer, for example. Most real-life problems of fluid flow depend greatly on viscosity, even when it would seem to have a small effect. The drag on a body moving in a fluid is mostly caused by the lower pressure in the turbulent wake behind it, and the drag is reduced by making this wake as small as possible.

The pressure on the front of the body has only a small effect. Trucks with the "streamlining" in front of the trailer would do better to put it behind!

We recall that the coefficient of viscosity μ is defined by Newton's equation $F/A = \mu(dv/dy)$, where F is the shearing force on an area A, and dv/dy is the velocity gradient normal to the surface. If μ is a constant independent of the velocity of shearing, the fluid is called *Newtonian*. Many, if not most, fluids met with in engineering are Newtonian. Some, like paint and blood, are not, and there is a variety of interesting behaviors in some other substances, especially those that are a mixture of phases. μ is measured in *poise*, dyne-s/cm^2.

STRESS AND STRAIN

To begin our study, we must review the nature of stress and strain. This will require the use of quantities with a number of components in three-dimensional space, such as vectors, as well as more complex quantities. These can be represented explicitly, and were indeed for many years, but this involves tedious repetitive writing.

Vector analysis provides an excellent shorthand, and is widely understood and used. For more complex quantities, this was extended to what were called *dyadics*, but these proved rather cumbersome, and seem to have disappeared.

What turns out to be very convenient is representation by Euclidean tensors using indices. This essentially writes all the components, as was originally done, but in a very compact and useful form. The vector a becomes the quantity a_i, where we understand that the index i takes the values 1, 2 and 3, or x, y and z, or whatever. The scalar produt of a and b is $a_i b_i$, where the repeated index implies summation over its range of values, called *contraction*.

Suppose that x_i is a point in space occupied by a solid body, and that ξ_i is the displacement of that point when the body is strained. Then, $\varepsilon_{ij} = \partial_i \xi_j$ is the *strain*, or ratio of displacement to distance.

The symbol ∂_i is short for ∂/∂_i. This is a new kind of quantity, depending on two directions. Suppose this region of the body isn't really strained, just rotated bodily by an angle θ.

Considering the displacements that result from such a rotation, it is clear that $\varepsilon_{ij} = -\varepsilon_{ji} = \theta$. We really don't want these if we are studying strains, and can get rid of them by forming $\Phi_{ij} = (\varepsilon_{ij} + \varepsilon_{ji})/2$, which is called *symmetrizing* Φ. What's left is just the *pure strain*, described by the *symmetric tensor* $\Phi_{ij} = \Phi_{ji}$.

There are six independent components of this tensor, three diagonal and three off- diagonal.

Among other things, this means that the order of the indices is immaterial, so we do not have to worry about it. In a fluid, instead of the displacements, we use the rate of displacement, the time derivative of ξ_i instead (which is just the fluid velocity v), so Φ becomes the *rate of strain* tensor.

Strain causes stress, so we must find some way to specify stress. If we imagine a plane cutting the solid, then the partA of the solid on one side of the plane exerts a force on the other side. The ratio of this force to the area considered is the *stress*. Given a normal unit vector to specify the orientation of the plane, then the stress is another vector that is a function of the direction of the normal vector, as well as of position.

We expect, then, that stress can be described by a quantity Ψ_{ij}. By considering the equilibrium of a small volume, the fact that it must not rotate is reflected in the *symmetry* of Ψ. Like Φ, then, Ψ is symmetric and has six independent components.

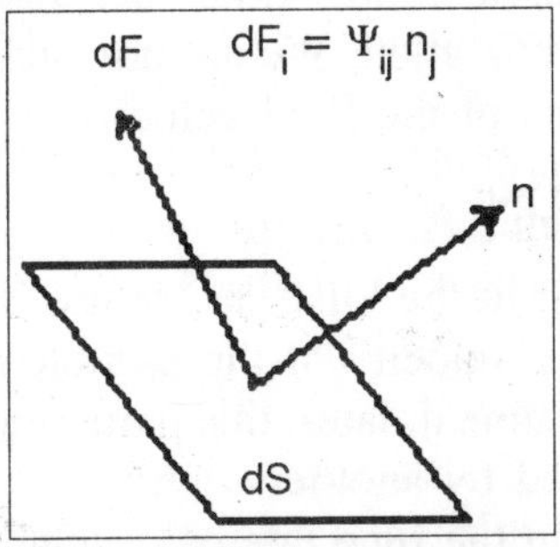

Fig. Stress Tensor

In a fluid at rest, the only stress that can exist is a pressure normal to any area, and this pressure is the same in any direction. This is, in fact, the definition of a fluid.

Therefore, considering areas normal to the coordinate axes, we find that $-pn_j = \Psi_{ij}n_j$, where n_i is a normal vector, and a pressure is considered a negative stress.

This means that $\Psi_{ij} = -p\delta_{ij}$, where δ_{ij} is the *Kronecker delta*, which is 1 when the two indices are equal, 0 otherwise. In solids or in viscous fluids, the stress may also have a component in the plane, called a *shear stress*. In this case, Ψ will acquire off-diagonal components.

In an isotropic solid, the relation between stress and strain, or between Ψ and Φ, takes the form $\Psi_{ij} = [k\Phi_{kk} - (2\mu/3)\Phi_{kk}]\delta_{ij} + 2\mu\Phi_{ij}$. The quantity Φ_{kk} is the sum of the diagonal elements of Φ, also called its *trace*. It is equal to $\Delta V/V$, the relative volume expansion of the material, and called the *dilatation*.

The trace does not depend on the orientation of the axes, which should be clear from its meaning.

In general, each of the six components of Ψ should depend on the six components of Φ, giving 36 constants of proportionality if we assume Hooke's Law, that stress is proportional to strain.

If the medium is isotropic, or the same in any direction, then these reduce to only two constants, the modulus of expansion k, $p = -k(\Delta V/V)$ (positive p causes a decrease in V), and the modulus of shear, μ. Young's modulus, for example, is $Y = 3k\mu/(k + \mu/3)$ and Poisson's ratio is

$\sigma = (3k - 2\mu)/(6k + 2\mu)$. This is much simpler than it appears. Try contracting Ψ and some normal vectors and see what stresses you get, for simple Φ's.

In the case of a fluid, we find analogously that $\Psi_{ij} = [-p - (2\mu/3)\Phi_{kk}]\delta_{ij} + 2\mu\Phi_{ij}$, in terms of the pressure p and the absolute viscosity μ, which relates a shear stress to the rate of shear strain. The analogy is clear. $\Phi_{kk} = \mathrm{Tr}\,\Phi = \partial_k v_k = \mathrm{div}\,v$ is the dilatation.

For an incompressible fluid, this quantity is zero, and the stress tensor become even simpler. Unlike solids, fluids are almost always isotropic, so all this is less of an approximation. We are now able to express the stresses in a viscous fluid in terms of the fluid velocity.

EQUATIONS OF MOTION

The motion of a fluid in the Eulerian picture is defined by the velocity field v or v_i, which is the velocity of the particle of fluid that happens to be at P(x,y,z) at time t. A time dt later, this particle has moved off a distance v_i t and has been replaced by another.

If we are interested in the time rate of change of some property of the particle of fluid, which we shall call F, then dF/dt must "follow the particle," or $dF/dt = \partial F/\partial t + v_i\partial_i F$.

This is called the *substantial derivative*, and is one of the first things explained in fluid mechanics.

The mass of fluid must be conserved in the motion, or $d\rho/dt + \rho\partial_i v_i = 0$, where the dilatation appears again. To find how ρ as a point function behaves, we use the substantial derivative to find $\partial\rho/\partial t + v_i\partial_i\rho + \rho\partial_i v_i = 0$, or $\partial\rho/\partial t + \partial_i(\rho v_i)$. This is the *equation of continuity* that must hold for any possible motion.

Newton's Second Law for a particle of fluid is $\rho(dv_i/dt) = \rho F_i + \partial_j\Psi_{ij}$, where we have used the divergence theorem to get the force per unit volume from the stress tensor. ρF_i is an external force per unit mass, such as gravity.

The total force on the particle is the sum of the external force and the force given by the stress tensor. Using the expression of Ψ in terms of Φ, and the substantial derivative, we find:

$$\rho(\text{“}v_i/\partial t) + \rho v_k\partial_k v_i = \rho F_i - \text{“}_i p + (\mu/3)\partial_i(\text{“}_k v_k) + \mu\partial_k\partial_k v_i.$$

This is called the Navier-Stokes equation, derived by Navier (1822), Poisson (1829), Saint-Venant (1843) and Stokes (1845) in successively more modern form. The second term on the left-hand side makes it nonlinear in v, which is extremely unpleasant if we are looking for analytical solutions.

If we specialize to the slow motion of an incompressible fluid, the Navier-Stokes equation becomes a bit simpler:

$$\rho(\partial v_i/\partial t) = \rho F_i - \partial_i p + \mu\partial_k \text{part;}_k v_i.$$

VORTICITY

As elements of a solid body can rotate, so can elements of a fluid. If we antisymmetrize the rate-of-strain tensor, that is form $\Omega_{ij} = (\varepsilon_{ij} - \varepsilon_{ji})/2$, we eliminate the pure strain and are left with the rotation.

This is an *antisymmetric tensor* with three independent (off-diagonal) components. These three components form a vector (1/2)ω = (1/2) curl v. curl *v* is the *vorticity* of the motion, a field quantity like any other.

From Stokes's Theorem, the line integral of the velocity around a closed curve is equal to the flux of vorticity through it. The diagram shows how the antisymmetric part of the strain is related to the angle of rotation.

Taking the time derivative gives the relation between vorticity and curl v. The lines of unit length and the strains in the diagram are actually multiplied by some small number so that the change in the lengths of the axes is infinitesimal, and the strains perpendicular to both.

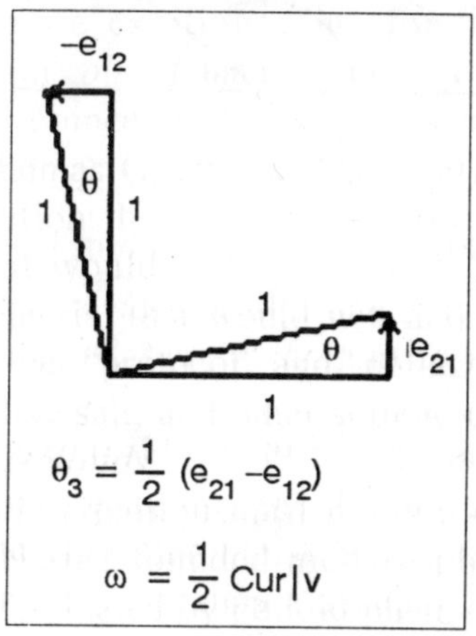

Fig. Rotation

If we take the curl of the Navier-Stokes equation (in its slow, incompressible form), we find that ∂(curl v)/∂t = μ div grad curl v, since curl commutes with div grad (write it out, if you don't believe it).

This is the same equation satisfied by the diffusion of heat for each component of the vorticity. We see that vorticity diffuses through the fluid with the diffusion coefficient equal to the viscosity. If the fluid is non-viscous, then ∂(curl v)/∂t = 0, and curl v = constant.

This is a simple form of Kelvin's Vorticity Theorem, stating that for an ideal fluid, the vorticity is a constant. If the motion began from rest, then this says that the vorticity will be zero everywhere ever after, and the flow will be *irrotational*. Even the smallest amount of viscosity will allow vorticity to be created, and it may dominate the flow pattern. In the present case, with Poiseuille's and Stokes's Laws, we expect vorticity.

FLOW AROUND A SPHERE

The problem is shown at the right. A rigid sphere of radius a is placed in a uniform flow of velocity V in the x-direction at large distances. The problem is symmetrical about the x-axis, so the velocity in the xy-plane has no z component, and conditions are the same in any plane through the x-axis.

We assume the flow to be steady, slow and incompressible. The velocity must satisfy the equation of continuity, div v = 0, Newton's Second Law, div grad v = $(1/\mu)$ grad p, and must vanish on the surface of the sphere. At great distances, its x-component must be V.

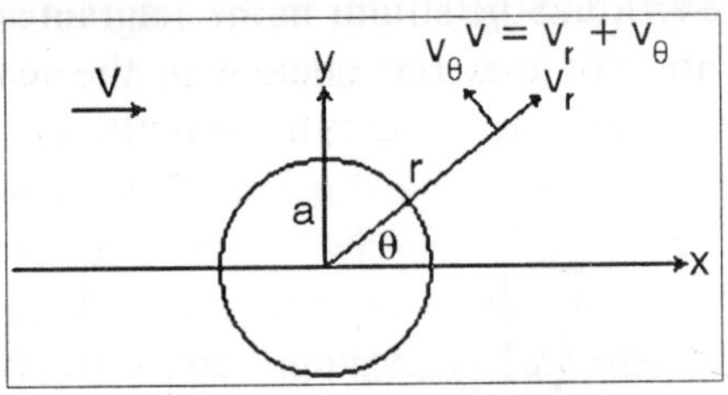

Fig. Sphere in a Uniform Flow

We'll use the vorticity, and a little luck, to solve this problem. Taking the curl of the equation of motion, we get div grad (curl v) = 0, so curl v is a solution of Laplace's equation.

From the axial symmetry of the problem, we can put curl v = $f(r,\theta)(-j\sin\varphi + k\cos\varphi)$, where j and k are unit vectors in the y and z (out of the page) directions, and φ is the azimuthal angle around the x-axis. Then, appropriate solutions of Laplace's equation are $r\sin\theta$ and $\sin\theta/r^2$, where the choice of $\sin\theta$ instead of $\cos\theta$ has been made so that the solution will vanish on the axis.

Substitution in Laplace's equation will verify that these are, indeed, solutions. The first solution must be thrown out because it is infinite at large distances. Then, curl v = $A(-jz/r^3 + ky/r^3)$ = curl (iA/r), where $z = r\sin\theta\sin\varphi$, $y = r\sin\theta\cos\varphi$, and i is a unit vector in the x-direction. Therefore, we must have v = $i(2A/r)$ + grad ψ, where ψ is some scalar function. The equation of continuity then gives div v = $-2A\cos\theta/r$ + div grad $\psi = 0$, or div grad $(A\cos\theta + \psi) = 0$.

This will be satisfied provided $\psi = -A\cos\theta + \varphi$, where div grad $\varphi = 0$. Since φ satisfies Laplace's equation, possible solutions are $B\,r\cos\theta$ and $C\cos\theta/r^2$, where solutions proportional to $\cos\theta$ have been selected to agree with the first term in angular dependence.

Finally, then, we have v = i(2A/r) – A grad cos θ + B grad (r cos θ) + C grad (cos θ/r^2). When we evaluate the gradients, we have an expression for the velocity, in terms of its radial and angular components.

This expression for v gives the required vorticity, satisfies the equation of continuity, and now if it can be made to satisfy the boundary condtions by proper choices of A, B and C, it will be the desired solution. The behaviour at large distances requires B = V.

The vanishing of the velocity on the surface of the sphere demands that A = -3aV/4 and C = $-a^3$ V/4.

For the components of the velocity, we have $v_r = V(1 - 3a/2r + a^3/2r^3) \cos\theta$ and $v_\theta = -V(1 - 3a/4r - a^3/4r^3) \sin\theta$. The z-component of curl v in the xy-plane is $-(3aV/4r^2) \sin\theta$.

Now we can use the stress tensor to find the stresses on the surface of the sphere, and by integrating them over the sphere to find the drag.

This is made much easier by choosing special x,y and z axes at a point on the sphere at angle θ with the x-axis, with the x-axis radially outward, the y axis in the direction of increasing θ, and the z-axis normal to the xy-plane.

Then, $\partial v_x/\partial x = \partial v_r/r\partial\theta$, $\partial v_y/\partial y = \partial v_r/r\partial\theta + v_r/r$, $\partial v_y/\partial x = \partial v_\theta/\partial r$, $\partial v_y/\partial y = \partial v_\theta/r\partial\theta - v_\theta/r$. Don't forget that in the second and fourth equation the unit vector also changes, and gives a contribution.

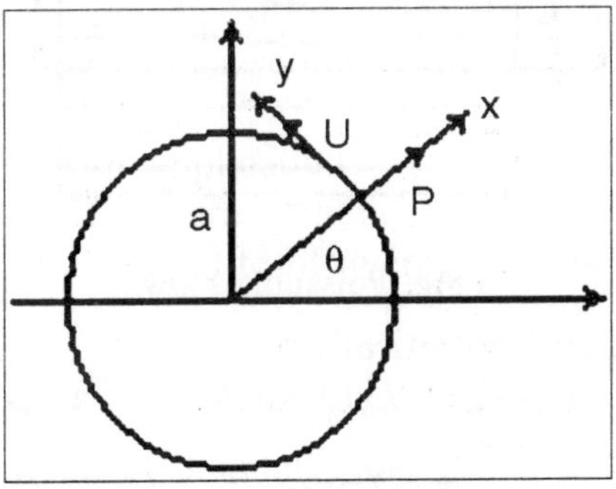

Fig. Temporary Axes

The only contributions that do not vanish on the surface of the sphere are the normal pressure, P = (3μV/2a) cos θ, and the shear stress U = $\mu(\partial v_y/\partial x)$ = -(3μV/2a) sin θ.

The net stress on unit area is P cos θ – U sin θ = 3μV/2a in the x-direction, so that the total drag becomes $4\pi a^2(3\mu V/2a) = 6\pi\mu aV$, the desired result. P and U can be separately integrated over the area, of course, and it is found that the pressure contributes 1/3, the shear 2/3 of the total drag.

Note that there is a positive pressure over the face of the sphere facing the flow, and a negative pressure over the other face. The shear force is a maximum at the equator, and is zero on the axis. Lamb's pictures of the stream functions for the irrotational flow around a sphere and for the viscous flow, and remark on how very different they are, repeating Lamb.

Indeed they are, but they refer to two different situations, the first where the sphere is moving, and the second where the fluid is moving. The actual difference is not as great as may appear from these diagrams, but, of course, there is a significant difference nonetheless.

FLOW THROUGH A TUBE

Poiseuille's problem is quite a bit easier. We can solve it using the definition of viscosity and a little statics. Consider a cylinder of fluid of thickness dr at a radius r, and of length L, and let this cylinder move at velocity $v = v(r)$.

The fluid on the side towards the wall moves a little slower, that towards the centre a little faster.

The velocity gradient causes the outer fluid to retard our cylinder, the inner fluid to urge it forward. In equilibrium, the velocity gradient will arrange itself so that the net force is balanced by the pressure difference at the ends of the cylinder.

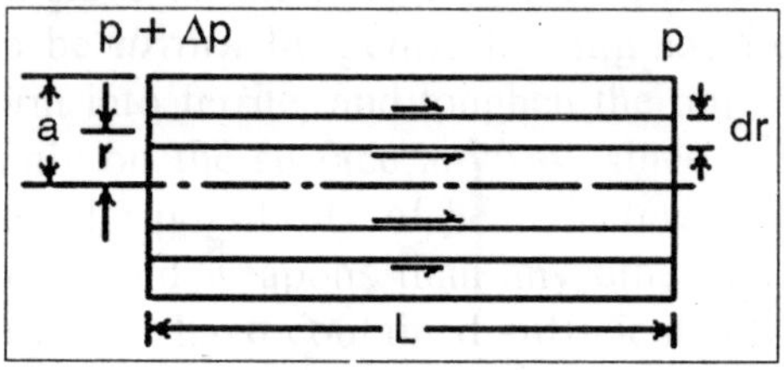

Fig. Poiseuille Flow

This is expressed mathematically by

$$(d/dr)[2\pi rL\mu(dv/dr)]dr = -2\pi r\Delta p\ dr$$

Integrating once, we find

$$2\pi rL\mu(dv/dr) = -2\pi(r^2/2)\Delta p, \text{ or } dv/dr = -(\Delta p/2L\mu)\ r + A,$$

where A is a constant. By symmetry, when $r = 0$, $dv/dr = 0$, so $A = 0$. Integrating a second time, $v(r) = -(\Delta p/4L\mu)r^2 + B = 0$, where B is a second constant of integration. Since $v = 0$ when $r = a$, the evaluation of B gives

$$v(r) = (\Delta p/4L\mu)(a^2 - r^2),$$

which is the parabolic velocity profile of Poiseuille flow. As with the sphere, the vorticity forms rings around the x-axis. It is easy to find the vorticity from the velocity profile.

The rate of discharge Q is found by integrating the velocity over the cross section,

$$Q = 2\pi\int v(r)r\ dr$$

The integral is easily performed after making the substitution $u = r^2$, with the result that

$$Q = (\pi a^4/8L\mu)\Delta p$$

The rate of discharge is proportional to the pressure difference Δp divided by the length L, or the pressure gradient, to the fourth power of the radius of the tube, and inversely proportional to the absolute viscosity. The mass discharge is proportional to the kinematic viscosity μ/ρ.

FLOW IN A PIPE

Flow in cylindrical pipes is a subject of great importance in engineering, and a great deal of empirical data has been built up that allows the solution of most practical problems with relative ease. For details, see Daugherty in the References. We will only make the connection with Poiseuille flow here.

In general, practical flow in pipes is turbulent, not laminar. The engineer speaks of "head loss" instead of pressure loss. Pressure and head are directly related through $p = \rho gh = \gamma h$, where p is pressure in dyne/cm^2, h is head in cm, ρ is the density in gm/cm^3 and γ is specific weight in dynes/cm^3. Of course, engineers use other units, such as lb/ft^2, ft, slug/ft^3 and lb/ft^3. The head can be measured directly by a manometer at any point, though actual manometers are seldom used these days. A certain volume rate of flow Q through a length L of pipe is associated with a loss in head of h/L, a dimensionless ratio. The engineer uses the equation

$$h = f(\mathrm{L/D})(\mathrm{V}^2/2g)$$

where D is the inside diameter of the pipe, V is the average velocity Q/A, where $A = \pi D^2/4$, h the head loss in a length L of pipe, and f the *Fanning friction factor*. The friction factor is a function of the Reynolds number $DV\mu/\rho$ and, for turbulent flow, of the relative roughness of the pipe, e/D, where e is the absolute roughness. The usual connection is given in the famous chart by L.F. Moody, which can be found in Daugherty. Turbulent flow is usual when the Reynolds number exceeds 2000. This is much larger than for a sphere, because the pipe tends to keep things smooth. Below 2000, Poiseuille flow governs. Poiseuille's equation, expressed in terms of head loss, is $h = 32\mu LV/\gamma D^2$. Comparing this with the Fanning equation, we find that $f = 64/R$, where R is the Reynolds number. This is represented by a line in the Moody chart showing that f decreases steadily to about 0.032, where the critical Reynolds number is reached, and then jumps abruptly to perhaps twice as much before it begins to decrease again as R increases.

TENSORS AND ELLIPSOIDS

An ellipsoid is a quadric surface described in canonical form by the equation $ax + by^2 + cz^2 = 1$. The lengths of the semiaxes of the ellipsoid are $1/a^{1/2}$, $1/b^{1/2}$ and $1/c^{1/2}$ along the x, y and z axes, respectively. If $b = c$, the ellipsoid degenerates into a spheroid, and if a = b = c, into a sphere. The centre of the ellipsoid is at the origin.

If the axes are rotated to a general position, then the equation for the ellipsoid can be written $ax^2 + by^2 + cz^2 + dyz + ezx + fxy = 1$, where a,b and c are new constants, as well as d, e, f.

The absence of linear terms means that we keep the origin at the centre of the ellipse, since displacing the centre of the ellipsoid will not add anything essential. There are six constants in this expression, and they specify the shape and size of the ellipse and the directions of the axes, which remain orthogonal. Any section of an ellipsoid by a plane is an ellipse, notably the principal sections that contain its axes.

A rank-2 symmetrical tensor Φ_{ij} has six components, three diagonal and three off-diagonal, which occur in symmetric pairs. The equation $x_i\Phi_{ij}x_j = 1$ is the equation of an ellipsoid associated with the tensor. A radius vector x is expressed as its components x_i, i = 1, 2, 3 instead of x, y, z. It is possible to find the rotation that transforms a symmetric rank-2 tensor into the *diagonal* form $\Phi_{ij} = a^{(i)}\delta_{ij}$. This will also transform the ellipsoid into canonical form and give the lengths of the semiaxes. To do this, let n_i be a unit vector (the components are direction cosines) and form the equation $\Phi_{ij}n_j = \lambda n_j$, where λ is some constant, which we may have to determine.

This expresses the condition that the vector resulting when the tensor Φ "operates" on n_i gives a vector in the same direction, but of possibly different magnitude. It is the fundamental duty of rank-2 tensors to operate in this way on one vector (by this is meant multiplication followed by contraction) to give another vector. The expression is really three simultaneous equations for the three direction cosines n_i.

Therefore, we have the three equations $(\Phi_{ij} - \lambda\delta_{ij})n_j = 0$ for the three values of i. These are *homogeneous* equations, and only have the solutions $n_i = 0$, unless the determinant of the coefficients vanishes. If it does, we can solve for the ratios of two of the direction cosines to the remaining one, and then determine them all by the condition that the sum of their squares must be unity.

The constant λ must be chosen so that the determinant of the coefficients vanishes, $|\Phi_{ij} - \lambda\delta_{ij}| = 0$. Since this equation is of third order, there are three roots for λ.

If the three roots are equal, then the tensor is already diagonal, and is the constant tensor $\lambda\delta_{ij}$.

The corresponding ellipsoid is just a sphere. If two roots are equal, then the remaining root gives the direction of an axis of symmetry, and the orientation of the axes perpendicular to this axis can be chosen any way one likes. The ellipsoid is then a spheroid, and

$$\Phi_{11} = \lambda,$$
$$\Phi_{22} = \Phi_{33}$$
$$= \lambda$$
$$= \lambda.$$

Finally, in the general case all three roots are unequal, and we have an ellipsoid with unequal axes, and the diagonal elements of Φ are the three values of λ. If the three values of λ are unequal, the resulting *principal axes* will all be *orthogonal*, that is, at right angles to each other.

If two values are equal, then the axis will be orthogonal to the plane of the other two axes.

The values of λ are called *eigenvalues* ("own values") and the unit vectors belonging to them are called *eigenvectors*. Working with principal axes simplifies the algebra greatly, and we usually assume that any symmetrical rank-2 tensor has been *diagonalized* in the way we have discussed. All of these matters are discussed at length in texts on linear algebra. We now have enough theoretical background to consider the most important rank-2 tensors met with in classical physics. We'll discuss the inertia tensor, the dielectric tensor, the strain tensor, and the stress tensor in this article.

Not in detail, of course, but enough to show how the tensors are defined and used, making their application easier. See the Examples for exercises in tensor diagonalization. There are good numerical methods for the diagonalization of matrices, and the basic theorems are proved in any quantum mechanics text.

THE INERTIA TENSOR AND RIGID BODY MOTION

The inertia tensor in the dynamics of rigid bodies is an excellent example of a rank-2 tensor where the associated ellipsoid aids in the visualization of the motion. We shall start from first principles, using index notation, and find the motion of a free rigid body—that is, a body under the action of no external forces.

Any rigid body can be considered as an assembly of mass elements m with fixed positions x_i relative to an orgin O, in a system of coordinates fixed in the body. The origin O can be a fixed point in space and the body, or the centre of mass of the body. In either case, the motion separates into translational and rotational parts. We are not concerned with the translational parts here, and will consider only the rotation.

The most general motion of the body is then a rotation with angular velocity ω_i. The direction of the vector specifies the direction of the axis of rotation. The angular velocity may change in magnitude and direction with time, in a quite general manner. The body coordinates rotate in the same way, and so do not form an inertial system.

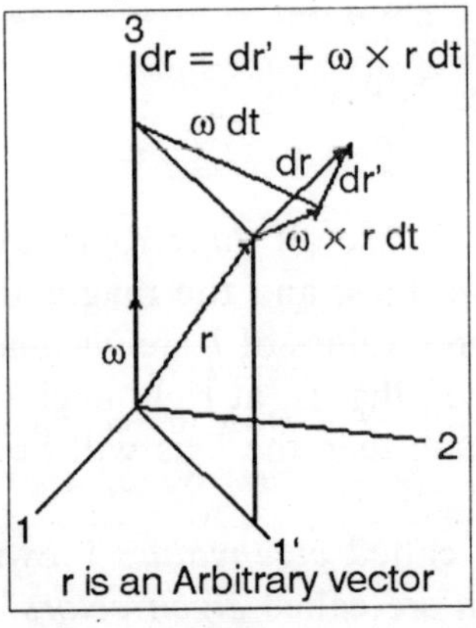

Fig. Increments in Fixed and Rotating Systems

The rate of change of any vector quantity v_i in a fixed coordinate system with origin O is $[dv_i/dt]_{fix} = [dv_i/dt]_{rot} + \varepsilon_{ijk}x_j v_k$. This is merely the relation that in vector notation is

$$[dv/dt]_{fix} = [dv/dt]_{rot} + r \times v,$$

which applies to any vector, not just a radius vector.

Accordingly, the velocity of an element of mass m in the inertial system is $v_i = \varepsilon_{ijk}\omega_j x_k$, since it is fixed in the rotating system. Summing over all mass elements, the angular momentum $H_i = \Sigma m\ \varepsilon_{ijk}x_j\ \varepsilon_{klm}\omega_l x_m$. Using the value of the contraction of two antisymmetric tensor densities, we find that $H_i = [\Sigma m(x_k x_k \delta_{ij} - x_i x_j]\ \omega_j$, The rank-2 symmetric tensor multiplying ω_j is the *inertia tensor* I_{ij} of the body.

The diagonal elements are called *moments of inertia*, and the off-diagonal elements *products of inertia*.

We note that the angular momentum is not necessarily in the direction of the angular velocity.

Now we find the kinetic energy T of the body by adding up the kinetic energies of each element of mass: $2T = \Sigma v_k v_k$. Using the expression for the space velocity, we again get a contraction of two antisymmetric tensor densities. In fact, $2T = \Sigma m\ \varepsilon_{ijk}\omega_j r_k\ \varepsilon_{irs}\omega_r r_s = I_{jr}\omega_j\omega_r$. This work is a good example of how much easier it is to use the index notation than vector notation in these things.

Newton's Second Law can be applied in the fixed system to each of the elements of mass, and we find that the net torque $L_i = \Sigma\varepsilon_{ijk}x_j F_k$ is the rate of change of angular momentum H_i in the fixed system. From the above equations, we have $L_i = I_{ij}(d\omega_j/dt) + I_{kl}\varepsilon_{ijk}\omega_j\omega_k$. These are three differential equations for the components of the angular velocity of the body relative to the body axes.

It is always possible to diagonalize the inertia tensor by a suitable choice of the orientation of the body axes. Let us assume this has been done, and the principal moments of inertia are $A^{(i)}$. The index is in parentheses because it is not a tensor index, merely an identifying label.

For ease of writing, we may also represent these quantities by A, B and C. Then, $I_{ij} = A^{(i)}\delta_{ij}$, and the equations of motion of the preceding paragraph become $L_i = A^{(i)}(d\omega_i/dt) + \varepsilon_{ijk}A^{(k)}\omega_j\omega_k$. These equations are known as *Euler's equations*.

Now we can consider the problem of a body moving freely with one point fixed, which can be taken as the centre of mass. An asteroid moving in space is a good example of such a body, since the torques exerted by unequal solar attraction are small and have little effect if the body is rotating rapidly. Such torques indeed cause the axis of the earth to precess, but this is a slow effect. We can safely neglect such torques with an asteroid rotating with periods of hours or less.

First of all, the kinetic energy T of the body must be constant, so $2T = H_i\omega_i = I_{ij}\omega_i\omega_j$ = constant.

The angular momentum is also constant, since no external torques act, so this says that the component of the angular velocity in the direction of H is also a constant. If we consider a space in which the components of the angular velocity are the coordinates, then this equation also defines an ellipsoid in this space, which is similar to the inertia ellipsoid. We shall call this the *energy ellipsoid*, since its size depends on the kinetic energy of the motion.

We shall choose the body axes so that the inertia tensor is diagonalized, so the body axes are also the symmetry axes of the energy tensor. The semiaxes of the energy ellipsoid are $\sqrt{(2T/A^{(k)})}$. The angular velocity is a vector from the origin O to a point on the energy ellipsoid P.

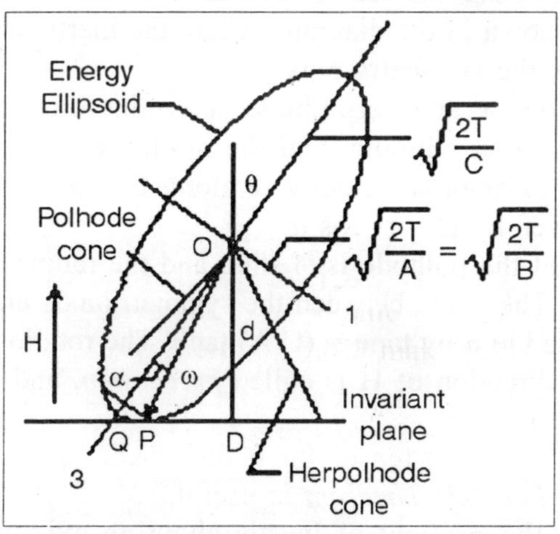

Fig. Free Motion of a Rigid Body

If the equation for the energy ellipsoid is differentiated, we find that

$I_{ij}\omega_j d\omega_i = 0$. Now $d\omega_i$ is a vector in the tangent plane at P, so this says that the angular momentum is perpendicular to the tangent plane. Because the angular momentum is constant, this means that the orientation of the tangent plane remains constant during the motion. Also, since the component of the angular velocity in the direction of the angular momentum is constant, the distance d from the origin O to the tangent plane is constant.

Therefore, we find the important result that the tangent plane to the energy ellipsoid is fixed in space. We call it the *invariant plane*. Since the angular velocity passes through the point of contact P, there is no relative motion between the energy ellipsoid and the invariant plane: the energy ellipsoid must roll on the invariant plane without slipping.

The angular velocity must terminate on the energy ellipsoid in any case. For a certain d, it must also lie on the surface obtained by squaring the relation Hd = 2T, which is $\Sigma(A^{(i)}d\omega_i)^2 = 4T^2$. Eliminating the constant terms between these two equations, we find that $\Sigma(A^{(i)}/2T)[(1/d^2) - (A^{(i)}/2T)]\omega_i^2 = 0$, which is the equation of a cone (not in general a circular one) with centre at O, on which the angular velocity must lie. This cone intersects the energy ellipsoid in a curve called the *polhode* (Greek: "path of the pole").

This cone rolls without slipping on a similar cone fixed in space determined by the path of the angular velocity in the invariant plane, which is called the *herpolhode*. We have now arrived at a complete description of the motion of the body, making use of the inertia ellipsoid.

Let's now specialize to the case frequently met when two of the moments of inertia are the same, say A = B, with C different. The case when C < A is shown in the diagram, where the inertia ellipsoid is prolate, and the 3-axis is the symmetry axis.

Now the polhode and herpolhode determine circular cones that roll on each other without slipping, and the motion can be easily visualized. The component of angular velocity Ω along the symmetry axis is called the *spin*. It is given by $\Omega = \omega \cos \alpha$.

The radius of the polhode is $\omega \sin\alpha$, and the radius of the herpolhode is $\omega \sin(\theta - \alpha)$. The angle between the symmetry axis and the direction of H is θ. It is related to α by $\tan\alpha = (C/A)\tan\theta$. The rotation of the symmetry axis around the direction of H is called *precession*, and its amount is $d\psi/dt = C\Omega/(A \cos\theta)$.

In finding the inertia tensor for bodies made up of simple geometric shapes, the *Parallel-Axis Theorem* is useful.

It says that the moment of inertia about an axis parallel to an axis through the centre of gravity is equal to the moment about the axis through the centre of gravity plus the total mass of the body times the square of the distance between the axes, or $I_{ii} = I_{ii}' + md^2$. For products of inertia (off-diagonal elements), the product of inertia is the product of inertia referred to the centre of gravity plus the mass times the product of the coordinate differences.

That is, $I_{ij} = I_{ij}' + mxy$. Tables of the moments of inertia of simple figures can be found in handbooks and in texts of engineering mechanics. The moment of inertia of a parallelepiped about an axis through the centroid perpendicular to a face is $(m/12)(a^2 + b^2)$, where a and b are the sides of the face.

The centroidal moment of inertia of a sphere is $(2/5)mr^2$, and of a disk relative to an axis perpendicular to its plane, like a wheel and axle, $(1/2)mr^2$.

THE STRAIN AND STRESS TENSORS AND ELASTIC CONSTANTS

Consider a solid elastic medium that is deformed by forces applied to its boundaries, or by forces exerted directly on the material by external influences. In response to these forces, a general point originally at x_i before the forces are applied moves to a point $x_i + \xi_i$.

The displacement ξ_i is not necessarily small, but its derivatives E_{ij} = "$_i\xi_j$ we presume are much less than unity. These quantities are called *strains*, and form a rank-2 tensor that is a function of position. It is convenient to separate E_{ij} into symmetric and antisymmetric parts, $E_{ij} = e_{ij} + \Omega_{ij}$.

The symmetric part e_{ij} is called the *pure strain* tensor, while the antisymmetric part Ω_{ij} represents the rotation due to the deformation. The vector associated with Ω_{ij} gives the rotational axis and the angle of rotation. We will be mainly interested in the pure strain, which represents the deformation of the medium.

If the medium returns to its initial state when the forces are removed, the deformation is called *elastic*, and all the strain energy put into the body is recovered. If the deformation is too large, the body does not return to the initial state, mechanical energy is dissipated, and the deformation is called *plastic*.

For solid bodies, deformation is usually elastic for sufficiently small strains. We shall deal with this case exclusively here.

Since the pure strain is symmetric, it can be diagonalized. Let the diagonal elements be a, b, and c.

Then, a small parallelepiped with sides dx, dy and dz and volume V = dxdydz becomes a small parallelepiped with sides (1 + a)dx, (1 + b)dy and (1 + c)dz after deformation, of volume V + dV = (1 + a + b + c)dxdydz, so that dV = (a + b + c)V.

Quantities of the order of squares of strains have been neglected. The sum a + b + c = dV/V is the relative increase in volume at the point, called *dilatation*. Now, $a + b + c$ is just the trace of the strain tensor, e_{ii} = "$_i\xi_i$, which we know is invariant under rotation.

As an example, consider a body strained as shown in the figure by a shearing force.

The only nonzero component of the strain tensor is $E_{12} = a$, where a is a small quantity equal to the angle of strain θ. Then the pure strain tensor has $e_{12} = e_{21} = a$, and the rotation tensor has $\Omega_{12} = -\Omega_{21} = a$. We note that the deformation is indeed equal to a rotation of a radians about the 3-axis plus an extension along the 2-axis and a contraction along the 1-axis. Indeed, the pure strain tensor has eigenvalues $\pm a/2$. The trace is zero, so there is no dilatation. This sort of deformation is called *pure shear*.

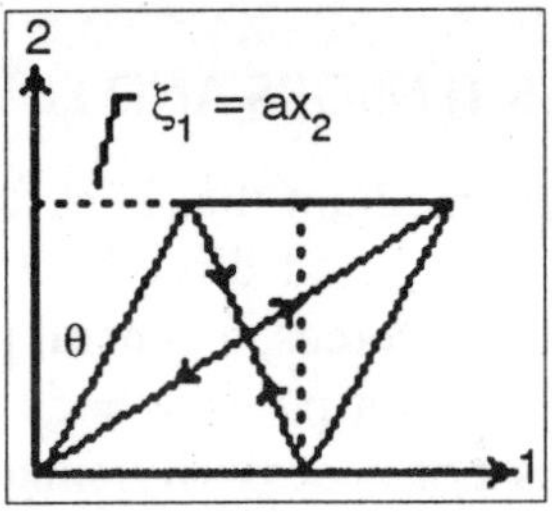

Fig. Deformation in Shear

Now we must consider the causes of deformation and rotation, stresses. If we imagine a plane surface in the body and that the material on one side of the plane surface is removed, the forces that it exerted on the material on the other side can be represented by forces acting on this plane surface.

Let the unit normal vector to the surface be taken as the normal pointing towards the removed material, or outwards.

The normal force is positive when it is a pull, or in the direction of the normal, or a *tension*. There may also be a force in the plane of the surface, perpendicular to the normal, called a *shear force*, and in any particular case the positive directions of its two components in the surface will be agreed upon. A force per unit area of surface is a *stress*. If a hydrostatic pressure p acts on the surface, then the normal stress is -p and the shear stresses are zero.

The stresses will, in general, depend on the direction of the normal. For any normal vector n_i, we will have stresses F_i that are a function of the n_i. This defines a tensor S_{ij} such that $F_i = S_{ij}n_j$, called the *stress tensor*.

That this is a satisfactory definition is can be seen by considering the equilibrium of a small tetrahedron. The three components of the force on the inclined face of area *dS* are balanced by equal and opposite forces on the other three sides.

Note that the area of each side is $n_i dS$, and the force on it will be in the opposite direction to the force on the face *dS*. To prove this, write out the forces in the 1, 2 and 3 directions separately and show that they balance.

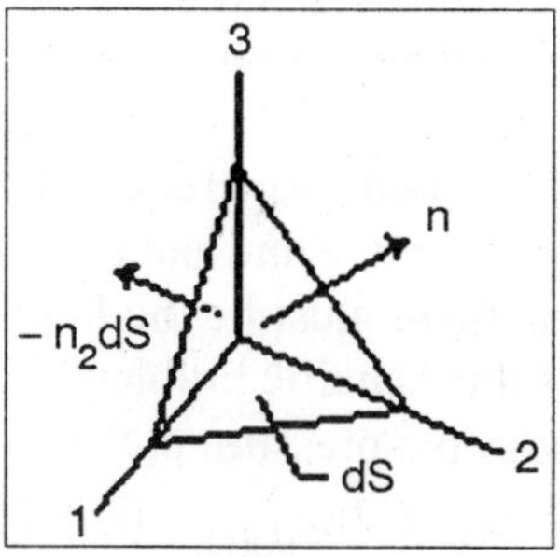

Fig. Tetrahedron

The stress tensor must be symmetric, or there would be unbalanced torques on small areas that would cause them to rotate. The absence of such rotation causes the stress tensor to be symmetric.

In the figure, we take the moments of the forces tending to rotate the cube about the 3-axis that come from the 1-2 component of the tensor, showing that $S_{12} = S_{21}$. The same holds for each of the other off-diagonal elements.

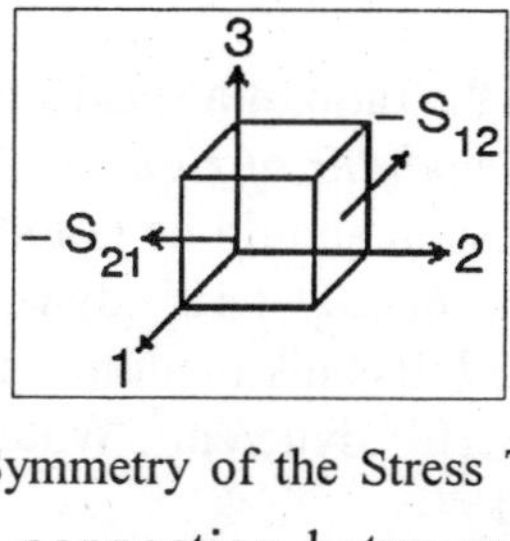

Fig. Symmetry of the Stress Tensor

We now look for the connection between the stress and pure strain tensors.

Since the strains are small compared to unity, we assume that Hooke's Law holds, and the stresses are linear functions of the strains. Actually, Hooke's Law is always valid for *elastic* deformations, since the strain is still small when inelasticity begins.

A systematic way to do this is to consider a deformation energy F defined by "F"$e_{ij} = S_{ij}$.

Hooke's Law then requires that *S* be proportional to *e*, or that *F* be a homogeneous quadratic function of the strains. The most general isotropic quadratic homogeneous function of the strains is $F = (1/2)\lambda(e_{kk})^2 + \mu(e_{ij})^2$, where the constants λ and μ are called the *Lamι* coefficients. If we differentiate *F*, we find $S_{ij} = \lambda e_{kk}\delta_{ij} + 2\mu e_{ij}$, which is the relation between stress and strain for an isotropic, elastic body.

Poisson took $\lambda = \mu$, as did Cauchy and others who thought there should be only one elastic constant for an isotropic medium. Their theories were based on central forces between supposed "molecules" that have little in common with our modern molecules. There was an extended debate between those who demanded more elastic constants and those who demanded fewer.

In the general case 21 constants were required by those who favored more, and only 18 by those favoring less. Time and experiment have come down on the side of those who think there must be the larger number.

Elastic constants useful in specific problems can be defined in terms of λ and μ. If we assume an isotropic pressure p, then

$$-p = \lambda e_{kk} + 2\mu e_{(k)(k)} = (\lambda + 2\mu/3)dV/V.$$

The *bulk modulus* k is defined by $dV/V = -p/k$, so $k = \lambda + 2\mu/3$. If an axial force F is applied to a rod of length L and cross-sectional area A, the rod lengthens by $\Delta L = FL/AY$, where Y is *Young's modulus*. At the same time, the rod contracts laterally, and the ratio of the lateral strain to the longitudinal strain is σ, *Poisson's Ratio*.

Under these conditions, we find that

$$\sigma = \lambda/(2\lambda + 2\mu) \text{ and } Y = 2(\sigma + 1)\mu$$

Poisson's ratio can be between 0 and 1/2, but is usually not far from 1/4, which Poisson assumed was the case. It is 1/2 for a body with no shear strength, such as a liquid.

The modulus μ itself is the ratio of a shearing stress to the shearing strain (an angle), so is called the *modulus of rigidity*. A quantity called a *modulus* has the dimensions of a stress, it should be noted.

Steel has $Y = 30 \times 10^6$ psi or 2.07×10^{12} dyne/cm^2, $\mu = 12 \times 10^6$ psi or 8.27×10^{11} dyne/cm^2, and $\sigma = 0.27$. Its bulk modulus, calculated from these figures, is $k = 20 \times 10^6$ psi or 1.38×10^{12} dyne/cm^2. Water, on the other hand, has $\mu = 0$, since it lacks rigidity, and

$$k = 3.16 \times 10^5 \text{ psi or } 2.18 \times 10^{10} \text{ dyne/cm}^2.$$

It is about 63 times more compressible than steel. As for gases,

$$dV/V = -dp/p$$

for an ideal gas, so $k = p$. At one atmosphere, this is

$$1.47 \text{ psi or } 1.013 \times 10^6 \text{ dyne/cm}^2,$$

about 22,000 times more compressible than water. Rubber is a curious material with σ *H*' 0.5, which implies that it has very low rigidity, a kind of "solid fluid." For rubber, $Y = 100$ to 600 psi, and $\mu = 30$ to 200 psi, so its rigidity is indeed small. Glass is on the opposite pole, with $Y = 10 \times 10^6$ psi and $\mu = 5 \times 10^6$ psi. Its Poisson's ratio is low, 0.20 to 0.27, which implies that it is quite rigid compared to its longitudinal elasticity.

A medium whose properties depend on direction is often called *anisotropic*, but a less barbarous word is *aelotropic* (Greek: change-turning). The above expression for F suggests the generalization to an aelotropic medium by setting F = $(1/2)\lambda_{ijkl}e_{ij}e_{kl}$, which gives $S_{1j} = \lambda_{ijkl}e_{kl}$. The rank-4 tensor λ_{ijkl} is called the *elastic constant tensor*, an operator connecting two rank-2 tensors just as a rank-2 tensor is an operator connecting two vectors.

In general, it has 81 components, but there is a great deal of symmetry here, since the energy is unchanged if *i, j* or *k, l* are interchanged (the strain tensors are symmetric), or if *i, k* and *j, l* are simultaneously interchanged.

As a result, there are only 21 independent components at worst. It is confusing to work this out, but all 81 components can be listed, and equal ones sorted out, as a last resort.

The symmetry of crystals will introduce more relations between the constants, reducing their number. Even in a triclinic crystal, the least symmetric, the possibility of choosing convenient coordinates reduces the number to 18.

A monoclinic crystal has 12, an orthorhombic crystal 9. A tetragonal crystal has 6, as does a trigonal (rhombohedral) crystal like calcite or quartz.

A hexagonal crystal has 5, and a cubic crystal 3. Cubic crystals are often considered to be isotropic in many respects, but they have an extra elastic constant that truly isotropic materials do not have. The constant corresponding to λ splits into two, one the coefficient of the squares of the diagonal elements of e, the other the coefficient of the cross products.

This reduction in the number of constants is found by applying the symmetry elements of the crystals, which turns some constants into their negatives, so they must vanish. This is an interesting point, but we cannot go into it further at this time.

An interesting application of this theory is to the propagation of mechanical waves in solids, which has application to seismology, nondestructive testing, and other fields. Such waves are almost always elastic, and have an interesting variety of properties that have been intensively studied.

The theory is generally more complicated than that of electromagnetic waves, and includes the subject of surface waves, as well as waves in an infinite medium.

THE DIELECTRIC TENSOR AND CRYSTAL OPTICS

The relation between the electric displacement D (statcoulombs/cm_2) and the electric intensity E (statvolts/cm), considered in the first approximation as a proportionality, will in general be given by $D_i = \varepsilon_{ij}E_j$, where the dielectric tensor satisfies $\varepsilon_{ij} = \varepsilon_{ji}$.

The requirement of symmetry comes from several sources, one of which is simply that the tensor should be diagonalizable by an ordinary rotation, which establishes the three orthogonal principal axes of polarization, and the three principal dielectric constants, its eigenvalues. This yields $D_{(k)} = \varepsilon_{(k)}E_{(k)}$, where k = 1, 2 and 3.

An electric intensity in a principal direction gives a displacement in the same direction. If all three directions are equivalent in the medium, then the three constants are equal: $\varepsilon_{(k)} = \varepsilon$, and an electric intensity in any direction creates a displacement in the same direction; the medium is then *isotropic.*

In crystals belonging to the cubic or isometric system, three equivalent axes can be found, so the medium is isotropic. In the trigonal (rhombohedral), tetragonal and hexagonal systems, two equivalent axes perpendicular to the third can be found.

If the unique axis is the 3-axis, then $\varepsilon_{(1)} = \varepsilon_2$, while ε_3 is different. In the orthorhombic, monoclinic and triclinic systems, all three principal dielectric constants may differ. In any case, dielectric constants may *accidentally* be closely the same, so the medium will resemble a more symmetric one.

Noncrystalline materials, such as polymer sheets, may be anisotropic as well. Nonisotropic transparent materials may show *double refraction*, discovered in 1669 by Bartholinus in Iceland spar, or transparent calcite, $CaCO_3$, and first explained by Huygens not long afterwards. Such materials are also called *birefringent.*

We must first discuss the propagation of plane electromagnetic waves in an anisotropic medium.

We shall assume throughout that B = H, or $\mu = 1$, which is true in transparent crystals. Assume that the wave depends on space and time through a factor $\exp[i\omega(t - nr\cdot s/c)]$, where ω is the angular frequency, v = c/n is the phase velocity, r the position vector and s is the unit vector normal to the wavefront. Then we can replace time derivatives by $i\omega$ and del by -$i\omega ns/c$ in Maxwell's equations. We find: $s\cdot D = 0$, $s\cdot H = 0$, ns x H = -D, and $ns \times E = H$. All the terms in these expressions are constant vector amplitudes.

The spatial relations between the vectors used to describe the wave are shown in the diagram at the right. The wavefront is the plane of H and D, with s the wavefront normal. E lies in the plane of s and D making an angle α with D.

Since browsers do not yet support the perpendicular symbol, the electric intensity along D will be represented in the text by E+. The Poynting vector S is in the direction of t, the ray unit vector. This gives the direction of energy flow, which is not normal to the wavefront. The phase velocity is vs, with v = c/n, and the ray velocity is $v_r t$, with $v_r = c/n_r$. The electric energy density $w_e = E\cdot D/8\pi = (n/8\pi)H\cdot(s \times E)$ and the magnetic energy density $w_m = H \cdot B/8\pi$ is given by exactly the same expression, so the total field energy $w = (n/c)s \cdot S$, where $S = (c/4\pi)(E \times H)$ is the Poynting vector. Now, $v_r = S/w$, so $v = v_r s \cdot t = v_r \cos \alpha$. The indices of refraction are then related by $n_r = n \cos \alpha$.

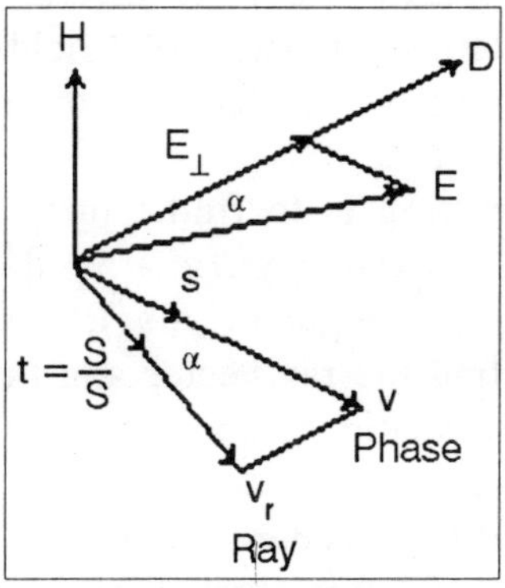

Fig. Wave Vectors

The magnetic field can be eliminated between the Maxwell curl equations with the important result that $D = n^2[E - s(s \cdot E)] = n^2E+$. This is a relation between E and D that must be satisfied by the wave, and is completely independent of the relation between them given by the dielectric tensor. Both relations must be satisfied in the wave. Since $E+ = (E \cdot D)D/d^2$, this relation can also be written as $n^2 = D^2/(E \cdot D)$. Similarly, it can be shown that $n_r^2 = (D \cdot E)/E^2$.

Applying the diagonalized dielectric tensor (which means everything must be referred to principal axes) we find $\varepsilon_k E_k = n^2[E_k - s_k(s \cdot E)]$. The k here is not a tensor index, merely an indentifier of one of the three components. Written out fully, we find three homogeneous equations for the field components E_k.

Such a system can have non-zero solutions only if the determinant of the coefficients vanishes. This condition will determine the value of n, or the phase velocity, for which both Maxwell's equations and the dielectric relation are satisfied. Fresnel discovered a cunning way to do this that yields a symmetrical equation for v in terms of the components of the wave normal vector. Rearrange the equation to $E_k = [n^2 s_k/(n^2{}_k)]$ $(s \cdot E)$. This is valid if the quantity in parentheses does not vanish.

Now multiply by s_k and add the equations for k = 1, 2 and 3. The dot product cancels, and we find a sum of three similar terms that adds to 1. Now the sum of the squares of s_k is also 1, so subtract this from both sides and combine the terms with the same s_k's. Now change from n's and ε's to the velocities. The final result is $s_1^2/(v^2 - v_1^2) + s_2^2/(v^2 - v_2^2) + s_3^2/(v^2 - v_3^2)$.

.This is *Fresnel's equation of wave normals*. To use it, multiply by the denominators to clear of fractions. The result is a quadratic equation in v^2, which gives two values for any value of s. The squares mean that the wave can travel in either direction. We will see that each pair of solutions for the same s gives us two plane-polarized waves polarized at right angles to each other which, in general, travel at different velocities.

A similar process in terms of t instead of s yields *Fresnel's ray equation*, which gives us the ray velocity v_r in terms of the ray unit vector. This can be found in Born and Wolf.

What is more interesting is to find t in terms of s and the phase velocity. The result is $t_k = s_k\{[v^2 + g^2/(v^2 - v_k^2)]/\sqrt{(v^4 + g^2)}\}$, where $g^2 = v^2(v_r^2 - v^2) = \{[s_1/(v_1^2 - v^2)]^2 + [s_2/(v_2^2 - v^2)]^2 + [s_3/(v_3^2 - v^2)]^2\}^{-1}$. From these formulas, we can find the ray vector and the ray (energy) velocity. When the phase and ray velocities re known, then the angle α can be found by a formula given above.

Now we can proceed to consider a geometric construction to find the directions of polarization of a wave travelling in an arbitrary direction in an anisotropic medium.

The field energy w is given by $8\pi w = D_1^2/\varepsilon_1 + D_2^2/\varepsilon_2 + D_3^2/\varepsilon_3$. Taking new variables $x_k = D_k/\sqrt{4\pi w}$, we have the equation $x_1^2/\varepsilon_1 + x_2^2/\varepsilon_2 + x_3^2/\varepsilon_3$, which is the equation of an ellipsoid whose semiaxes are the square roots of the principal dielectric constants, or the indices of refraction for waves polarized along a principal axis.

This ellipsoid has been called the *optical indicatrix* or the *index ellipsoid*. Consider the intersection of the plane $x_1s_1 + x_2s_2 + x_3s_3 = 0$ normal to a wave vector s with the ellipsoid. This intersection will be an ellipse. The axes of this ellipse are the directions of polarization of the two waves with the given wave vector.

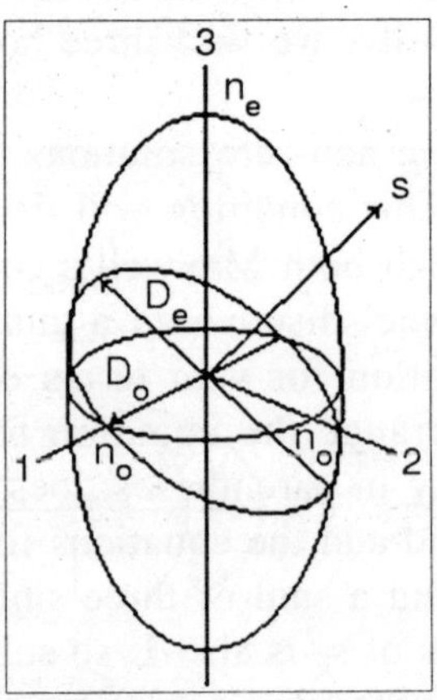

Fig. Index Ellipsoid

It should be clear that once we assume a direction for the wave, the polarization in the perpendicular plane must be carefully chosen so that E and D are coplanar with s, which in general they will not be. This happens for two polarizations that differ by 90°, and the two polarizations will travel with different velocities.

This allows the two polarizations to be separated by refraction, as in a Nicol prism. This is proved by finding the extrema of the vector from the origin to the ellipse, and showing that the result is the same condition that we found above from the combination of Maxwell's equations and the dielectric tensor. Any direction s for which the intersection is a circle is a direction along which a wave may have any polarization, and all polarizations travel at the same velocity.

Such a direction is called an *optic axis* of the crystal. For an ellipsoid with axes all of different lengths, there will be two such directions symmetrically located in the plane of the two axes with the largest and smallest indices on either side of the axis of greatest index. Such media are called *biaxial.*

If two indices are equal, then a plane perpendicular to the axis of the third index cuts the ellipsoid, which is now a spheroid, in a circle. Therefore, this axis is the optic axis, which is the only one. Such media are called *uniaxial*. If the third index is greater than the other two, the medium is called *positive*, and *negative* otherwise.

The two equal indices, or velocities, are called *ordinary* and denoted by a subscript "o" while the third is called *extraordinary* and denoted by a subscript "e." If we write Fresnel's equation of wave normals for this case, for a wave travelling in a direction making an angle θ with the optic axis, we easily find the two solutions $v^2 = v_o^2$ and $v^2 = v_o^2 \cos^2 \theta + v_e^2 \sin^2 \theta$. This gives two surfaces, one a sphere of radius v_o for the "ordinary" wave, and one a surface of fourth order, an ovoid, for the "extraordinary" wave. For a positive uniaxial medium, the ovoid is inside the sphere, while for a negative uniaxial medium, it is outside. These figures are not the index ellipsoid.

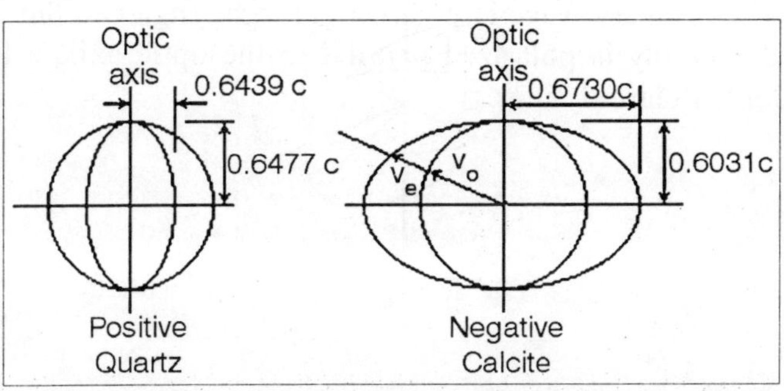

Fig. Uniaxial Wave Normal Surfaces

The distance between the ordinary and extraordinary wave normal surfaces is greatly exaggerated in the diagrams, as can be seen from the actual figures for calcite and quartz that are given.

The extraordinary velocity surface is drawn as an ellipse for convenience, but it is not an ellipse, being blunter at the ends of the major and minor axes than an ellipse. Huygens took it for an ellipse, which indeed it closely resembles, and elementary Optics texts have followed him in this, but this is an error, although it gives the ray direction quite well using Huygen's construction.

This is reminiscent of the Bohr atom, in which erroneous concepts give the correct result. In quartz, the two surfaces do not exactly meet on the optic axis, and this is the source of *optical activity* in quartz, but the effect can be neglected here.

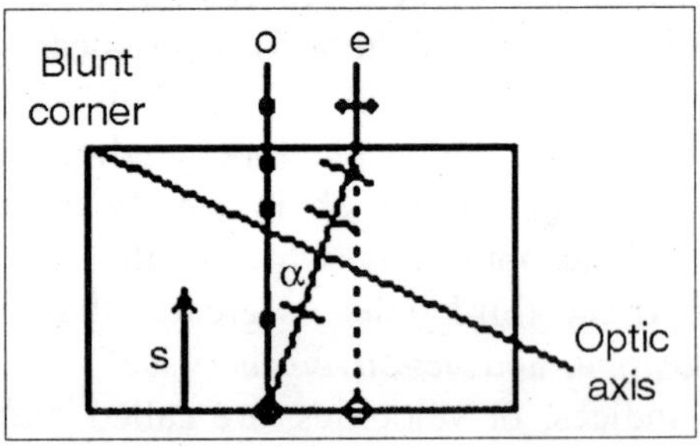

Fig. Double Refraction in Calcite

The optic axis of a cleavage rhomb of calcite passes through a blunt corner making equal angles with the sides and faces. If you look at a dot through the crystal, two images will be seen. When you rotate the crystal, the ordinary image will remain fixed while the extraordinary image rotates around it.

The reason is that the ray velocity of the extraordinary wave is inclined to the wavefronts, lying in the plane containing the wave normal and the optic axis. The ordinary wave is polarized at right angles to the optic axis, the extraordinary ray is polarized parallel to the optic axis, a fact easily checked with a Polaroid filter.

Chapter 5

Boltzmann's Factor

Boltzmann's factor is $e^{-E/kT}$, which expresses the "probability" of a state of energy E relative to the probability of a state of zero energy. This factor can be used to introduce temperature into a wide variety of physical problems, and is often taken as a starting point. In this article, the origin of this factor, which is surprisingly simple, is discussed. To do this, it is necessary to understand entropy on a fundamental basis, and this also is really fairly easy to do.

Then the calculation of thermodynamic functions through the use of the partition function Z is explained, and finally this is applied to the "ideal gas" of wide applicability.

The reader will follow the argument better if prepared with a basic knowledge of classical thermodynamics, which the statistical thermodynamics presented here greatly extends, especially in understanding.

Classical thermodynamics is a deductive theory of remarkable power and accuracy, which deals with the thermal behaviour of macroscopic bodies. It is based on two "laws" or postulates, the first of which is that heat is a form of energy and energy is conserved, and the second is that heat cannot be abstracted from a body and converted into mechanical energy with no other effect.

There are many equivalent ways to state these postulates, and it is entertaining to show that they are all equivalent. Temperature is considered a fundamental notion, abstracted from sensation, that indicates the direction of heat flow and the state of thermal equilibrium. It is often included as a "zeroth" postulate.

In working out the consequences of these postulates, it is very useful to define a function of the state of a system called the *entropy*, with dimensions of energy divided by temperature.

This temperature is a very special one, the *absolute temperature*, for which classical thermodyamics can give a definition, which is a very great accomplishment. Entropy is of great use, and in fact is essential, in engineering and chemical problems. However, what entropy "is," and what are the limits of its application, are not explained at all by the classical theory. English-speaking physicists, in particular, have had a very difficult time understanding it, though they have been able to apply it with great power and ingenuity.

Engineers have been very foggy on entropy, and to all appearances remain so, while engineering students have learned to just follow the rules and think of more pleasant things.

The application of classical mechanics to the explanation of thermodynamics resulted in a complex and disgusting mess that explains nothing, though great efforts were expended along this line.

This is not surprising, since thermodynamics depends sensitively on mechanics at an atomic scale, for which classical mechanics fails spectacularly. Nevertheless, Ludwig Boltzmann (1844-1906) and Josiah Willard Gibbs (1839-1903) came very close to a reasonable theory, which was completely clarified with the appearance of quantum mechanics.

Gibbs seems to be little remembered these days, but he was the most important physicist that the United States has ever produced, and of world quality.

Since the introduction of quantum mechanics, it has been possible to understand entropy clearly and completely on a very simple basis, so that all the clouds of mystery dissolve into nothingness.

Entropy and the Second Law have been applied outside of thermodynamics to such things as economics and sociology, but these applications reflect only the ignorance of those who propose them, and have no basis. Thermodynamics is very simple, resting only on the existence of quantum-mechanical stationary states that can be counted.

ENTROPY

Thermodynamics applies to systems with a very large number of possible states available to them, and in which transitions between the available states are continual and rapid. Not simply a large number of states is required, but a *very* large number, as argued by Kittel. Very large numbers have properties different from those of merely large numbers, and even their logarithms are large numbers.

Many such systems, and particularly those used as examples, are composed of numerous subsystems, which may be called *particles*, though we do not mean the structureless dots usually called by this name. Our particles here may be molecules, for example, with internal degrees of freedom in addition to states of motion.

Even simple mass points may have a large number of states available in a macroscopic volume.

Let's take a hydrogen molecule as such a particle, confined in a cube a millimetre on a side. Suppose it has available all the states of kinetic energy less than, say, 4kT, a reasonable thermal excitation energy, where $k = 1.38 \times 10^{-23}$ J/K is Boltzmann's constant. At room temperature, this is about 1.62×10^{-20} J.

The quantum number n for a state with energy E is given by $n^2 = 8mL^2E/h^2$. With a mass of 3.32×10^{-27} kg, we find $n = 9.136 \times 10^5$. The total number of states with smaller energy is 1/8 the volume of a sphere of radius n, or 4.00×10^{17} states.

Our particle will have at least this many available to it as it jostles with other hydrogen molecules in the cubic millimetre, since there will be about 10^{16} of these. We need have no concern about the rapidity at which transitions between states are made in this system.

Our single particle has more than 10^{17} states available to it. This is already a pretty large number.

There are only about 10^7 seconds in a year, and the universe itself may be only about 10^{18} seconds old. Now we can make another leap and consider the number of states available to the system of hydrogen gas in a cubic millimetre at STP.

This will be the product of the number of states available to one particle raised to the power of the total number of particles, or 10 to the power 17×10^{16}. This is, indeed, a very large number. Stop and consider its breathtaking impudence for a minute. Write it out; 10 to the power of 17 followed by sixteen zeros.

It is so large that multiplying it by, say, 10,000 hardly changes it. Let us denote the number of states available to a system as W, and to a particle as w, so that $W = w^N$, and is a number like the one above. This counting of states is by no means arbitrary or indefinite, but very clearly defined. This is the thing that classical mechanics could not do, and it is fundamental. Matter has a very definite structure, determined by the sizes and properties of atoms, that is beyond sensory perception and the contemplation of fools.

It must be mentioned here that in simply multiplying the number of states of different, but identical, particles we are making an error. In fact, we are counting many states more than once, and have to divide by N!, where N is the number of particles.

We shall take this up later, but at the moment only notice that it will make our estimate of the entropy somewhat too large. Also, hydrogen has rotational energy in addition to translational, and these degrees of freedom contribute to the entropy as well.

In this article, we are concerned mainly with the translational degrees of freedom.

The entropy of a system is defined, following Boltzmann, as S = k ln

W. This equation is carved on his monument in the Zentralfriedhof zu Wien, and is as remarkable as $E = mc^2$.

The "k" is simply an arbitrary constant that gives S its dimensions, which for temperature in K is more precisely 1.380658×10^{-23} J/K. Kittel represents the entropy as $\sigma = \ln W$ to show that it is essentially a dimensionless counting.

This means that W is an exponential function of S, $W = e^{S/k}$. For our cubic millimetre of hydrogen, $S = k\,(2.3034)(17 \times 10^{16}) = 5.4 \times 10^{-6}$ J/K. Since a mole of hydrogen occupies about 22.4 litre at STP, this gives S = 121 J/mol-K, a not unreasonable result.

Of course, this is not an accurate evaluation of the entropy of hydrogen at STP, but shows the principles, and the sizes of numbers, involved.

Note that if we had made an error in estimating W by a factor of 10,000, the answer for the cubic millimetre would only be changed by k ln 10,000 = 1.27×10^{-22} J/K, an unnoticable difference. Entropy is just a logarithmic measure of the number of states available to a system, nothing more mysterious than that.

Suppose now that we have two systems, 1 and 2, with available states W_1 and W_2. The number of states available to the system formed by putting 1 and 2 into thermal contact will then be $W = W_1W_2$. Therefore, the entropy of 1 + 2 will be $S = k \ln W = k \ln W_1 + k \ln W_2 = S_1 + S_2$. The entropy, then, is an *additive* property, like mass.

This certainly does not mean that entropy is any kind of substance, as we see from the definition. It cannot be thought of as a fluid (neither can energy, but people persist in thinking this way).

When we put 1 and 2 into contact, they may not be in equilibrium, and energy may pass from one to the other, which we call a flow of heat. Let's suppose that system 1 gives up energy dU to system 2.

The loss of dU causes a restriction in the number of available states (some become energetically unreachable), while the gain of dU causes a gain in the number of available states. Since W is a function of U, so is S, and we can express the net change in entropy of 1 + 2 in the form $dS = (\partial S_1/\partial U)(-dU) + (\partial S_2/\partial U)(dU)$.

A more convenient way to write the partial derivatives is to set "S/ "U = 1/T. T will have the dimensions of a temperature because of the definition of k, and in fact is just the absolute temperature of the system. Now we have $dS = (1/T_2 - 1/T_1)dU$.

Energy will pass back and forth between the two systems until they have thoroughly explored the available states, and they will come into equilibrium when the two systems jointly have the maximum number of states available to them. Then dS = 0 (since any energy interchange away from the optimum value will lead to fewer available states), and this can be guaranteed for any dU only when $1/T_2 = 1/T_1$, or $T_1 = T_2$.

More than this, if $T_2 > T_1$ initially, dS will be positive for a positive dU, or energy flow from 1 to 2. We have explained not only thermal equilibrium, but the direction of approach to equilibrium by considering numbers of available states.

If we know the entropy of a system, we know the temperature of the system. Temperature exists whenever the derivative is well-defined (even for a system of two states only), not just for large systems that can be used as thermal reservoirs, as is important in this example.

We can now briefly indicate the role of time in entropy. A system can be prepared in a non-equilibrium state in which only a limited number of states are available to it. As time passes, more and more states become available, until finally all the states possible given the restrictions on the system (total volume, etc) are available to all the particles.

In terms of entropy, this means that the entropy will increase steadily to a maximum value that is the equilibrium entropy of the system. Think of a system that is created with temperature differences between its parts that only slowly become equalized.

Every step towards equalization of temperature gives the particles more and more freedom to occupy different states. The initial temperature differences could be used to drive a heat engine to extract useful work while this takes place.

Alternatively, we can just let heat conduction take its course, and lose the possiblility of extracting useful work. The effect on the system is the same (provided we replace the energy extracted) in either case, and the entropy increases to its final maximum value.

Suppose now that the system consists of a large part at a temperature T, and a small part, perhaps one particle, on which we focus our attention. We are interested in the relative probabilities of the small part's being in some state 1 and another single state 2.

To form these states, we have to extract energies E_1 and E_2, respectively, from the large part, which decreases the number of states available to the large part. Let W_1 and W_2 be the available states of the large part in the two cases. Then it is clear that the probability we want is just W_1/W_2. Now $W_1 = e^{S/k}$, and $S_1 = S' - E_1/T$, keeping only the first term in the series, with dU – $-E_1$ and S' the entropy of the large part before any energy is taken away.

Therefore, $W_1 = e^{S'/k}\exp(-E_1/kT)$, and we have a similar expression for W_2. Therefore, our desired probability is $p_1/p_2 = \exp[-(E_1 - E_2)/kT]$, which is the familiar *Boltzmann factor*.

The reason for the exponential in the Boltzmann factor should be clear. The ratio of probabilities that we want is the ratio of numbers of states, just as if we were working with dice in elementary probability, and the numbers of states are exponential functions of the entropy.

Also, our ratio is for individual states; if they come in bunches (degeneracy), then the factors must be multiplied by the numbers of states in each level. The Boltzmann factor is perhaps the most useful result of statistical mechanics, applicable to many problems, and which can even be carried over to problems expressed in classical mechanics, though it is fundamentally a counting of states.

Kittel carries through this argument allowing the number of particles N in the system to be an independent variable as well. The entropy is then a function of N, and its partial derivative with respect to N is $-\mu/kT$, where μ is another very important quantity, the *chemical potential.*

In classical thermodynamics, this quantity is even more mysterious than the entropy, but quantum mechanics gives it a clear meaning, by means of the argument presented here. It is remarkable that the only quantum mechanical result we use directly is simply the existence of stationary states that can be counted. The thermal behaviour of matter is, indeed, conclusive evidence of the existence of atoms. The Second Law is little more than a statement that more probable things are more probable than less probable things.

PARTITION FUNCTION

The probability that state E_i is occupied is $P_i = C\exp(-E_i/kT)$, which is the meaning of the Boltzmann factor.

The constant C can be found by summing over all P_i, which should give unity—the probability that *some* state is occupied. Thus, $1 = C\ \Sigma P_i$, or $C = 1/Z$, where $Z = \Sigma P_i$ is called the *partition function.* Each state of the system is represented in Z by its Boltzmann factor.

The internal energy U of the system can be expressed as $U = \Sigma E_i P_i$, the *ensemble average* of E_i.

The *ensemble* is an (imaginary) collection of systems, with one in each of the possible states, associated with its corresponding probability. In this case, we are dealing with the Gibbs *canonical ensemble.*

A change in U can be expressed as $dU = \Sigma P_i dE_i + \Sigma E_i dP_i$. The second term is the change in U due to a change in the probabilities, as, for example, caused by a change in temperature. This implies some transitions between particle states if the system is considered as composed of particles. We have already seen this when systems are placed in thermal contact, and heat flows between them. From the definition of temperature, $dU = TdS$ in this case, so we identify this term with TdS.

The first term results from a change in the energy levels without a change in probabilities. For example, if the volume V occupied by the system changes, so do the translational energy levels.

In fact, if the quantum number n remains constant, then the energy of a state is proportional to L^{-2}, if it is confined in a cube of side L. If we consider the state as exerting an outward pressure p_i on each face of its cube, then the work done by the state if L is increased by dL is $3p_iL^2dL = p_idV$, so $dE_i = -p_idV$. The term ΣP_idE_i is the ensemble average of the work done, or -pdV. Now we have dU = -pdV + TdS, which is called the *thermodynamic identity*.

Changes in U can take place due to changes in occupation of states, or by changes in the energies of the states, and these changes are called heat and work, respectively.

We assume that the changes associated with pdV take place with no change of entropy—that is, they are "reversible" in the usual terminology of thermodynamics. Work expressed as pdV is only one kind of work that can be done by a system; the energy of a state may depend on other parameters besides the volume, such as the magnetic field, which can be treated analogously.

We have just seen that $(\partial U/\partial V)_S = -p$. This is merely another way to write that the change in energy dU caused by a change in volume dV at constant entropy S is the negative of the pressure, or dU – –pdV (S = constant), which we have just discussed.

A change in S can be expressed in terms of partial derivatives by dS = $(\partial S/\partial U)_V$ dU + $(\partial S/\partial V)_U$ dV, where S is considered as a function of U and V, as we have assumed above. Now set dS = 0, divide by dV, and replace dU/dV by $(\partial U/\partial V)_S$ by -p. Since $(\partial S/\text{''}U)_V = 1/T$, we then have $(\partial S/\text{''}V)_U = -p/T$. One might be tempted to use this expression as a definition of p, as we defined T by a partial derivative of S, but that would not be legal, since p is already defined in terms of the forces exerted by the system.

What we have to do is show that this externally defined p also appears in the partial derivative expression, and we have just done that.

Just as in the case of temperature, we can now show that $p_1 = p_2$ in equilibrium, and that if $p_1 > p_2$, V_1 increases at the expense of V_2, so that system 1 does work on system 2. The systems must also be in thermal equilibrium for this reasoning to be valid. If they are not, we must have recourse to mechanical arguments, since dS will be zero in any case.

Since $P_i = \exp(-E_i/kT)/Z$, $\ln P_i = -E_i/kT - \ln Z$, or $E_i = -kT \ln P_i - kT\ln Z$. Since $\Sigma P_i = 1$, $\Sigma dP_i = 0$, so multiplying the previous equation by dP_i and summing over i, we find $\Sigma E_idP_i = -kT\ \Sigma dP_i\ln P_i$. The sum on the left is, as we have seen, just TdS, and $d[\Sigma P_i\ln P_i] = \Sigma\ dP_i\ln P_i$ since $\Sigma dP_i = 0$, so $dS = d[-kT\ \Sigma P_i \ln P_i]$, and thus $S = -kT\ \Sigma P_i \ln P_i$.

The constant of integration is zero, as we can see by considering a system occupying a single state. This is Boltzmann's expression for the entropy in terms of the probabilities.

Now we go back to the first equation in the preceding paragraph, where $-kT \ln Z = E_i + kT \ln P_i$. Multiplying by P_i and summing, which is taking the ensemble average, we find $-kT \ln Z = U - TS$, using the expression for S that we have just derived.

The function $U - TS$ is the *free energy* F. It has the property that $dF = dU - TdS - SdT = -pdV - SdT$, so that $dF = 0$ in a process where $dV = dT = 0$, or an isothermal process at constant volume, such as a chemical reaction taking place in a fixed volume at a fixed temperature.

This means that F is an extremum (usually a minimum) at equilibrium, and can be used as a criterion of equilibrium in an isothermal process as entropy is used in a process at constant energy. The fact that $F = -kT \ln Z$ makes the partition function very useful for finding the thermodynamic functions of a system.

From F, we can find $p = -(\partial F/\partial V)_T$, and $S = -(\partial F/\partial T)_V$ by differentiation, and then $U = F + TS$, and so on. These properties come out as functions of T and V, which is much more convenient than the expressions in terms of U and V that we have found up to this point.

As an example, consider a linear harmonic oscillator whose states have energies $(n + 1/2)hf$, where n is the quantum number, f the frequency, and h Planck's constant. We have included the zero-point energy hf/2.

Then, $Z = \Sigma \exp[-(n + 1/2)(hf/kT)] = \Sigma \exp[-(n + 1/2)x]$, where $x = hf/kT$. This series can be summed to $Z = \exp(-x/2)/(1 - e^{-x}) = e^{x/2}/(e^x - 1)$. Therefore, $F = -kT[(x/2) - \ln(e^x - 1) = -hf/2 + kT \ln(e^x - 1)$. The entropy is $S = -k \ln (e^x - 1) + (hf/T)e^x/(e^x - 1)$.

Finally, $U = F + TS = hf/(1 - e^{-x}) - hf/2$. For $kT >> hf$, $U \rightarrow kT$, while for $kT << hf$, $U \rightarrow hf/2$, which are the correct limits.

The specific heat is $C = dU/dT = kx^2e^{-x}/(1 - e^{-x})^2$, the Einstein expression. For $kT >> hf$ (large x) $C \rightarrow k$. The Einstein temperature is $T_E = hf/k$, a measure of the quantum size. When $T = T_E$, then $C = 0.92k$, only 8% below its high-temperature limit.

When $T = T_E/10$, $C = 0.0045k$, and the oscillator is "frozen out." This has an important application to the vibration of molecules. The Einstein temperature of O_2 is 2260K, and of H_2 6300K. The vibration of hydrogen is frozen out at room temperature, while that of oxygen is only a little excited.

If the zero-point energy is omitted, then $Z = \Sigma e^{-nx} = 1/(1 - e^{-x}$. We find $F = kT \ln (1 - e^{-x})$ and $S = -k \ln (1 - e^{-x}) + (hf/T)/(e^x - 1)$.

Then, $U = F + TS = hf/(e^x - 1)$. This is appropriate for photons, which are not conserved and so cannot have a zero-point energy. We recognize here the Bose-Einstein distribution for zero chemical potential, which appears in the theory of black-body radiation. The spectrum of black-body radiation (the energy density) is this quantity times the number of modes (oscillators) in the frequency interval df.

A two-state system with a ground state of energy 0 and an excited state of energy E can also be easily handled, since the partition function is the simple expression $Z = 1 + e^{-E/kT}$. Then, $U = Ee^{-x}/(1 + e^{-x}) = E/(e^x + 1)$, where $x = E/kT$. The specific heat is $C = kx^2e^x/(e^x + 1)^2$. This is the same as for the harmonic oscillator, except that a +1 replaces the -1 in the denominator, which makes a considerable difference. For $x \to \infty$ (low temperatures), $C \to 0$, while for $x \to 0$ (high temperatures), C also approaches zero.

The value of x for the maximum C can be found by setting the derivative of C with respect to x equal to zero. The result is the solution of $2/x = \tanh(x/2)$, which is about $x = 2.40$. The peak of the C versus T curve is at $T = 2.40E/k$, from which E can be estimated experimentally. This peak is called a *Schottky anomaly.*

The entropy can be found in the usual way from $F = -kT \ln(1 + e^{-x})$. If we set $E = 1$ for simplicity, we find $S = -k[U \ln U + (1 - U)\ln(1 - U)]$. This is zero for $U = 0$ and $U = 1$ (note that $\lim(x \to 0)\ x \ln x = 0$), and symmetrical about a maximum at $U = 1/2$, for which $S = k \ln 2$.

The temperature is given by $1/T = dS/dU = k \ln[(1 - U)/U]$. We must replace k by k/E if the temperature is to come out in K. For $U = 0$, or the upper state unoccupied, we find $T = 0$, which is not unexpected. For $U < 1/2$, the temperature is positive, approaching $\int$ for $U = 1/2$, where the two states are equally occupied on the average.

For still greater U, the temperature becomes *negative*, approaching 0 from below as $U \to E$, when the lower state is unoccupied. Negative temperatures are higher than positive temperatures, with -0K the highest temperature of all. Whenever the temperature is negative, the upper state is more populated than the lower, and there is a *population inversion.*

If we put systems with temperatures +T and -T in contact, the result at equilibrium will be $T = \infty$, not zero. This is quite reasonable, since in the combined system the upper and lower levels will be equally occupied, which corresponds to infinite temperature. Population inversions are created in lasers so that the system will radiate spontaneously.

THE IDEAL GAS

An ideal gas is a system consisting of N identical particles in a volume V. The particles are assumed to move entirely independently. Of course, in any real case they collide very frequently, and these collisions are necessary to permit the system to occupy all of its available states.

Nevertheless, the particles feel each other's presence very rarely, and are *almost* independent. We shall totally neglect the collisions, and trust that the gas can reach equilibrium quickly when disturbed. We will consider only the translational motion of the particles, and not any internal degrees of freedom they may have. This restriction is purely for simplicity.

Let's begin by considering one molecule in a volume V, thought of as a cube of side L, such that $L^3 = V$.

The states are labelled by three quantum numbers, and the energy E depends only on the sum of squares n^2 of the quantum numbers. In fact, $E = h^2n^2/8mL^2$, where m is the mass of the particle. In n-space, where the quantum numbers are plotted along three orthogonal axes, there is one state per unit volume, and only points in the first octant correspond to different states. This means that the number of states in (n,n+dn) is D(n)dn $= \pi^2n^2dn/2$. Now it is easy to find Z by replacing the sum with an integral. Setting $x = (n/L)(h/8mkT)^{1/2}$, we have $Z = [L(2mkT/h^2)]^3\ (\pi/2)\ \int(0, \infty)\ x^2\exp(-x^2)dx$.

The value of the definite integral is $\pi^{1/2}/4$, so $Z = V[2\pi mkT/h^2]^{3/2}$. This is the partition function for a single particle.

The energy can be found from Z by the formula $U = kT^2\partial(\ln Z)/\partial T$ (this is just $-\partial F/\partial T$). The only temperature-dependent part of ln Z is $T^{3/2}$, so we find $U = 3kT/2$. This is an expression of the famous *principle of equipartition*, which states that to each degree of mechanical freedom an energy kT/2 corresponds in thermal equilibrium.

In an ideal gas we have three degrees of translational freedom. A harmonic oscillator has two degrees of freedom, one for the kinetic energy and one for the potential energy, and we have already seen that its average energy is kT at a high enough temperature.

The pressure is given by $-\partial F/\partial V$, or $p = kT\ \partial(\ln Z)/\partial V$. The only volume-dependent part of Z is simply ln V, so we find $p = kT/V$ or, $pV = kT$, the ideal gas law for a single particle.

If we have N particles instead of just one, it would not take much argument to convince one that $U = 3NkT/2$, and $pV = NkT$, instead. These are, indeed, correct results, and would be obtained if we simply raised the single-particle Z to the Nth power.

It happens, however, that if we do this, the free energy and the entropy do not come out correctly! The problem is quantum-mechanical. In the microscopic world, two electrons, or two hydrogen atoms, are not just similar, they are identical. If we have a state containing any number of them, interchanging any pair gives back exactly the same state, not a different one. That is, the number of states calculated on the basis that the particles are all distinguishable (raising Z to the power N, for example) is too large by a factor N!.

This is the number of ways we can permute the particles in the state and get back the same state. It is actually a little worse than this. The states of particles like electrons, that have half-integral spins, must be antisymmetric on the exchange of any two particles (i.e., it must change sign), while states of particles of integral spin must be symmetric under exchange. We will not go into the complexities of state-counting that these considerations require, but stick with the simple N! factor.

The partition function for N particles is then $Z = (V^N/N!) [2\pi mkT/h^2]^{3N/2}$. In handling the N!, the Stirling approximation $\ln N! = N \ln N - N$ will be accurate enough. The internal energy U and the pressure p come out as above, but F and $S = -\partial F/\partial T$ are different.

Most important is the expression for the translational entropy, $S = (3Nk/2)\ln(2\pi mk/h^2) + (5Nk/2) + Nk \ln(V/N) + (3Nk/2)\ln T$, which differs from the naive result by $-k(N \ln N - N)$, the contribution of the N!. This is the famous *Sakur-Tetrode* (Tetrode has three syllables) equation.

For the cubic millimetre of hydrogen we used as an example above, the first two terms, the constant terms, contribute $+8.842 \times 10^{-6}$ J/K to the entropy; the third term gives -7.63×10^{-6} J/K. Part is volume-dependent, but the rest is constant. The final term, the temperature dependence, adds 1.180×10^{-6} J/K, for a total of 2.392×10^{-6} J/K.

This is a much more accurate value than our early rough estimate, but it is of the same order of magnitude. The Sakur-Tetrode equation has been thorougly confirmed by experiment. For a monatomic gas, such as helium or argon, it gives the total entropy.

We have seen that the usual ideal gas constant $R = N_A k = 8.31451$ J/mol-K. Note that this is for a gram-mole, for which $N_A = 6.02 \times 10^{23}$. It is sometimes given for a kg-mole, a unit which is never used by the sane. The Boltzmann constant is just the gas constant per molecule, and is the same constant used to get entropy from probability in $S = k \ln W$.

Since 1 cal = 4.186 J, R = 1.986 cal/mol-K, quite close to the easily-remembered round value of 2 cal/mol-K. Measurements on ideal gases is one way to determine k experimentally. As a practical matter, most gases under normal conditions are very well approximated as ideal gases, consisting of independent, classical particles, and their thermodynamic properties are easily determined.

Linear molecules have an additional two rotational degrees of freedom, and nonlinear molecules have three. Rotational energies are small, so they are not frozen out at usual temperatures as vibration is.

The principle of equipartition then applies, so the internal energy is increased by RT and 3RT/2 per mole, respectively, giving U = 5RT/2 and U = 3RT for the two cases.

The specific heat at constant volume is dU/dT, so $C_V = 3R/2$ for a monatomic gas, 5R/2 for a gas of linear molecules (such as a diatomic gas) and 3R for a gas of nonlinear molecules (such as a polyatomic gas). These predictions are well confirmed by experiment.

An *adiabatic* process is one that takes place without heat exchange. If it is *reversible*, then it will be *isentropic*.

Such processes are usually fairly rapid ones, such as the compressions involved in the propagation of sound, or the convective rise of bodies of air in the atmosphere, and are of great practical importance. For a monatomic gas, S = (3R/2)ln T + R ln V + constant terms J/mol-K. Therefore, S = constant implies that $T^{3R/2}V^R$ = constant, or $VT^{3/2}$ = constant. Since pV = RT, this can be cast into the form $V(pV/R)^{3/2}$ = constant, or $p^{3/2}V^{5/2}$ = constant.

As C_V = 3R/2 and $C_p = C_V + R$ = 5R/2 (this latter relation is proved in the Appendix; C_p is the specific heat at constant pressure, when the gas must be allowed to expand and do work).

In view of this, an isentropic process for an ideal monatomic gas is characterized by $p^{C_V}V^{C_p}$ = constant, or $pV^{Cp/Cv}$ = constant. It is shown in the Appendix that this is a general relation.

The ratio of specific heats Cp/Cv is usually denoted by γ, so finally we have the familiar pV^γ = constant. For a monatomic gas, $\gamma = 5/3 = 1.667$, for a gas of linear molecules $\gamma = 7/5 = 1.400$ and for a gas of nonlinear molecules, $\gamma = 8/6 = 1.333$. The ratio of specific heats can be determined quite accurately by measuring the speed of sound, and these predictions are fully confirmed.

DIFFERENCE IN THE SPECIFIC HEATS

We begin with the thermodynamic identity, dU = TdS – pdV, and consider one mole of substance. In a change of temperature dT at constant pressure, dU = C_VdT for an ideal gas, while the heat absorbed TdS = C_pdT. Hence, $C_p = C_V + p(\partial V/\partial T)_p$, where the partial derivative is the ratio of dV to dT at constant p. Since the equation of state of one mole of an ideal gas is pV = RT, $p(\partial V/\partial T)_p$ = R. Therefore, $C_p = C_V + R$. Q.E.D.

In general, the internal energy can depend on the volume as well as on the temperature. In fact, $(\partial U/\partial V)_T = T(\partial p/\partial T)_V - p$, which is zero for an ideal gas. In the more general case, $C_p - C_V = -T(\partial V/\partial T)^2(\partial p/\partial V)_T$. The necessary derivatives can be found from the equation of state.

RELATION BETWEEN P AND V IN AN ISENTROPIC PROCESS

The thermodynamic identity is again the starting point, dU = TdS – pdV. In an isentropic process, dS = 0, so 0 = C_VdT + (RT/V)dV, where we have used the equation of state for one mole of an ideal gas, and have made the same assumption about the internal energy as above. Then

$C_V(dT/T) + R(dV/V) = 0$. This integrates easily to $C_V \ln T + R \ln V =$ constant, or $T^{Cv}V^R =$ constant, or $T^{Cv/R}V =$ constant. From the equation of state, $T = pV/R$, so $p^{Cv/R}V^{Cv/R+1} =$ constant. Raising to the power R/Cv, we find $pV^\gamma =$ constant, where $\gamma = (Cv/R + 1)/Cv = Cp/Cv$. this result is exact for an ideal gas only. Q.E.D.

RELATION BETWEEN ADIABATIC AND ISOTHERMAL BULK MODULI

The bulk modulus is defined by $k = -V(dp/dV)$, where an increase in pressure dp causes a decrease in volume -dV. If the temperature is held constant, then $pV = RT =$ constant, and the isothermal bulk modulus is $k_T = p$ for an ideal gas. If the process is adiabatic, then $pV^\gamma =$ constant.

The derivative with respect to p is $V^\gamma + p\gamma V^{\gamma-1}(dV/dp)_s$. The adiabatic (isentropic) bulk modulus is then $k_s = \gamma p$. The speed of sound c is given by $c^2 = k/\rho$. For an ideal gas, this becomes $c^2 = \gamma p/\rho = \gamma RT/M$. The adiabatic process is "stiffer" than the isothermal. Although we have proved it only for an ideal gas, the relation $k_s = \gamma k_T$ is true in general.

THE GAS CONSTANT IN OTHER UNITS

We gave two values for the gas constant above, 8.314 J/mol-K and 2 cal/mol-K, which are convenient in some problems, but not in all. The gas constant for one mole in any units can be found from $R = pV/T$, where corresponding values for p, V and T are inserted. For example, p = 1 atm, V = 22.4 l and T = 273.15K are well-known corresponding values, which give R = (1)(22.4)/(273.15) = 0.082006 l-atm/mol-K.

Note that 0.0820 is a value good to better than 1 part in 10,000, and not too hard to remember. If you are converting to U.S. Engineering Units, you need to know that a pound-mole is 453.59237 times more than a gram-mole, and a °R (rankine) is 9/5 of a kelvin (0°C is 491.67R). At STP the pound-molar volume is 358.8 cubic feet, so R = 0.73 cuft-atm/lb-mol-R. Since 1 atm is 14.696 psi, R = 10.73 psi-cuft/lb-mol-R as well.

Chapter 6

Thermodynamics

From its Greek etymology, "thermodynamics" could be the study of the forces associated with beans, or with heat. We'll consider heat here, leaving beans for another article.

Heat was long considered something like a substance, the Empedoclean element fire, with the ill-defined concept of temperature as a sort of driving force. Heat even got its own units, of which the calorie survives even today, and was the object of the practical science and technique of calorimetry. The heat substance was the subtle fluid *caloric*, which combined with matter in obscure ways, and could be latent (hidden) or evident. Relics of these ideas still survive in the terminology of thermodynamics.

In the early 19th century, it slowly became evident that heat was not a conserved quantity, as it should have been if it were material, but that work and heat were interconvertible. Count Rumford showed that limitless amounts of caloric were created in the boring of cannon, and Watt showed that heat could be converted into mechanical work. It is not an easy practical matter to show that the work done is proportional to the quantity of heat destroyed, but the converse is readily observed.

It was a matter of experiment to find that 1 calorie of heat was the equivalent of 4.186 J of work, or 1 Btu the equivalent of 778 ft-lb. Heat and work were indentified as forms of energy, which was a conserved quantity. This was the fundamental concept of thermodynamics, the new science of heat and force. Things were not as simple as they seemed in the days of caloric.

The deductive science of classical thermodynamics rests on simple foundations, which were dignified as the Laws of Thermodynamics. Much effort has been rather uselessly expended in arguing over the best way of expressing these fundamental postulates.

Let's restrict ourselves here to a pure substance whose state can be specified in terms of its pressure p, volume V and temperature T. Actual substances are extremely varied, and it would be tedious to attempt to handle all the variety in our discussion. All the important matters can be discussed on the basis of such a pure substance. The fixed amount of substance under consideration is called the *system*.

We assume, first of all, that the variables p, V and T satisfy an equation f(p,V,T) = 0 which we call the *equation of state*. The variables p and V are easy to define and measure, but the temperature T presents a great mystery. One way of proceeding would be to choose a definite substance, such as mercury, and arbitrarily define temperature as proportional to the volume of a certain amount of mercury at a standard pressure.

Thermodynamics can limp along on such an unsatisfying definition, but it was found possible to make a better definition that does not depend on the properties of a substance.

Carnot showed that such a definition could be based on a cyclical thermodynamic process in which Q/Q' = T/T', where Q,Q' are the amounts of heat transferred and T,T' are the corresponding *absolute temperatures*. This result did not depend on the particular substance used to execute the cycle. Carnot himself actually was working with caloric, but his results were interpreted later in this manner.

This shows that there is a natural zero of temperature, where Q = 0, and that absolute temperatures are determined up to a constant, which can be chosen to suit one's whim.

If we decide that the absolute temperature of ice in equilibrium with air-saturated water at 1 atm is 273.15, we get the Kelvin scale, in which the difference in temperature of the steam and ice points is 100°, characteristic of the Celsius scale. If we take 9/5 of this, the difference is 180°, characteristic of the Fahrenheit scale (212 – 32 = 180), and the absolute temperatures are on the Rankine scale.

Neither scale is any more metric or absolute than the other, but the Kelvin scale is standard.

The practical method of finding absolute temperature uses an ideal-gas thermometer. The equation of state for one gram-mole of an ideal gas is pV = RT, so that T = pV/R.

This looks like basing the scale on a particular substance, but really is not, since the ideal gas is a theoretical concept that can be approached more or less closely by an actual gas, but is a universal concept independent of any particular substance.

Taking an ideal gas through a Carnot cycle shows that T in the equation of state is the absolute temperature. In practice, the temperature is extrapolated to zero pressure, where the gas will be ideal in fact. Thermodynamics is much neater when the absolute temperature is used.

Now that the temperature T is properly defined, and the equation of state f(p,V,T) = 0 of a substance has been determined by experiment, thermodynamics postulates that two functions, the *internal energy* U =

U(p,T) and the *entropy* S = S(p,T), exist. These are written as functions of p,T but really they are functions of p,V,T with the restriction f(p,V,T) = 0 that makes them functions of two independent variables. Any two variables can be used to specify U and S: p,T; p,V; or V,T. Moreover, U or S can be solved for either of the independent variables, giving p = p(U,T) or even S = S(U,V).

It turns out that most of classical thermodynamics is merely the working out of mathematical relations between p,V,T,U,S and other functions defined in terms of them for special purposes. In particular, there are lots of the partial derivatives that are a familar feature of classical thermodynamics. The validity of these mathematical deductions gives classical thermodynamics an impressive accuracy as a physical theory.

Classical thermodynamics is haughtily independent of the structure of matter, as if it existed in an ideal Platonic realm of pure reason. This is, alas, only an illusion, and thermodynamics is a creature of the quantum-mechanical atomic properties of matter.

The fundamental concepts of entropy and temperature are given a rigorous basis by considering the mechanics of matter, and rest on the huge numbers of quantum states available to a macroscopic system, among which the system continually and rapidly wanders.

Classical thermodynamics is an expression of the atomic structure of matter, just as Chemistry is.

The atomic theory of thermodynamics is called *Statistical Mechanics,* for historical reasons. A macroscopic system is one that we can see with our eyes, if only in a microscope, while a microscopic system consists of only a few atoms.

There is no sharp distinction; as we go to smaller and smaller systems, fluctuations around the values specified by thermodynamics increase, and concepts such as entropy have less and less significance. Thermodyamics just gradually evaporates.

The most important result of statistical mechanics is that the entropy of an isolated system never decreases, and is a maximum at equilibrium. This condition, which can be expressed as $dS > 0$, is called the *Second Law of thermodynamics*. It applies to an *isolated* system. If two systems are brought into thermal contact, the entropy of one may decrease, and the entropy of the other increase, but the net entropy change will be zero or positive.

The existence of the functions U (and S as well) which depend only on the state of the system, constitutes the *First Law*. Statistical mechanics shows that the change in internal energy of a system $dU = TdS - pdV$ (called the Thermodynamic Identity), where TdS is the heat transferred to the system, and pdV the work done by the system, in a reversible process.

A reversible process is one in which the net increase (system + surroundings) in entropy is zero. Of course, dS may be greater than or less than zero, depending on whether the system absorbs or gives up heat. Unlike the pressure, volume and temperature, the energy and entropy of a system cannot be directly measured experimentally.

Now that we have covered the fundamentals of thermodynamics, we can proceed to make some mathematical deductions from them. The notation for partial derivatives that appears in the reasoning may appear formidable, but the meaning is clear and simple, specifying the ratio of the change in one quantity due to a change in another, with what other variables are held constant.

In $(+x/+y)_z$, x is considered as a function of y and z, $x = x(y,z)$, and the partial derivative is the ratio of dx to dy when z is held constant. Most of the functions we consider are functions of two independent variables, so this notation is very apt.

For example, if $x = 2yz + z^2$, then the above partial derivative is 2z, while $(+x/+z)_y = 2y + 2z$. Also, $(+y/+x)_z = 1/(+x/+y)_z$.

PARTIAL DERIVATIVES AND POTENTIALS

First of all, let's consider all the partial derivatives that flow from the equation of state, $f(p,V,T) = 0$.

We can write $df = (+f/+p)dp + (+f/+V)dV + (+f/+T)dT = 0$, which tells us how the differentials of the three variables are interrelated. For simplicity, we write $(+f/+p) = f_p$, and so on.

There are six partials between pairs of variables, which can be expressed in terms of the partials of f. For example, if $dT = 0$, then $f_p dp + f_V dV = 0$, or $(+V/+p)_T = -f_p/f_V$. From symmetry, the general pattern is easy to see. If we multiply the above result by $1 = f_T/f_T$, we find that $(+V/+p)_T = -(f_p/f_T)/(f_V/f_T) = -(+V/+T)_p/(+p/+T)_V$.

This is a relation between the three partials that is difficult to see by inspection, but which we shall find useful.

Since $dU = TdS - pdV$, $(+U/+S)_V = T$, and $(+U/+V)_S = -p$. The cross second partials must be equal, so $(+T/+S)_V = -(+p/+V)_S$. This is one of *Maxwell's Relations*, which can prove very useful. We also say that the "natural variables" of U are S and V.

Also, in a process where $dS = dV = 0$, we have $dU = 0$, which is the condition for an extremum of U, either a maximum or a minimum. Therefore, the equation also gives a *criterion of equilibrium*.

The function $H = U + pV$, called the *enthalpy*, can be handled the same way. $dH = dU + pdV + Vdp = TdS + Vdp$.

The first partials give $(+H/+S)_p = T$ and $(+H/+p)_S = V$. Equality of the second partials gives a second Maxwell relation: $(+T/+p)_S = (+V/+S)_p$.

In a process at constant entropy and pressure, the enthalpy is an extremum. The *free energy* F = U – TS. dF = dU – TdS – SdT = -pdV – SdT.

In a process at constant temperature, which describes many actual processes, the work done is the decrease in the free energy, which is the source of the name. At constant temperature, the free energy F is a minimum at equilibrium, not the energy U.

This is a very important result with many applications. As for partials, $(+F/+V)_T = -p$, and $(+F/+T)_V = –S$. The Maxwell relation is $(+p/+T)_V = (+S/+V)_T$. Finally, the *Gibbs function* G = U – TS + pV, and dG = -SdT + Vdp. In a process at constant temperature and pressure, the Gibbs function is constant. When temperature and pressure are held constant, the Gibbs function is a minimum at equilibrium.

If g is the Gibbs function per mole for the reactants in a chemical reaction, and g' the Gibbs function per mole for the products, then at equilibrium g = g'. A similar relation also holds for a liquid and its vapour in equilibrium at a temperature T and a pressure equal to the vapour pressure. $(\partial G/\partial T)_p = –S$ and $(\partial G/\partial p)_T = V$. The Maxwell relation is $(\partial S/\partial p)_T = -(\partial V/\partial T)_p$.

The internal energy, enthalpy, free energy and Gibbs function are known as *thermodynamic potentials*, because their derivatives are of so much use. The free energy is also called the Helmholtz function, and the symbol A is often used in place of F. The Gibbs function per mole is called the *chemical potential*, which is of great importance. The meaning of the chemical potential is greatly clarified in statistical mechanics.

SPECIFIC HEAT AND THE TDS EQUATIONS

Although the internal energy cannot be measured directly, heat transfers affecting it can be. At constant volume for an ideal gas, dU = TdS = C_VdT, which defines the *specific heat* at constant volume, C_V. Specific heat is also called *heat capacity*. It is proportional to the amount of substance, so is usually specified per gram or per mole, in cal/g/K or J/mol/K. As a partial derivative, it is $C_V = T(\partial S/\partial T)_V$.

At constant pressure, we have dH = TdS = C_pdT, or $C_p = T(\text{“}S/\text{”}T)_p$. C_p is the specific heat at constant pressure, much easier to measure experimentally than the specific heat at constant volume. We can find U and H at any temperature T by integrating the corresponding specific heat from T = 0 to T.

The knowledge of specific heats at low temperatures is very useful in the calculation of thermodynamic functions.

The entropy is a particularly important function, so it is useful to look at how it depends on the specific heats.

Let's begin by considering S a function of T and V, S = S(T,V). Then,

$dS = (\partial S/\partial T)_V dT + (\partial S/\partial V)_T$. The first partial is C_V/T, as we have just seen, and a Maxwell relation says the second is $(\partial p/\partial T)_V$.

This gives $TdS = C_V dT + T(\partial p/\partial T)_V dV$. Now dS is given in terms of experimentally measurable quantities only. This is called the *First TdS Equation*. If S is now considered as a function of p and T, $S = S(p,T)$, we find $dS = (\partial S/\partial T)_p dT + (\partial S/\partial p)_T dp$.

The first partial is C_p/T, while a Maxwell relation gives the second as $-(\partial V/\partial T)_p$. Therefore, $TdS = C_p dT - T(\partial V/\partial T)_p dp$. This is the *Second TdS Equation*. The TdS equations relate the specific heats to the entropy and to derivatives of the equation of state.

If we subtract the first TdS equation from the second, we get $0 = (C_p - C_V)dT - T(\partial V/\partial T)_p dp - T(\partial p/\partial T)_V dV$. If this is solved for dT, we find that $dT = T[(\partial V/\partial T)_p/(C_p - C_V)]dp + T[(\partial p/\partial T)_V/(C_p - C_V)]dV$. Now this is a relation between the three differentials dT, dp and dV, which we know are related by the equation of state. Considering T as a function of p and V, we have $(\partial T/\partial p)_V dp + (\partial T/\partial V)_p dV$.

The coefficients of dp and dV in the two expressions for dT must be equal. The coefficients of dp give $C_p - C_V = T(\partial V/\partial T)_p/(\partial T/\partial p)_V = T(\partial V/\partial T)_p(\partial p/\partial T)_V$. At the top of the preceding section, we showed that $(\partial p/\partial T)_V = -(\partial V/\partial T)_p(\partial p/\partial V)_T$, so our result can be cast into the form $C_p - C_V = -T(\partial V/\partial T)_p^2(\partial p/\partial V)_T$. Since $(\partial p/\partial V)_T$ for all substances is negative, this shows that C_p is always greater than C_V. The difference is expressed entirely in derivatives of the equation of state.

This equation has several interesting consequences. At $T = 0$, $C_p = C_V$, and they are also equal when $(\partial V/\partial T)_p = 0$, when there is a minimum or maximum of the density, as with water at 4°C. For an ideal gas, $(\partial V/\partial T)_p = R/p$ and $(\partial p/\partial V)_t = -RT/V^2$, so that $C_p - C_V = -T(R/p)^2(-RT/V^2) = R$.

If we set $dS = 0$, then $C_p/C_V = -[(\partial V/\partial T)_p/(\partial p/\partial T)_V)] (\partial p/\partial V)_S$. The last partial is just the ratio of dp to dV in this case, when $dS = 0$. Again we use our relation between three partials of the equation of state to find that $C_p/C_V = (\partial V/\partial p)_T(\partial p/\partial V)_S$.

Now, the bulk modulus $k = V(\partial p/\partial V)$, where isothermal or adiabatic condtions must be specified. Denoting C_p/C_V by γ, we have $\gamma k_T = k_S$. The ratio of C_p to C_V is the ratio of the adiabatic (isentropic) bulk modulus to the isothermal bulk modulus.

From this, we also conclude that the adiabatic bulk modulus is always greater than the isothermal bulk modulus, and that they are equal at $T = 0$ or extrema of the density as a function of temperature. The reciprocal of the bulk modulus, the *compressibility*, is often used instead.

These results for the difference and the ratio of the specific heats at constant pressure and constant volume are very important and useful. They are borne out to great precision by experiment, and show the power of the mathematical arguments of classical thermodynamics. The volume thermal expansivity of a substance is $\beta = (1/V)(\partial V/\partial T)_p$. For an ideal gas, it is simply 1/T, so that absolute zero can be determined by measuring the thermal expansivity of a gas. At 0°C, $\beta = 1/273.15$, as we know.

The isothermal bulk modulus of an ideal gas is the pressure. If you wish to try the equations on a non-gaseous substance, for mercury at T = 273K, $C_p = 6.69$ cal/gmol-K, specific volume 14.72 cm^3/gmol, $\beta = 1.81 \times 10^{-4}K^{-1}$ and $k_T = 2.58 \times 10^{11}$ dyne/cm^2. From this data, it is easy to find that $C_V = 5.88$ cal/gmol/K, and $\gamma = 1.14$. A gmol of mercury is 200.59 g. In terms of β and k, $C_p - C_V = -TV\beta^2k$. If we are dealing with specific heats per mole, then V is the volume per mole.

Water is an unusual liquid, with a high specific heat of $C_p = 18$.cal/mol/K, and V = 18 cm^3/mol. At 0°C, $\beta = -6.814 \times 10^{-5}$ per K, but at 20°C it is $+2.066 \times 10^{-4}$ per K. At about 4°C, it is zero. k decreases steadily from 1.99×10^{10} dy/cm^2 at 0°C to 2.21×10^{10} at 20°C.

At 0°C, it is close to 2.04×10^{10}. At 20°C, $C_p - C_V = 0.111$ cal/mol/K, so $\gamma = 1.006$. The speed of sound in a liquid is $c = \sqrt{(\gamma k/\rho)}$. In water, this amounts to 1491 m/s at 20°C. In mercury, the figure is 1474 m/s.

Iron, or steel, has density 7.86 g/cc, molecular weight 55.847, so its specific volume is 7.11 cc/mol. β is three times the linear expansion coefficient, or 3.6×10^{-5} per K. k, calculated from Young's modulus E and the shear modulus μ by $k = E\mu/(9\mu - 3E)$, is 1.51×10^{12} dyne/cm^2. C_p is 0.1075 cal/g/K or 6.00 cal/mol/K. From these figures, $C_p - C_V = 0.097$ cal/mol/K. Therefore, $C_V = 5.90$ cal/mol/K and $\gamma = 1.017$. For solids and liquids, we see that the difference in the specific heats is small, and γ is about 1.0.

You may have noticed that C_V for mercury or for iron is not far from 6 cal/mol/K (that is, 3R). This is the rule of Dulong and Petit, reflecting the fact that a mole contains the same number of atoms of any substance. For a gram-mole this number is Avogadro's number, $N_A = 6.02 \times 10^{23}$. N atoms have 6N vibrational degrees of freedom (3N for kinetic energy and 3N for potential energy), and at sufficiently high temperatures equipartition of energy gives each degree of freedom kT/2 of energy, or $3N_A\,kT = 3RT$ in all, which means a specific heat of 3R.

This is evidence of the atomic structure of matter from classical thermodynamics, and is of the same kind as similar evidence in classical Chemistry. A more detailed investigation, such as Debye's theory, gives a better result and allows the behaviour of the specific heat at low temperatures to be described, where it is proportional to T^3.

CLAPEYRON'S EQUATION

The change of ice to water, and water to steam, as a sample of H_2O is heated, provides a familiar example of *phase change*. The molecules are arranged differently in the three different manifestations, although the chemical composition remains the same. Ice is a rigid crystalline lattice, really a large molecule. Water is a transient association of molecular clusters locally with dodecahedral symmetry.

Steam is a gas of mostly independent H_2O molecules, and perhaps some dimers. There are many ways to arrange the molecules, but these are the most stable of each general type.

All are bound largely by strong hydrogen bonds between the protruding protons and the unbonded pairs of electrons on the oxygens. The strength of the hydrogen bonds makes H_2O an atypical substance. There needs to be a word for all appearances of H_2O, so that ice, water and steam can be restricted to the individual structures. Each of the distinct structures is a *phase*, with distinctive thermodynamic properties.

When water is heated at atmospheric pressure, it *boils* at about 100°C. At this temperature, the *vapour pressure* of water is equal to the atmospheric pressure, so bubbles of steam can form and the water can rapidly turn into steam as quickly as sufficient energy can be supplied. The water has been turning into steam all along, which has been leaving the surface of the water and diffusing into the atmosphere.

What makes boiling distinct is the formation of the bubbles when the vapour pressure equals the atmospheric pressure. If the atmospheric pressure is changed, then so is the boiling point.

In Denver, where the atmospheric pressure is about 850 mb, water boils around 95°C. It is noticed that the temperature remains constant at 100°C or whatever as long as there is any water left. Water can be superheated, and then may burst explosively into steam, and there are various kinetic phenomena associated with boiling, but when treated gently, it behaves pretty much as described.

If water is heated in a closed vessel in the absence of air, it does not boil in this way. At 0°C, say, the water lies at the bottom of the vessel, and above it is vapour (usually not called steam below 100°C) at a pressure of 4.579 mmHg. This is about the atmospheric pressure on Mars, so any water there would boil.

If we increase the temperature to 100°C, the vapour pressure rises to 760 mmHg, roughly exponentially. Nothing particular happens here, and when we increase the temperature to 200°C, the pressure has risen to 11 659 mmHg, or 225 psi. That is, for every temperature there is a corresponding pressure, the vapour pressure, at which the vapour is in equilibrium with the liquid.

If we allow some of the steam to escape, it is quickly replaced by more by evaporation of water to maintain the appropriate temperature. If the vessel is a cylinder fitted with a piston, and we press on the steam as if to increase its pressure, this does not happen, but some of the steam condenses to water so the pressure does not increase.

It is easily perceived that if some water changes to steam, heat must be added to maintain the temperature. At 100°C, the amount of heat is about 540 cal/g or 9720 cal/mol. This supplies the energy to break the hydrogen bonds and to give the liberated molecules a kinetic energy appropriate to the temperature (3kT/2 each).

When heat was thought to be the fluid caloric, this represented caloric that was hidden in the water but suddenly became evident in the steam, where it lubricated the water into a gas. It was called the *latent heat* L, and the term is still used. A phase change of this type, where there is a latent heat, is called a *first-order* phase change. Since the temperature remains constant during the change, the entropy changes by $\Delta S = L/T$.

An abrupt change in entropy is characteristic of a first-order phase transition. An abrupt change in the density, or volume V, practically always accompanies the change in entropy. In the case of water at 100°C, the specific volume of the water is 1.0435 cc/g, and of the steam, 1673 cc/g. The entropy changes from 1.3071 J/g/K to 7.3546 J/g/K.

The phase change from water to steam is an excellent example of the competition between energy and entropy to determine the state of a substance. Energy says be water, since water molecules attract one another.

Entropy says be steam, since steam molecules have so much more freedom. It is the minimum of $F = U - TS$ that determines what happens (at constant V and T). At constant pressure, which is the case for the equilibrium between water and vapour, it is the Gibbs function that governs, and what the pressure is for any temperature is the point at which $g_s = g_w$, where g is the Gibbs function per gram or mole, and "s" refers to steam, "w" to water. If a small amount of water dm evaporates, the net change in Gibbs function will be $g_s dm - g_w dm = 0$.

If the temperature is changed by dT, then the Gibbs functions will change by amounts $dG = -SdT + Vdp$. For equilibrium, then, $dg_s = dg_w$, or $-S_s dT + V_s dp = -S_w dT + V_w dp$. Of course, dT and dp are the same for the water and the steam. Therefore, $(S_w - S_s)dT = (V_w - V_s)dp$, and $dp/dT = (S_s - S_w)/(V_s - V_w)$, or $dp/dT = L/(T\Delta V)$, which is Clapeyron's equation.

Clausius later demonstrated the role of the entropy, so the equation is also known as the Clausius-Clapeyron equation. This equation gives the slope of the curve of vapour pressure versus temperature.

For water at 100°C, we have $dp/dT = (7.3546 - 1.3071)/(1673 - 1.0435)$ J/cc/K $= 3.617 \times 10^4$ (dy/cm^2)/K $= 27.13$ mmHg/K. Experiment gives 787.57 mmHg at 101°C and 733.24 mmHg at 99°C, or $dp/dT = 27.165$

mmHg/K, which is good agreement. The reciprocal of this is 0.0368 K/ mmHg, from which we can calculate the boiling point at Denver when the pressure is 850 mb or 638 mmHg. The result is 100 – (760 – 638)(0.0368) = 100 – 4.49 = 95.51°C.

The ideal gas law gives 1701 cc/g as the specific volume of water vapour at 100°C, which is not too far from 1673 cc/g. If the ideal gas law is used in Clapeyron's equation, we find $dp/dT = MLp/RT^2$ if L is per gram, and $dp/dT = Lp/RT^2$ if per mole.

Trouton's Rule says that the entropy change at the normal boiling point at 1 atm is about 20 cal/mol/K, or 4.65 J/g/K, which is at least of the same order of magnitude as the actual 6.05 for water (water does not obey Trouton's Rule very well). The approximate equation $dp/dT = 10p/T^2$ then results. At 1 atm and 373K, it gives 0.055 mmHg/K, which is not too bad. It should work for any liquid near its boiling point as a rough estimate.

Clapeyron's equation is useful in many situations. When applied to the ice-water transition, the fact that water is denser than ice means that dp/dT is negative, so as the pressure increases, the transition temperature decreases. This is an unusual situation, which occurs rarely, though antimony acts in the same way.

Expansion on solidification is advantageous in casting type, since the solid must fill the mold completely. The heat of fusion of ice at 0°C is 1436 cal/mol or 79.72 cal/g. The specific volume of water is 1.000160 cc/g and the specific volume of ice is 1.09086 cc/g.

This is common ice, hexagonal ice or ice I, where the oxygens are tetrahedrally stacked and there are tunnels through the crystal in the c-direction. The properties of ice are not listed in the Handbook of Chemistry and Physics, a strange thing for such an important substance. Ice also sublimates directly into vapour. At –10°C, the vapour pressure of ice is 1.950 mmHg, and at -20°C it is 0.776 mmHg. Of course, dp/dT is positive for the sublimation of ice. The entropy of sublimation is 44.674 cal/mol/ K at 0°C, corresponding to a latent heat of 678 cal/g.

The coefficient of cubical expansion, β is about 2×10^{-4} per K at 0°. At -15°C, the coefficients of linear expansion are 46 perpendicular to the c-axis, and 63 parallel to it, with $\beta = 156$, all times 10^{-6} per K. The adiabatic bulk modulus at about the same temperature is 7.81×10^{10} dyne/cm^2. The specific heat C_p of ice at 0°C is 9.09 cal/mol/K, or 0.505 cal/g/K, half that of water.

Another interesting example of the competition between energy and entropy is the existence of lattice vacancies in crystals above 0K.

It requires energy to make a vacancy, but the entropy increase is so favorable (because the atoms can now be shuffled) that some vacancies always exist at positive temperatures.

It is good to remember that this competition is taken care of by reasoning with the free energy F instead of with the energy U.

Water vapour and air form a nearly ideal gas mixture under normal conditions, with each gas acting as if the other did not exist. If the partial pressure of water vapour at 20°C is 17.535 mmHg, then it will be in equilibrium with water of that temperature.

If we cool the air slightly, some water vapour must condense so that the pressure is appropriate to the lower temperature. This will appear as a fog, or as condensation on surfaces, which is dew. Therefore, the temperature at which the water vapour will be in equilibrium with liquid water is called the *dew-point temperature*.

The *relative humidity* of a sample of air is the actual pressure of water vapour divided by the equilibrium vapour pressure at that temperature, time 100%. Air that has a relative humidity of 100% at 0°C, when brought into the house and heated to 20°C, will have a relative humidity of 4.579/17.535 × 100 = 26%.

The fact that dew forms when moist air is cooled is sometimes faultily expressed as "cool air holds less moisture than warm air." Well, air doesn't "hold" any moisture at all—it behaves quite independently of the water vapour. It's just that the equilibrium vapour pressure of water decreases with temperature.

THE CONCEPT OF TEMPERATURE

When school physics turns to Heat, the first item of business is the definition of temperature as a parameter determining whether a body will warm or cool when placed in contact with another. The superfluous 'Zeroth' Law of Thermodyamics is asserted, the first step in the axiomatic thermodynamics that ignores the structure of matter.

The inadequacy of this description is soon evident, in that it gives no hint of how to measure temperature, or the meaning of a temperature interval, or how to proceed in certain difficult cases that come up later, for example with interacting systems or negative temperatures. This approach is actually quite difficult and ambiguous, leading to much inconclusive argument and mental confusion.

A much more profitable procedure begins with the recognition of bodies, or systems, that can be isolated, and only weakly interact with other systems. This is true of electrically neutral samples of matter where gravity can be neglected, and which interact with a neighbouring system only through the interfaces or surfaces.

Since the volume scales as L^3, and the surface as L^2, for a sufficiently large, compact system the volume energy dominates the surface energy. For such a body, the internal energy E is well-defined. We postulate that the body consists of many parts (atoms), in interaction with one another. All degrees of freedom are included, not just translation, and energy is conserved in all interactions.

Many possible distributions of energy among the atoms gives the same total energy, and the system is constantly undergoing redistribution of this energy among its parts. The logarithm of the number of possible states W for a given energy E and volume V of the system is a quantity called the entropy S.

The fact that the logarithm is used, and the factor of proportionality is taken as unity, are only for later convenience, and have no fundamental significance. We can, therefore, conceive of a function S = S(E,V) that is essentially the consequence of the atomic nature of matter. Without the atoms, there is no S. This is not a unique function, but depends on the history of the system. We can make it definite by saying that we take it to be the S that results when the system is allowed to come into macroscopic equilibrium. For example, if we consider a gas of energy E and volume V suddenly allowed to expand into a larger volume V', S' = S(E,V') will be different from S = S(E,V). We now assert that in any such process in an isolated system S' > S. This means only that the system has taken advantage of access to a greater number of states.

Of course, this is the Second Law of Thermodynamics, but now it follows from principle, not as a postulate. The entropy of an isolated system can only increase. Entropy has the deciding role in determining the direction of a spontaneous process, such as a chemical reaction, not energy.

One must be careful to regard the energy E and the entropy S as functions of the whole system, not as some fluid quantities distributed around the system. One commentator has likened it to the beauty B of a painting, which cannot be considered as the summation of infinitesimal bits of beauty distributed over the canvas. The fact that most macroscopic bodies can be subdivided, and the whole considered as the sum of its parts merely obscures the essence of E and S. At some point, this procedure fails, and the unwary is trapped.

Nevertheless, it is useful to distinguish E, S and V as *extensive* quantities that are proportional to the amount of matter, in a sense. The continual increase in entropy does not mean that it is piling up somewhere as a giant grey mountain.

Let us now solve S = S(E,V) for E: E = E(S,V). Incidentally, for many systems this function has the useful property that E(λS,λV) = λE, or E is a homogeneous function of first order. However, now we are chiefly concerned with how E varies with S at constant V.

This corresponds to an exchange of heat, as contrasted with an exchange of work. For small changes, $dE = (dE/dS)dS$, where for normal systems $dE/dS > 0$. Now, suppose 1 and 2 are isolated bodies that are brought into contact, so they can exchange energy, perhaps through their surfaces. The interaction is assumed sufficiently weak that the separate identities of the bodies is not compromised.

The total change in energy $dE = dE_1 + dE_2 = (dE/dS)_1 dS_1 + (dE/dS)_2 dS_2 = 0$. $(dE/dS)_1 = < (dE/dS)_2(-dS_2/dS_1)$. Let us suppose that $dS_1 > 0$, which in normal cases means that system 1 receives energy, and system 2 loses an equal amount. dS_2 must be negative, and may not be less than $-dS_1$, by the Second Law, so $(dE/dS)_1 < (dE/dS)_2$. In equilibrium, the two quantites will be equal.

The quantity $kT = dE/dS$ can be called the Temperature, since it has all the necessary properties of this magnitude as defined in school physics. However, we now have a definition that is independent of the properties of any substance, that gives meaning to temperature intervals and even to the absolute value of the temperature, and which can be extended to unusual cases (such as when $dE/dS < 0$). The constant k has the dimensions of energy per temperature unit. If the temperature unit is the kelvin K, and the energy unit the joule J, then $k = 1.381 \times 10^{-23}$ J/K, called *Boltzmann's constant*. It is usual to include k in the definition of the entropy, $S = k \ln W$, where W is the *thermodynamic probability*. The kelvin is defined as 1/100 of the absolute temperature difference between the ice point and the steam point, incidentally.

The derivative $dE/dV = p$, where p is the pressure exerted by the system on its boundary. There are other ways in which work can be done by or on the system, but this is a good example. We now have $dE = TdS + pdV$, which is the differential form of the First Law of Thermodynamics. By considering the atomic structure of matter, we can bypass completely the sterile postulatory basis of axiomatic thermodynamics, and gain a deeper insight into what is going on, as well as an appreciation of the limitations of our analysis.Judging from a good school physics reference (Whelan and Hodgson), all material on entropy is beyond "A" level. Even that material does not contain the argument given here, and does not show that a statistical definition of temperature is related to the direction of heat flow.

Only the concept of entropy is needed above as one of the variables determining the internal energy, not its explicit calculation by statistical mechanics. The Second Law of Thermodynamics is actually much more easily understood than the First Law. It is a shame that thermodynamics is made so difficult and inaccessible to the student.

Chapter 7

Surface Tension and Phase Rules

It is wonderful the way the free surface of a liquid seems to be covered by a thin film in tension, that makes drops spherical, jets cylindrical, and climbs up the wall of a container.

Soap films are a related wonder, forming iridescent spheres so light they float in the air, and thick foams when agitated. The way it is presented in school physics often reinforces this view, though the film has properties unlike those of a material substance.

We are dealing with a surface phenomenon, one of the rare ones that is mentioned in school physics, and one with a great number of interesting applications. The molecules on the surface have water on one side, and something else on the other. Let us say we are dealing with familiar water, though it could be any fluid.

These molecules are less well nestled among their kind, and so there is a deficit in attractive energy. If we make new surface, we must put some energy in when we bring molecules from the comfort of the interior to the challenges of the borders. This is not just individual molecules, but in the case of water a change of structure, since the order of bulk water cannot apply on the frontiers. Liquid water has a curious structure, in which dodecahedrons probably play a role.

Free energy $F = U - TS$, where U is the internal energy, T the absolute temperature, and S the entropy. Then $dF = dU - TdS - SdT = pdV + TdS - TdS - SdT = pdV - SdT$.

At constant temperature, $dT = 0$, $dF = pdV$, where pdV represents the work done on the system (it is only one kind of work, but stands here for them all, everything but heat).

In the present case, the work in making new surface of area dA at constant temperature is γdA, so $dF = \gamma dA$. γ is the free energy per unit area of new surface. Dimensionally, this is erg/cm^2 or, what is the same thing, dyne/cm. Indeed, if $dA = Ldx$, then $dF = (\gamma L)dx$, as if γL were a tension in dynes. This is the usual definition of surface tension, though even school physics makes it clear that it is really a free energy per unit area. When a surface is extended, it cools, and heat must flow in to maintain isothermal conditions. However, the free energy F is something quite different from the usual energy U. In the present case, $\gamma = U + T(d\gamma/dT)$. Surface tension decreases with temperature, so F is less than U.

In fact, U and γ are quite different, though both become zero at the critical temperature.

The Figure shows how U and F vary with temperature. If you are considering what happens to molecules on the surface, U is probably closer to what you need.

U is not too far from constant, until the surface itself disappears. Under normal conditions, we do not get near the critical temperature, of course, since the water boils first.

Water molecules are not only attracted to other water molecules. If they are more strongly attracted to the molecules of another substance, that substance is said to be wetted, or hydrophilic, and surface molecules tend to cuddle up to it.

If they are less attracted, the substance is hydrophobic, and surface molecules tend to shy away from it. The water-air surface meets another at an angle of contact θ. This angle is small for a hydrophilic surface, large for a hydrophobic surface. It is 90° if the attraction is the same as that of water itself. All of this is strongly affected by the purity of the surfaces involved.

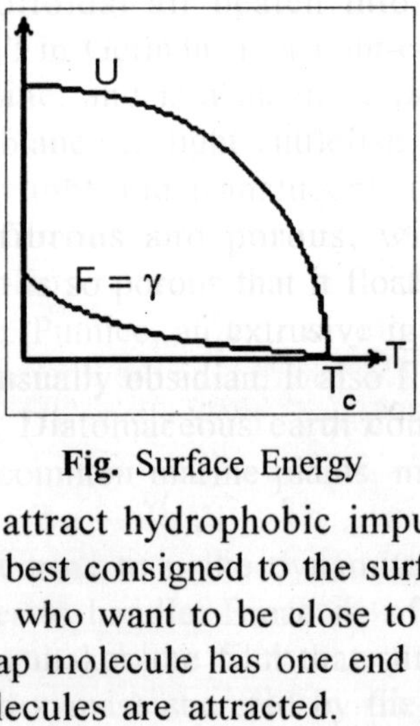

Fig. Surface Energy

Surfaces postively attract hydrophobic impurities. If a molecule or body is not wetted, it is best consigned to the surface so that it cannot get in the way of molecules who want to be close to one another. Soap is an interesting case. The soap molecule has one end that is polar, like water, and to which water molecules are attracted.

The other end is fatty and saturated, strongly hydrophobic (but equally strongly fatophilic!). Soap molecules in water tend to make a carpet on the surface, fatty ends outward, polar ends inward.

This lessens the expense of making a surface, so the surface tension falls. In addition, the surface is now rendered somewhat permanent, and the evaporation of water through it is discouraged. The surface will now attract fats and oils, which is why the soap is used.

Consider a spherical ball of water. This occurs when the water is free from gravity, as when falling freely (even at terminal velocity in air, small drops are quite spherical). Cut the sphere neatly in half in your imagination. The tension at the surface is $2\pi r\gamma$, and this must be balanced by a higher pressure p over the diametral plane.

Thus $2\pi r\gamma = \pi r^2 p$, from which $p = 2\gamma/r$. This is the difference in pressure inside and outside the sphere, due to the surface energy. The surface energy is made a minimum by the spherical shape, which has the least surface area for a given volume. In terms of the principal radii of curvature R and R', $p = \gamma(1/R + 1/R')$.

If we have a soap bubble instead, the film is an inner and an outer surface carpeted with soap molecules, with water in between. Hence, the pressure is $4\gamma/r$ higher inside the soap bubble than outside.

The γ is certainly not the value for pure water, 73 dyne/cm (290 K), but something less. One reference gives 25 dyne/cm. Cgs units have a very pleasant size for this quantity.

Mercury has the high surface tension of 470 dyne/cm. The surface tension of water falls with increasing temperature, to 59 dyne/cm at the boiling point. Rayleigh obtained 74 dyne/cm for clean water, 53 for greasy water, 41 with an olive oil surface, and 25 for a surface of sodium oleate (soap).

The capillary rise in a narrow tube can be estimated by using the angle of contact. We have $2\pi r\gamma \cos\theta = \pi r^2 h\rho g$, so the capillary elevation is $h = 2\gamma \cos\theta/\rho g r$. With a zero contact angle, the rise in a 1 mm diameter capillary is 3 cm. This is, of course, an upper limit.

Surface tension encourages the evaporation of small droplets (the added pressure inside a convex surface), and the collapse of small bubbles, for the same reason.

This can lead to supersaturation on condensation in the one case, and to superheating on boiling in the second.

Surface tension can be measured by capillary rise (θ must also be measured), or by the pressure necessary to produce a small bubble (Jaeger's method).

A method not as simple as it looks measures the weight of a droplet detached from a thick-walled capillary of radius r. If you write $mg = 2\pi r\gamma$, you will be in error about 100%, as a better balance of forces will show. The correct relation turns out to be about $mg = 3.8r\gamma$ instead, as determined empirically by Lord Rayleigh.

If the surface of a body of water is deformed in a wavelike manner, the higher parts will create a positive pressure from surface tension and their curvature, and the lower parts a negative pressure for the same reason. This will add to the gravitational force acting against the inertia of the water, and wave propagation will be the result. Therefore, we are led to an investigation of waves on water. We describe the motion by its velocity potential φ, whose negative derivative with respect to distance (gradient) is the particle velocity. The motion is irrotational, and φ satisfies Laplace's equation. Consider plane waves travelling in the direction x, with z the depth from the surface of the water, $z = 0$. A suitable solution of Laplace's equation is $\varphi = e^{i(\omega t - kx)}(Ae^{kz} + Be^{-kz})$. Because $d\varphi/dz = 0$ at the bottom, $z = L$, A and B are chosen so that $\varphi = Ce^{i(\omega t - kx)} \cosh k(z - L)$.

The relation between ω and k depends on the dynamics. If p is the overpressure, $p/\rho = gz - d\varphi/dt$. Therefore, at $z = -h$, the surface, $(T/\rho)d^2h/dx^2 = gh + d\varphi/dt$. Now, $dh/dt = -d\varphi/dz$ (the vertical velocity), so $(T/\rho)d^3\varphi/dx^2dz = g\, d\varphi/dz - d^2\varphi/dt^2$. If we substitute our form for φ into this equation, we find the dispersion relation

$$c^2 = (\omega/k)^2 = (g/k + Tk/\rho) \tan h\, kL.$$

The g/k applies to pure gravity waves, and the Tk/ρ to pure capillary waves. In deep water, $\tanh kL = 1$, and the dispersion relation is Kelvin's formula, published in 1871.

The phase velocity of gravity waves is proportional to the square root of the wavelength, and the phase velocity of capillary waves is inversely proportional to the wavelength. There is a minimum phase velocity $(4Tg/\rho)^{1/4}$. For water, with $T = 76$ dyne/cm, this velocity is 23.1 cm/s, the wavelength 1.71 cm, and the frequency 13.6 Hz. Capillary waves were studied by Faraday in 1831, who termed them crispations. Scott Russell observed capillary waves made by an obstacle in flowing water in 1844.

These are called stationary waves, but are different from the usual stationary waves formed by wave trains moving in opposite directions. Their velocities are fixed by the velocity of the stream and the angle of their wavefronts. The stream velocity must be greater than 23 cm/s (1/2 mph) for such waves to form, since this is their minimum phase velocity. Capillary waves form on the upstream side, gravity waves on the downstream.

PHASE RULES

SYSTEMS, PHASES, CONSTITUENTS AND VARIANCE

The *system* is simply the matter that we are studying, collected in one place and with definite boundaries.

Our systems will generally not exchange matter with the surroundings, and so will be called *closed.* They may be *isolated* and not exchange energy with the environment, or sometimes in contact with a *thermal reservoir* of temperature *T*.

A system may absorb a quantity of *heat Q* or do *work W* on the environment. Heat is the transfer of energy of disordered molecular motion, while work is a macroscopic, ordered, transfer of energy, such as the movement of a piston or an electric current. The *state* of a system is a specification of conditions sufficient to determine its macroscopic characteristics completely. A system is in *equilibrium* if there are no changes of its state with time.

Phase is derived from the Greek verb fasis, "appearance, assertion, accusation" which, as indicated, has important figurative meanings. Its use for the phases of the moon, which repeat in a regular sequence, gave rise to its use for the time sequence of a signal, particularly a sinusoidal signal, where the phase is represented as an angle that steadily increases with time. In this article, its meaning is quite different, referring to matter that appears in a homogeneous form: its "appearance."

A material phase is a homogeneous substance at an atomic level, with the same conditions existing everywhere within it. The conditions may be expressed as *intensive* parameters that do not depend on the amount of substance, such as the pressure *P* and the temperature *T*, or by *extensive* parameters that are proportional to the amount of substance, such as the entropy *S* or the volume *V*.

Extensive parameters may be referred to a fixed amount of substance, either a certain mass, or, better to a gram-mole. In this case, we shall refer to them with small letters, as *s* or *v*, and call them "specific" quantities. A phase may be a pure substance or a mixture, provided the mixture is like a solution or a solid solution, where the mixing is at an atomic level. A mixture of sand and salt is not a phase, but two phases each of which can have its own macroscopic characteristics.

The *system* that we study may consist of one or more phases. The system also has a certain number of chemical constituents. If they can react with one another to form new substances, the new substances are not independent constituents so long as their amounts can be determined by the conditions existing in the system. For example, if Na^+ and Cl^- ions are present in a system, then

$$NaCl \rightarrow Na^+ + Cl^-$$

is not a new constituent.

However, H and He are two constituents of a system, even though He can be formed by fusion of four hydrogen atoms.

The fusion reaction will not take place under the conditions of any reasonable system, so the amounts of each will be constant. If the system is solar matter, on the other hand, we may want to talk about only one constituent.

Let us suppose our system consists of only one phase, say water vapour. We also have only one constituent, water. We can describe the state of the system uniquely by specifying a pressure P and a temperature T, which can be established independently.

If the vapour is considered to be an ideal gas, then $PV = NkT$, where k is Boltzmann's constant and N is the number of molecules. From this, it is clear that the *molar volume* is $v = RT/P$, where $R = N_A k$. It cannot be specified at will, but is determined once the pressure and temperature are fixed. A system consisting of a single phase of one component is said to have *variance* $f = 2$. This is not the same variance as in statistics, and is, in fact, a bad choice of words, but a conventional one.

This conclusion is valid even if the ideal gas assumption is dropped, and an arbitrary *equation of state* is assumed instead. The *Phase Rule* will give the variance for more complicated systems.

SOME THERMODYNAMICS

Thermodynamics is unavoidable when studying phase changes, equilibrium, chemical reactions, electrolytic cells, and similar subjects. Here is a brief synopsis of the subject to serve as a handy reference that includes the most-used relations and concepts.

The *internal energy* U of a system is the total energy of its molecules, kinetic and potential, referred to an arbitrary zero. The *specific internal energy* is the internal energy per unit mass: $u = U/M$.

The modifier "specific" will mean "per unit mass." For a simple system, u may be a function of the macroscopic variables s and v, the specific entropy and specific volume. We write $u = u(s,v)$.

Changes to U can be divided into heat and work, so that $\Delta U = Q - W$, or, for convenience, $dU = dQ - dW$, or $du = dq - dw$. This is the *First Law of Thermodynamics*. dq and dw represent infinitesimal changes, not the differentials of functions. du is, of course, the differential of a function. The First Law says that u is a function of state that depends on energy transfers with the system.

For a system whose volume can change, $dw = Pdv$, or $P = (\partial u/\partial v)_q$. This partial derivative means the same thing, and the reader should not be put off by the notation. The subscript on the partial derivative means that $dq = 0$. By analogy, we can write $dq = Tds$, or $T = (\partial u/\partial s)_v$.

A new function S or s is introduced, called the *entropy*. It can be found in principle by integrating $ds = dq/T$ at constant volume, and in fact this is done in elementary thermodynamics.

However, entropy cannot be properly understood on a macroscopic basis alone (though many texts make a valiant attempt to do so), and we shall say more about it below.

Let us assume that entropy s is a function of the state of a system that can be determined independently the same way that v can be.

The *Second Law of Thermodynamics* is that s is a function of state, that the heat absorbed in a reversible process is Tds, and that for any process in an isolated system, $ds \geq 0$.

There is a lot in this statement, and it is the core of thermodynamics. Note that entropy can increase without the absorption of heat in an irreversible process. An example is the free expansion of a gas into a vacuum. More randomness is given to the gas, but no energy is absorbed or lost.

In a small reversible change of the system, $dU = TdS - PdV$. This is a concise statement of both the First and the Second laws in one equation. If we consider the function $H = U + PV$, then $dH = TdS - PdV + PdV + VdP$, or $dH = TdS + VdP$. If we consider an isobaric process, $dP = 0$, then dH is the heat transferred to the system, or $dH = C_P dT$.

Therefore, $dS = C_P dT/T - (V/T)dP$, from which we can find S as a function of P and T. For an ideal gas with constant specific heat, $s = c_p \ln T - R \ln P + s_o$. The constant of integration s_o should be noted.

Energy is arbitrary up to a constant. Adding a constant to the energy will not change any results. Entropy is different.

Unlike energy, entropy has an absolute value, and the constant s_o is not arbitrary. A serious fault of thermodynamics based only on macroscopic considerations is that this constant cannot be determined by its means.

The entropy has just been introduced in the most mysterious way possible, as the S in $dQ = TdS$. It is very difficult to understand entropy on this basis, if indeed it can be understood at all.

To know what entropy means, we must consider the individual molecules of the substance in their chaotic environment. The entropy of a substance of fixed energy and volume, $S = S(U,V)$, is $S = k \ln W$, where W is the number of states of equal probability open to the system.

For a macroscopic system with an entropy of only 1 cal/K, W is about 10^{10}, an extremely large number. At the other extreme, a system with only two states available to it has the unbelievably low entropy of about 0.25×10^{-23} cal/K. Entropy is a help only when the system has a large W, like all macroscopic systems made up of molecules.

It is not a substance, does not flow, and cannot be isolated. The expression for the molar entropy of an ideal gas shows that it increases with the molar volume, as then the molecules can be spread over an even larger volume with greater randomness.

It should be clear that we could just as well take s and v as parameters as P and T. Which two parameters we choose is unimportant, just so we choose two independent parameters that can be varied in any way. Whichever pair we choose, the other two are uniquely determined.

We can now write the Phase Rule for systems of one phase and one component as $f = 2$. There is not much of a surprise here, but we will be able to generalize this expression.

Let's make a little excursion here to see what the thermodynamic conditions are for equilibrium. We are specially interested in a system held at a fixed pressure P and temperature T, typical of laboratory measurements and easy to imagine. The processes that could take place might be a chemical reaction in which the components rearrange themselves chemically, electrolysis with work done electrically, or a *change of phase*, say from liquid to vapour.

We shall assume that Pdv alone represents the work done by the system, for simplicity. Define a function $g = u + Pv - Ts$, where we have used molar quantities. This is called the *Gibbs function*. Then, $dg = du + Pdv + vdP - Tds - sdT = vdP - sdT$.

In any process at constant P and T, then, we must have $dg = 0$. The Gibbs functions of the initial and final states when such a change is made reversibly must be equal, or $g(f) = g(i)$. Equations like this can be written for each pair of phases in equilibrium, and these were some of the equations used by Gibbs in his deduction of the phase rule.

We also find that $(\partial v/\partial T)_P = -(\partial s/\partial P)_T$. This is one of *Maxwell's relations* that appear to drop out of the sky on us in thermodynamics. They are no more than the mathematical consequences of our definition of certain functions.

If we wish to consider equilibrium at constant temperature and volume instead, then the function we need is $F = U - TS$, or $f = u - Ts$, per mole. F is the *free energy*. Then, $df = -sdT - Pdv$, and the free energies must be equal in final and intial states, $f(f) = f(i)$.

If we go a little farther and consider irreversible processes, we find that the conditions on the Gibbs function and the free energy are that they assume a *minimum* in equilibrium. The Maxwell relation that falls on us here is $(\partial s/\partial v)_T$ = (part;$P/\partial T)_v$.

The functions F and G both contain the term TS, which is the reason why absolute entropies are important. Any arbitrary addition S_0 to the energy of any component would change F and G by $-TS_0$, which would not be a constant (and thereby unimportant to F and G) but would depend on T, and affect the equilibrium.

Arbitrary additions to the internal energies of the various phases would have no effect, because of the conservation of energy and the absence of the factor T. While we are on a roll, we might as well consider the internal energy U, where $du = Tds - Pdv$, an expression of the First and Second Laws of Thermodynamics, all wrapped up in one package.

In a process at constant entropy and volume, the energy is conserved: u(f) = u(i). Furthermore, when dv = 0, du = c_vdT = Tds = du, or ds = c_v/T, and cv = $(\partial u/\partial T)_v$. Maxwell's relation for this function is (part;T/∂v)$_s$ = (part;P/∂s)$_v$.

That leaves us with only H = U + PV, called the *enthalpy*. In this case, dh = du + Pdv + vdP = Tds + vdP. This is conserved for processes at constant entropy and pressure, h(f) = h(i).

If we heat the system at constant pressure, then dq = c_PdT = Tds = dh, and $c_P = (\partial h/\partial T)_P$. The Maxwell's relation is (part;T/∂P)$_s$ = $(\partial v/\partial s)_P$.

Also, ds = $(\partial s/dT)_v$dT + $(\partial s/\partial v)_T$dv. Using the definition for c_v and Maxwell's relation from the free energy, we find Tds = c_vdT + T$(\partial P/\partial T)_v$dv. Doing the same thing for T and P instead of T and v, we get ds = $(\partial s/\partial T)_P$dT + $(\partial s/\partial P)_T$dP, or Tds = c_PdT – T$(\partial v/\partial T)_P$dP.

In both equations, the second terms involve only derivatives of the equation of state.

Much use can be made of the functions and formulas we have just derived in relating the thermodynamic functions of a substance. They may seem complicated, but really are just mathematical deductions with little physics in them.

SYSTEMS WITH MULTIPLE PHASES

Now let us consider a more general system of c components and φ phases. In 1875, Josiah Willard Gibbs, by considering the conditions for equilibrium in such a system, and the number of indpendent variables necessary to specify the state of every phase, found the Phase Rule to be f = c – φ + 2. This agrees with our result, but is capable of extension to system where we could not guess the result. We should mention here that we consider systems that can do work only by PdV work—pressing on a piston.

There are many other ways, such as electrical or magnetic, and the phase rule can be extended to them. Here, PdV work will serve as representing all such forms of work.

We know that H_2O can exist as a vapour, steam, a liquid, water, and a solid, ice. Each substance forms a phase with a single constituent, and when it alone is present, the system has variance 2, and the state can be specified by two parameters. Let's use T and P for the sake of argument, and draw a diagram with these parameters as axes.

It is easy to identify regions where each type of phase exists, and there must be conditions under which both exist at the same time. When this is the case, the Phase Rule says that f = 1 – 2 + 2 = 1. Therefore, we can at most specify one parameter. Suppose it is T. Then once we have fixed T,

P and v are also determined for each phase. This means that two phases can only exist along a line in the diagram.

This is no small prediction! It means that there are no *areas* where two phases can exist.

On the line where they both exist, the pressure P is common to both, but the specific volumes v can be (and usually are) different. We also know that when water changes to steam, a large amount of energy is required (around 540 cal/g), the *latent heat* L.

This heat is transferred into the steam at a constant temperature, raising its entropy by $\Delta s = L/T$ while its specific volume increases by $\Delta v = v(s) - v(w)$. A phase transition with a jump in entropy and specific volume is called a *first-order* phase transition.

The same thing takes place for all three of the possible changes, water→steam (vaporization), water→ice (freezing), and ice→steam (sublimation), and so there are three corresponding lines on the phase diagram. Now, should all three be present, the Phase Rule gives $f = 1 - 3 + 2 = 0$. This means that this can occur only at some one point, called the *triple point*. For water, this is at T = 0.01°C and P = 4.58 mmHg, and we have no choice in the matter.

Furthermore, this point must lie on all three of the lines where two phases are present, so these three lines must meet at the triple point. There are no exceptions to this prediction by the Phase Rule.

The slopes dP/dT of these lines where phase changes take place are expressed by Clapeyron's Equation.

One derivation uses the first TdS equation we found above. For $dT = 0$, it gives $Tds = T(\partial P/\partial T)_v dv$.

In the present case, $(\partial P/\partial T)_v = dP/dT$ for both phases, $Tds = T\Delta s = L$, and $dv = \Delta v$, so we have $dP/dT = L/T\Delta v$.

In going from water to steam, both L and Δv are positive, while from steam to water, they are both negative.

Therefore, $dP/dT > 0$. In going from water to ice, L is negative, but Δv is positive for water, so $dP/dT < 0$. Normally, dP/dT is positive for freezing as well. Another, possibly better, derivation uses the fact that $g(f) = g(i)$ at both T and T + dT (two neighboring points on the curve), so that $dg(f) = dg(i)$ as well. Using the expression for dg when T changes by dT and P by dP gives Clapeyron's Equation as well.

The specific volume v can be considered as out of the page, and each of the lines is an edge-on view of a surface. Areas are regions of $f = 2$, lines of $f = 1$, and the triple point of $f = 0$. The critical point is just the place where the difference of specific volumes of water and steam becomes zero, as does the latent heat. Beyond this point, the two phases cannot be distinguished and become a single phase.

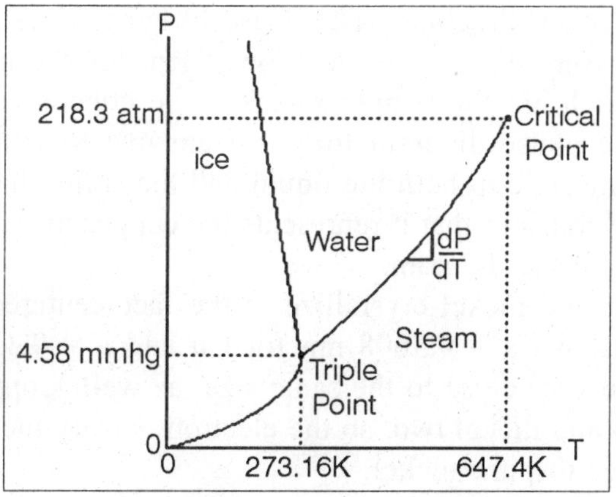

Fig. Phase Diagram for water

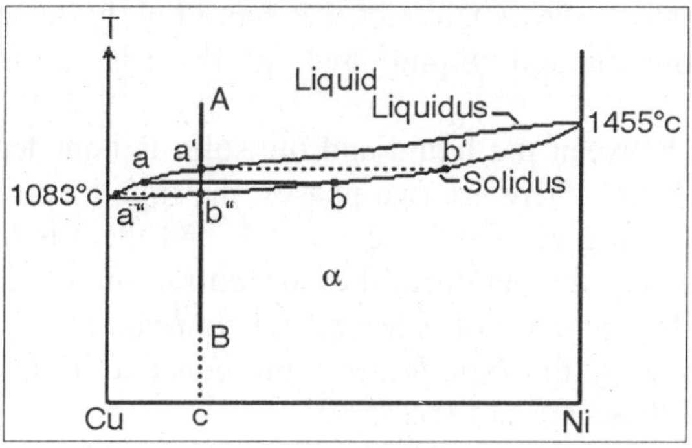

Fig. Phase Diagram for the Copper-Nicket Ally

BINARY SYSTEMS

Now we consider systems with two constituents. If these are mixtures of two metals, they are called *binary alloys*. The concentration of the constituents can be specified by one number, the mole or weight fraction of one component. If the constituents do not react, then their concentrations remain constant for any given system.

We still have the pressure P and temperature T, which are common to all phases. Therefore, a phase has variance 3 in general, since the Phase Rule gives $f = 2 - 1 + 2 = 3$. A three-dimensional model would be required to depict the state, and one phase would correspond to a volume. When we are dealing with metal alloys, the pressure is of little importance and is usually constant, so we can omit it in the diagram.

The result is a two-dimensional phase diagram, as for a single substance, where the coordinates are the temperature T and the concentration c. We must remember that the pressure and the specific volume point in other directions. The Phase Rule becomes $f = c - \varphi + 1$ if we leave the pressure out of consideration. Then, for a single phase, $f = 2 - 1 + 1 = 2$, which are the temperature and the concentration.

A schematic phase diagram for a system in which the constituents are completely miscible in both the liquid and the solid phases is shown at the left. We may consider that it represents the copper-nickel system, which has this kind of phase diagram.

Both copper and nickel crystallize in the face-centered cubic system, with lattice constants a = 0.3608 nm for Cu and a = 0.3517 nm for Ni. The two ions are very close to the same size, as well. Copper donates one electron per ion, and nickel two, so the electron density increases as we go from pure copper to pure nickel.

Therefore, copper and nickel can replace one another in the lattice quite freely, and the energy difference is reflected in the difference in melting points, which shows that nickel is more tightly bound, as expected.

In the regions marked "liquid" and "α" there is one phase, and so a variance of 2.

The region between the liquid and the solid is bounded by two lines, and within this region there are two phases, the liquid metal and the solid. The Phase Rule then gives $f = 2 - 2 + 1 = 1$, and there is only one degree of freedom. For any temperature, the concentrations of liquid and solid are determined by the ends of a horizontal *tie line*. The concentration in the liquid is given by the *liquidus*, and the concentration in the solid by the *solidus*, and these are not the same.

Let us suppose we have a liquid of concentration c, and cool it slowly from point A to point B, where it has solidified.

The overall concentration of the system cannot change, of course, and must follow the vertical line. When we reach point a', that is, of course, the concentration of the liquid.

The first solid to separate is at point b', much richer in nickel than the liquid. If we cool the system a little more, the liquid is now at concentration a, while the solid that is separating is at concentration b. Further cooling gives points a" and b", connected by a tie line.

Note that solid b is poorer in nickel than solid b' that was deposited earlier, while liquid a is richer in copper than the original liquid at a'. In order for this to happen, the solid must have gained some copper and lost some nickel, while the opposite has happened to the liquid. Usually, this must happen by diffusion, which can be slow in a solid. If insufficient time is taken to reach equilibrium, then the deposited metal becomes successively richer in copper, as does the liquid, since nickel solidified preferentially at first.

The average concentration remains constant, but the metals have been separated somewhat. The solid shows cores rich in nickel, the outsides rich in copper. If diffusion is allowed to do its work, the composition in the solid phase becomes uniform.

Note that every time a phase boundary is crossed, the number of phases changes by 1. On cooling, the phases increase from 1 to 2 at the liquidus, and back from 2 to 1 at the solidus.

The difference in concentrations in the regions of two phases is like the difference in specific volume in the regions of two phases in the PVT surface of a single substance. In our PT diagram for water we did not see any such areas, but they would appear in a Pv diagram, where the surface is seen from another direction.

It is more common to have systems of two metals that are miscible in the liquid state, but crystallize in different systems in the solid state, or have ions significantly different in size, so they cannot occupy the same lattice sites without considerable strain. Usually each crystal lattice can accommodate a few foreign ions and make a homogeneous solid solution phase, but when the concentration of the foreign ions reaches a certain limit that depends on the temperature, the phase may refuse to accept any more.

Then we have a liquid state that is one phase, while the solid state is a two-phase mixture of the *terminal* phases.

This happens in the tin-lead system. The terminal tin phase, α, can hold less that 3% lead, while the terminal lead phase, β, can hold no more than 20%, and that at an elevated temperature. The solid must, in general, be a mixture of the two phases, nearly pure tin, and nearly pure lead. After a while, they may diffuse into one another to approach homogeneity.

The phase diagram for the tin-lead system is shown schematically at the right. Lead melts at 327°C, and tin at 232°C. Adding one to the other *lowers* the melting point in either case.

This is largely a result of the increase in entropy produced by the mixing. There are three possible phases, α, β and the liquid. The Phase Rule says that $f = 2 - 3 + 1 = 0$, so they can all three be present only at one point, which is point E. This point is at 183°C and a lead concentration of 38% by weight.

It is called a *eutectic*, and represents the minimum melting point of the system. Three large regions correspond to areas with two phases: liquid + α, liquid + β, and α + β.

The liquidus is composed of the two upper lines meeting at E. The solidus is represented by the horizontal line through E, that is terminated by triple points at each end, together with the lower lines from the melting points. The final two lines represent the limits of stability of the terminal phases.

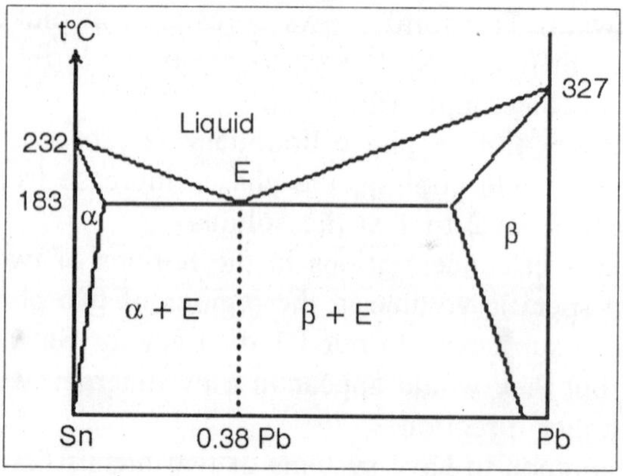

Fig. Tin-Lead Equilibrium Diagram

Suppose a tin-rich liquid solidifies rather quickly. The first metal to solidify is nearly pure tin α phase. This makes the liquid richer in lead, so its concentration moves to the right.

This continues until 183°C is reached. Then we have α phase solid, and a liquid of the eutectic composition. Now this liquid solidifies, depositing characteristic thin layers of alternating α and β phases as it departs from equilibrium one way and then the other, correcting itself every time.

If we annealed the solid and gave time for diffusion to be complete, everything would be uniform at the original composition.

The liquid begins to solidify as soon as the liquidus is reached. Past this point, we have some solid, and some liquid, which are mixed like ice and water. If we are close to pure tin or pure lead, the liquid solidifies to the corresponding terminal phase, plus a little of the other.

We can reach a phase supersaturated in the other constituent, which may then *precipitate* as small crystals. At the eutectic, nothing happens until we reach 183°C. Then this temperature remains constant until all the metal has solidified, as the characteristic eutectic mixture.

Another behaviour that is seen in some important alloys is that the crystal structure of the solid solution changes with composition. This is seen in brass, the Cu-Zn alloy, and bronze, the Cu-Sn alloy. For discussions of this important case. In brass, we also have the interesting phenomenon of an order-disorder transition, which is a *second-order* transition without a discontinuity in the entropy and specific volume.

ICE AND SALT

Alloys are not the only binary systems for which phase diagrams are useful. A portion of the water-salt phase diagram is shown at the right below. At the left is pure water, in this case ice below 0°C. On the right, but not shown, is pure salt, NaCl above 0.15°C and $NaCl \cdot H_2O$ below.

Below the horizontal line at -21°C is solid ice + salt. We presume that ice and salt are immiscible, so that in the solid state neither includes any of the other, which is a good approximation.

Above the horizontal line, there are three regions, the middle of which represents a homogeneous solution phase, the other two mixtures of solution with solid ice and solid salt. The point E is a eutectic, where three phases are present, so that f = 0. The vertical line above point B represents a saturated solution of salt in water. The solubility of salt does not depend much on the temperature.

More accurate figures from a handbook give a solubility of 35.7% at 0°C, 39.8% at 100°C. The heat of solution of salt is -1180 cal/mole or -20.2 cal/g. The minus sign indicates that heat is absorbed when salt dissolves. Since this happens at constant pressure here, it is an enthalpy change, ΔH.

The latent heat of fusion of ice, another enthalpy change, is 79.67 cal/g. This means that when a gram of ice melts to water, the heat required could cool that water to -80°C.

In equilibrium, of course, the water cannot cool and the heat must come from outside, say from the drink the ice cube floats in. This is the familiar property of ice in maintaining a low temperature, which it does effectively by the large latent heat.

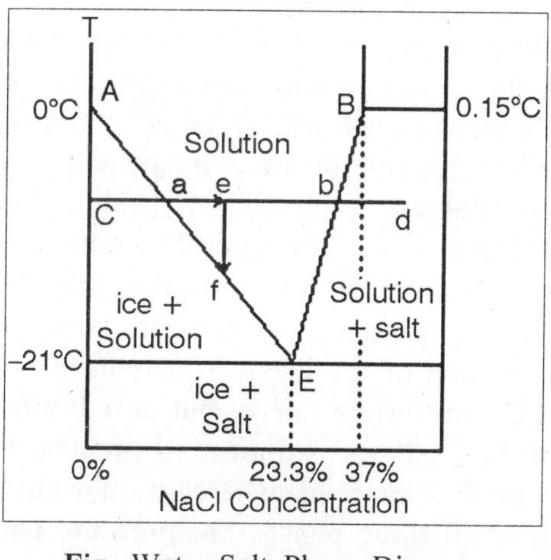

Fig. Water-Salt Phase Diagram

The ice will get your drink down to 0°C or die trying. Usually, it dies trying and melts continuously no matter how well you have insulated the glass. A mixture of crushed ice and water can usually get down to 0°C and provide a fixed reference temperature for a thermocouple.

This has to be done properly, for water has maximum density at 3.8°C, and the ice is usually floating on 3.8° water (the increase in density is very slight, however, only about 1 part in 10,000). The reference junction must be kept surrounded by ice. Incidentally, the equivalent of a reference junction can be supplied in a fairly accurate and drier fashion with electronics.

So we see that ice is a powerful cooling engine because of its latent heat, but can only go to 0°C, which is not enough to freeze things. Also, ice is not always welcome in the winter, and it would be good to have some way to melt it so the water would flow off and not make things slippery. Both demands can be met with the help of salt, NaCl.

Suppose the temperature corresponds to point c, and we sprinkle salt on the ice. The ice melts, by means of the heat supplied by the environment, and some salt solution is formed in the puddle, because the ice is not in equilibrium with all that salt.

The salt solution becomes more concentrated as the salt lying around dissolves in the water from the melting ice.

The ice tries to dilute the solution, while the salt wants to concentrate it, and when point a is reached, all the ice has melted and we have a salt solution only. If more salt is present, the solution becomes more concentrated and may reach point b, where the solution is saturated, and any more salt would just precipitate out. Therefore, the system ends up beyond b, with saturated solution and some solid salt, waiting for more ice.

Now suppose we mix ice and salt on purpose, to get a freezing mixture. The environment is not a heat reservoir now. In fact, we make insulate it so that it must supply all the latent heat if the ice is to melt. Ice and solution are both present, so we must be in the corresponding region to the left of line AE in the phase diagram.

Suppose the solution has a concentration corresponding to point a. If a little more salt dissolves, the concentration may correspond to point e, to the right of the equilibrium concentration, so ice dissolves to dilute the solution. The provision of the latent heat required cools the system to point f, again on the equilibrium curve, but at a lower temperature.

Therefore, with ice, salt and solution all present, there cannot be an equilibrium until point E is reached, where equilibrium can be established and maintained while all three phases are present. The concentration at this point is 23% (some sources say 29%), from which the amount of salt required can be estimated. The temperature is -21°C or -5.8°F, which is sufficient to freeze ice cream.

The melting of ice extracts most of the heat, but the dissolving of salt contributes its part as well. Magnesium chloride forms a eutectic at -33.6°C (-28°F) and 21.6%, and calcium chloride at -55°C (-67°F) and 29.8%.

Calcium chloride can be used to freeze mercury at -38°C. The hydrated salts should be used for making freezing mixtures, and everything should be kept cold, insulated, and mixed.

Aqueous solutions of alcohol, and other substances, such as ethylene glycol, depress the freezing points of the solutions, as is well-known from motor-vehicle antifreezes. These can make freezing mixtures and melt ice as well.

A skull-and-crossbones Martini from denatured alcohol and ice cubes, and got it down to –6°C in a few minutes.

A 30% ethyl alcohol solution, with a gravity of 0.954, gets you down to –19°C (–2°F). Glycerine gives a minimum at –37.8°C, in a 70% glycerine solution.

Carbon dioxide sublimates at –78.5°C. Carbon dioxide snow, compressed into blocks, is "dry ic≥ which is drier than freezing mixtures. It is often mixed with liquids like alcohol, acetone, chloroform or ether to make a cooling bath.

The temperature, however is not lower than the sublimation temperature of CO_2, and no eutectic is formed. Liquid nitrogen boils at -196°C and is convenient to use, leaving no clean-up. With all these means at one's disposal, temperatures down to 77K are available without machinery.

IRON

Sometimes the constituents of a mixture may react to form a new substance, which has significant effects on the phase diagram. When carbon is dissolved in iron, the hard, brittle, white compound *cementite*, Fe_3C, may be formed.

From the atomic weights of Fe and C, 55.85 and 12.00, respectively, we can find that cementite corresponds to 6.7% C by weight. Cementite melts at 1837°C and has a density of 7.4 g/cc.

On the phase diagram, cementite corresponds to a vertical line at 6.7% C, and divides the phase diagrams into two independent sections, one with less carbon, the other with more. Engineering is interested only in the section with less carbon.

Pure iron has a body-centered cubic crystal structure at room temperature, called α-iron or *ferrite*. At 910°C the structure changes to face-centered cubic, and this material is called γ-iron or *austenite*.

At 1403°C, the structure reverts to bcc, and the material is called δ-iron, though it is really the same as α-iron.

At 1535°C, the iron melts. Austenite is much softer and more easily worked than ferrite, which explains the blacksmith's heating of iron to red heat for working it.

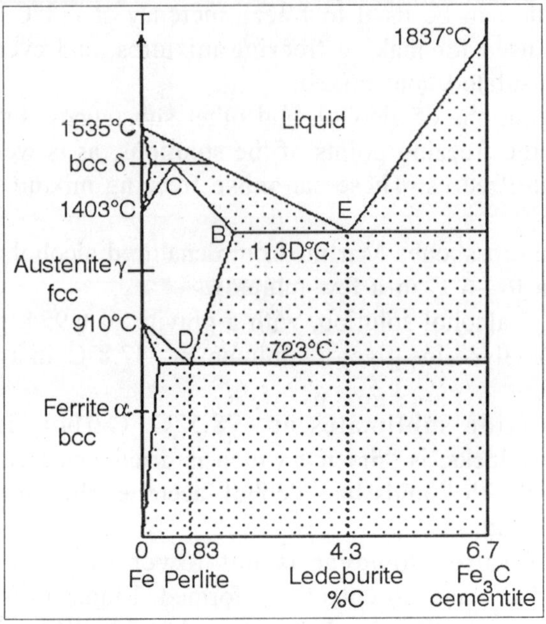

Fig. Iron-Carabon Phase Diagram

These things are reflected at the left and right boundaries of the iron-cementite phase diagram, which is sketched at the left. The actual lines are more or less curved, not straight, and some of the areas have been magnified for clarity. Only the part of the iron-carbon phase diagram to the left of cementite has been shown, since it is the only area of technical interest.

The plain areas are those in which there is only one phase, for which the Phase Rule gives f = 2 – 1 + 1 = 2, and the dotted areas are those with two phases, where f = 2 – 2 + 1 = 1.

For any temperature, the concentrations of the two phases present can be read off at the ends of a horizontal tie line. The names of the phases are given in all cases. They are ferrite, austenite, δ-iron, perlite, ledeburite and cementite. Only the first three occupy regions with f = 2; the final three are of definite composition.

Point E is easily recognized as a eutectic. Its temperature is the lowest fusing temperature of an iron-carbon mixture, 1130°C. A eutectic liquid will solidify to ledeburite, a typical layered eutectic solid of austenite and cementite. A liquid to the left of E will solidify to austenite + ledeburite, while one to the right will solidify to cementite + ledeburite. At 723°C, there will be a further change to ferrite + cementite.

These changes in the solid state are not rapid. Any material whose composition is to the right of point B (greater than 1.7% carbon) is called *cast iron*. When iron is reduced in the liquid state with carbon, cast iron is the usual result. In early times, when iron men were trying to get a low-carbon iron by directly reducing the iron ore with charcoal, their work would be ruined if the furnace temperature rose above 1130°C, and the iron absorbed carbon and melted.

The cast iron could not be hammered or worked; it was just a hard, disappointing, inedible, useless lump. Using cast iron to make castings was a much later development, since for several reasons iron is not as easy to cast, as bronze is.

The blast furnace just goes ahead and melts the iron in the interests of rapid mass production, so it comes out with 3% or 4% carbon, which forms cementite and graphite on cooling. This is obviously not an equilibrium phase, but is quite stable.

The iron was once led through channels into ingot moulds dug in the cinders of the casting floor, which suggested a sow and her piglets, and gave the name "pig iron" to the product.

It is produced because methods of removing the excess carbon are available, first the puddling furnace that made wrought iron, and later the Bessemer converter, open-hearth furnace, and basic oxygen converter. There is quite an interesting history, for which the reader is referred to books on the history of technology.

Any material with a composition to the left of B is called a *steel*. For such a material, heating will cause the steel to become a uniform austenite phase. This is a lucky circumstance that is responsible for the ability to heat-treat steel to suit its application.

A steel of composition corresponding to point D will become a layered mixture of ferrite and cementite of average composition 0.83% carbon, called *perlite*. It is so-called because it can have a pearly lustre in the microscope, a play of colors produced by the interference of light in its regular layers, which act as a diffraction grating.

This part of the phase diagram looks just like a eutectic, except that the upper material is solid austenite, not a liquid. It is called a *eutectoid*, "like a eutectic" though there is no melting involved.

A steel that at room temperature is a structural mess of different phases and crystals, but of the average composition of the eutectoid, will, on heating above 723°C, become a uniform solid solution after a while, that will cool to a nice pearlite structure throughout. This *normalization* can be applied to any steel, and does not require melting. The steel must be heated to a high enough temperature that it all becomes austenite, however—a temperature that is certainly below 1130°C.

There is a funny bit at high temperatures near the δ-iron region. This looks like an inverted eutectoid, but it isn't a eutectoid. If we consider a melt with a composition just to the left of point C, δ-iron will freeze out and the liquid will become richer in carbon. When the temperature of C is reached, the carbon-rich liquid will dissolve some of the carbon-poor δ-iron until all the liquid freezes to a mixture of δ-iron and austenite of the composition of point C. Further cooling will decrease the amount of δ-iron and increase the amount of austenite until the solid phase is uniform austenite.

What happens at point C is called a *peritectic reaction*, as δ-iron is dissolved and austenite formed. The carbon-cementite phase diagram illustrates most of the interesting things that can happen.

Austenite cooled quickly does not become a tough ferrite-pearlite mixture, but a difficult structure called *martensite* that is very hard and brittle. It is as if the carbon, trying to escape as the bcc crystals of austenite shift to fcc, are trapped in the attempt, and hinder the structure change by wedging themselves in and preventing the formation of a stress-free crystal. If you are making a sword or a knife, this may be just what you want.

The steel is first cooled slowly, so ferrite and pearlite are formed, and the item can be shaped and worked. Then it is heated above the transition temperature to austenite, and finally *quenched* by plunging it into water, or into the body of a slave on the advice of local priests.

The temper can be *drawn* by gentle heating to allow some of the martensite to transform into ferrite, and toughen the object. In most cases, the hard temper is only on the surface anyway, since the interior cannot be quickly quenched. The possibility of heat treatment makes steel a far superior metal for tools and weapons than any other available alloy.

Ancient iron may well have contained sufficient dissolved carbon to be considered at least very mild steel, perhaps 0.1% carbon, without any special handling. Austenite does dissolve carbon, though the process would be slow. Such steel would harden and temper. Pure iron, which was seldom produced or desired except for ornaments, would be too soft for tools and weapons. Higher concentrations of carbon can be attained by packing the iron with charcoal in a closed crucible, and heating for a good while.

This, apparently, was done in India at an early date, and the resulting "wootz" was obtained by the West in trade and used for making superior weapons, as at Damascus.

The Huntsman crucible process, developed in Sheffield, was similar. Steel with 1% carbon, good for making cutlery, was an expensive material that could only be used for high-value objects. Cheap steel did not arrive until the Bessemer process of burning the carbon out of cast iron with an air blast was invented around 1855.

The phase diagram had no region for martensite, because martensite is not an equilibrium phase. This indicates that many important properties may not be described by the phase diagram. Another example are the effects of small amounts of impurities. Bessemer iron contained amounts of oxygen and nitrogen that were detrimental. The oxygen forms oxides that concentrate on the grain boundaries and weaken them. The iron had to be deoxidized, by adding silicon or manganese, before it was satisfactory.

These impurities are beneficial even in larger amounts, silicon in magnetic alloys and manganese in steel that will work-harden and become more durable. Sulphur and phosphorus, often found in ores or coke, are extremely detrimental in any quantity.

Phosphorus forms a brittle compound with iron that collects on grain boundaries and makes the steel brittle and subject to cracking when cold. Sulphur forms compounds that are fusible, and lubricate the grains so the steel crumbles when an attempt is made to hot-roll it, called *hot shortness*. The control of impurities governs most iron and steel processes.

COLLOIDS

There should be a more exciting name for the phenomena to be discussed here. The dull term *colloid* that reminds us of glue is, nevertheless, the accepted word. It was coined in the early 19th century by the Father of Physical Chemistry, Thomas Graham (1805-1869), to distinguish those materials in aqueous solution that would not pass through a parchment membrane from those that would.

Glue was indeed a material that would not, and the Greek for glue is *kolla*, from which we also get "protocol" and "collagen." Those that would pass through were things like salt, and other soluble crystalline substances, which Graham called *crystalloids*.

Colloids received little attention until the end of the century, when van't Hoff, Oswald and Nernst founded modern physical chemistry and they, and others, became fascinated by colloid phenomena. There had been famous observations by Tyndall and others in the meantime, but chemists could not get excited over glue.

Then, in the 1920's and 1930's, the importance of colloids to industrial processes and biochemistry changed everything. Colloids became a hot field, and soon every elementary textbook said something about them.

An interesting philosophical point was suggested by this study. Colloids are often called a "fourth state of matter," We shall find that it is very difficult to encapsulate any concept concerning colloids in a word, though heaven knows chemists have tried, and many words have been coined. It is necessary to name things to think about them efficiently, and one thing scientists have done assiduously is to assign names.

Biology comes to mind, with endless terms and names based only on surface appearances, at least until recently.

Naming gives the appearance of knowledge, where there is no real knowledge at all. The antithesis to mere naming is mathematical analysis, which gives real conclusions and effective knowledge. The danger of names comes when they are regarded as real things and are used to delimit instead of simply to denote and describe.

It is easy to recognize the three conventional states of matter in ice, water and steam. The names solid, liquid and gas can be attached to certain suites of properties, and makes a useful distinction. In a gas, particles of the substance move freely and have to be stopped by walls.

In a liquid, the particles are sometimes associated, sometimes not, but always occupy a certain volume. In a solid, the particles cannot move far relatively, and can only vibrate.

Many substances can be classified by these properties, but the terms do not separate matter into three mutually exclusive boxes, and may not be descriptive enough. Where is tar, for example, or jelly, or a substance above its critical point?

Colloids will give many examples of substances for which the simple classification into three states is wholly inadequate.

The properties of matter depend almost completely on its structure. All metals are alike, to a good approximation. They are shiny, soft, tough substances that conduct electricity. All ionic crystals are alike (granted differences in crystal symmetry).

They are hard, transparent and do not conduct electricity in the solid state. They can usually be crushed into white powders. The variations between metals, or between ionic crystals, are very much less important than their similarities. Saying that a substance is a metal, or an ionic crystal, says much more than simply that it is a solid. Solidity is only a macroscopic appearance, of no fundamental significance, like being green.

The definition of matter as "that which occupies space." But what about gases? They occupy space, of course, but two or more gases can occupy the *same* space, as far as appearances go.

The important thing is to use terms like solid, liquid, gas only as far as they are useful descriptions, and not consider them as exclusive classifications into which everything must fit. To see that this is not trivial, consider the many sciences (not generally chemistry or physics) in which there are bitter controversies about which named category to assign to some object or process. We should not be limited by the arbitrary names we give our concepts.

PROPERTIES OF COLLOIDS

Colloids were particles larger than molecules, but smaller than grains of sand. This is true, and colloidal dimensions can be considered to be from about 10 nm up to 1000 nm, or 1 μm, but mere size is not the important thing about colloids. The overwhelmingly important property of colloids is that they have very large surface area.

To some degree, they are *all surface* and their properties are those of their surfaces. Incidentally, 1 μm is about the size of a bacterium. We'll use the word "colloid" to refer to a substance of colloidal dimensions, or to a colloidal system, indifferently.

To see the significance of this observation, consider the cubic centimeter in the diagram at the right. In this form, it has an area of 0.0006 m^2. We could say that it is almost all volume. Most of its molecules are safely resident behind its surface, secure from disturbance or attack.

Let us now divide it into thin laminae, 10 nm thick, a colloidal distance. The cube becomes a million laminae, with a total surface area of 200 m^2. Every molecule is now only a short distance from the cold outdoors, and the material is all surface. We have turned the mass cube into a *laminated colloid* by this delicate slicing alone.

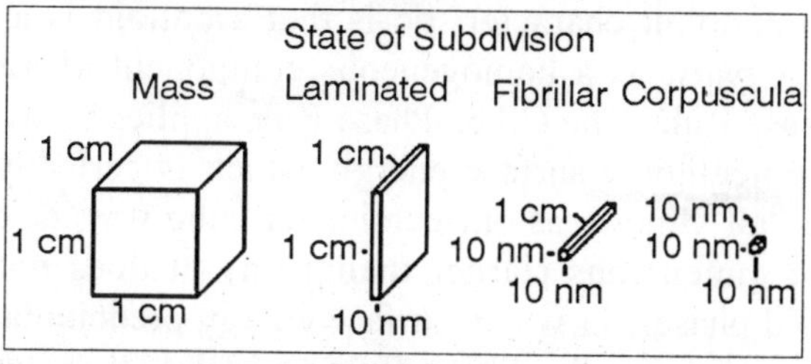

Fig. State of Subdivision and Total Area

Continuing, we now slice each of the million laminae into a million fibers, and the surface area doubles. We still have a colloid, of course, with two dimensions colloidal, but have not increased the area greatly, not as we did in the first slicing.

We can expect *fibrillar* colloids as well as laminar ones. Finally, each fibre is chopped into a million bits, giving a *corpuscular* colloid. This increases the surface area only by 50%, to 600 m^2. From the mass to the corpuscle, the surface area has been increased by a factor of a million, which is typical of a colloid. Note that most of this increase came with the first dimension to "go colloidal," so we can call anything with any least dimension of colloidal size to be a colloid.

The large area emphasizes surface effects relative to volume effects, giving colloids different properties than those of bulk matter. The surface tension of a liquid is the free energy required per unit area to create new surface. For water against air, its value at 20°C is 72.75 dyne/cm. For mercury against air, it is 450 dyne/cm. For mercury against water, it is 375 dyne/cm, not surprisingly close to the difference of the other two values. The ratio of surface to volume for a sphere of diameter d, or a cube of side d, is 6/d.

The surface energy of small drops will strongly affect their properties. If a cubic centimeter of water were divided into 10 nm cubes as above, the coalescence of the cubes would release enough surface energy to heat the water by about 10°C. A cubic centimeter of mercury would be warmed by 143°C by the same procedure.

It is better to define a colloid as a system in which the surface area is large and in which surface effects are predominant, rather than simply in terms of particle size. Indeed, in foams there are no colloidal particles at all—it is the thinness of the films that creates the colloidal behaviour. Similarly, in a gel the fibrous structure is what is colloidal. In any colloidal system, there must be at least one structural dimension of colloidal size in order for the large surface area to exist in a limited volume, however. This broadened definition of colloid is not only reasonable, it is useful. A colloid is a material system that is mainly surface.

The next important characteristic is that a colloid is a *two-phase* (at least) system. A *phase* is a homogeneous component of a system, in the sense of the Phase Rule. The Gibbs Phase Rule applies to systems in which the phases have negligible surface energy, which is perfectly applicable to phases that are "all volum≥ as our centimeter cube was, or even to phases of microscopic dimensions (larger than 1 μm). It does not apply to the system of colloid phases, in which surface energy predominates. Therefore, be careful when applying the Phase Rule to colloidal systems.

A colloidal system consists of an *internal phase*, which is the material of colloidal dimensions, and an *external phase*, which is the material in which the colloid is dispersed. These designations are analagous to the terms *solute* and *solvent* used for simple solutions (which form a single Gibbs phase). As the particles of a corpuscular colloid become smaller and smaller, we go over imperceptibly from a two-phase colloid to a single-phase solution, and there is no definite boundary.

The colloidal system that is most similar to a simple solution is a dispersion of corpuscles, or particles, in a liquid. This is called a *sol*, and the liquid is the external phase.

This is the classical colloid as described by Graham. If the external phase is a solid instead of a liquid, the system is called a *solid sol*. The only difference is the mobility of the molecules. In a solid sol, they can move only by diffusion. If the external phase is a gas, usually air, instead of a liquid, we have an *aerosol*. There is no definite boundary between a sol and a solution, but still they are significantly different. There is also no definite boundary between a sol and a coarse suspension. A coarse suspension will settle out rapidly, while a sol may be permanent.

The particles that appear in a sol may be wetted by the liquid, or may not. Wetting is a typical surface effect, and so is of paramount importance in a colloid system. In the first case, the liquid is *adsorbed* on the surface of the particle. The terms *adsorb* and *absorb* sound alike, but are quite different. A substance that is absorbed is taken into the volume of the absorbing substance, like water into sand.

If is is adsorbed, it attaches itself only to the surface. Since colloids are all surface, as we have pointed out, adsorption is what is important with them. If the particle adsorbs the external phase, it is called *lyophilic*, or *hydrophilic*, if the external phase is water. The Greek verb "luo" means to dissolve or destroy, and philic is from "philos," love.

A lyophilic colloid "loves the external phase." On the other hand, if the particle does not adsorb the external phase, is is said to be *lyophobic*, or "fears the external phase."

Sols that contain inorganic particles, such as metals, are mostly lyophobic, as are most aerosols and solid sols. Lyophobic hydrosols are a very common kind of colloid, and deserve detailed description. For example, consider the hydrosol of gold with particles about 4 nm in size. This was one of the first sols studied extensively, and has interesting properties.

With about 0.1% gold, the sol is a rich ruby red. The similar solid sol in glass makes ruby glass. The gold particles absorb strongly in the green and blue, so the transmitted light is red.

There is a little yellow-green scattered light, but mostly it is a case of absorption by the gold metal. If the gold particles clump together, which they may do as time passes, the colour of the solution changes. When the particles are about 40 nm in diameter, the solution is blue, with considerable scattered light. If the particles agglomerate further, the colour disappears and gold flakes settle out.

Bacteria, which are about 1 μm in diameter, can be suspended in water to form a sol, which has all the classic properties. The Brownian motion, the Tyndall effect (turbidity), and even electrophoresis are seen. The bacteria act as a hydrophobic sol, peptized by their electrostatic charge. The properties of a sol are largely independent of the nature of the internal phase.

There are two interesting questions here about the stability of the sol. First, what keeps the gold suspended in the red solution so that the tiny particles do not settle out? Second, how are the particles kept from agglomerating? Let's take the first question first.

The gold particles fall under gravity through the water. The terminal velocity v of their fall is given by Stokes's equation, $(mg - m'g) = 6\pi\eta av$, where m is the mass of the particle, a its radius, m' the mass of the water displaced, and η is the viscosity of water (1.002 centipoise at 20°C). A correction factor $(1 + K\lambda/a)$ must be applied for the small particles, where K is a constant, λ is the mean free path of the liquid molecules (this factor really applies better to aerosols), and a is the radius of the particle.

On the other hand, the particles are subject to the bombardment of the molecules of the liquid, which produces the Brownian movement. For a sufficiently small particle, the upward diffusion produced by the Brownian movement overcomes the gravitational fall, and an equilibrium is reached, much like the equilibrium of gases in the atmosphere. Perrin first verified this effect, although it is quite complicated in this case, the gas-like effect only occurring close to the upper surface of the sol. There is a *critical size* for a particle, below which it will not settle out. At any rate, this is why colloidal-size particles do not settle out of a sol.

Now for the second question. Generally, if two colloidal particles collide, they will stick together and make a bigger particle, because it is usually favored by energy.

Eventually, the particles get larger than the critical size to be suspended by the Brownian movement, and they settle out. There must be some good reason if this is not to happen. In most lyophobic colloids, the particles are electrically charged with the same sign, and this keeps them apart, since they repel one another.

The particles are charged mainly because they adsorb certain ions in the environment. In water, they may be OH^- ions, which are generally present, and give the particles a negative charge. The H^+ ions are hydrated, so are not as easy to adsorb, but apparently some particles like them, and become positively charged. If you try to make a hydrosol with particles of opposite charges, they neutralize each other and the sol collapses.

Since lyophobic sols are stabilized by electric charge, adding electrolytes generally destroys the sol. When rivers reach the sea with their loads of colloidal sediment, the ions in sea water coagulate the sol, and the load is deposited in the delta. There are other ways for a particle to become charged.

A zinc particle, for example, may become negative when some zinc dissolves according to $Zn \rightarrow Zn^{++} = 2e^-$, and the electrons remain on the particle.

How the charges are distributed is an interesting question. The sol appears electrically neutral on the large scale.

The particle with its adsorbed charges is called a *granule*. The charges are in a thin layer on the surface. They attract an atmosphere of opposite charges from the external phase, just as an electron in a plasma surrounds itself with a shielding positive charge by attracting positive ions and repelling electrons. The whole neutral structure, granule plus mobile external charge, is called a *micelle*. This, then, is what moves around, the charged particle and its cloud of opposite charge. When we apply an electric field to the sol, the cloud of charge is moved in one direction, and the granule moves in the other.

There is a local viscous flow about the granule, and the micelle moves toward the anode, if the granule is positive, or toward the cathode, if it is negative. This movement is called *electrophoresis*, and can be practically useful. The particles of a sol do not repel one another until they come quite close, and their micelles overlap, because of the shielding.

The *mobility* of the colloidal particle is its velocity in a unit field. Helmholtz developed a formula for the mobility, $M = \zeta\kappa/4\pi\eta$. Here, κ is the dielectric constant of the external phase, η its viscosity, and ζ is the potential difference across the micelle from the outside to the adsorbed charges on the particle, that is, through the fluid that is sliding around the granule. From measurements of M, it is possible to find ζ, which becomes a kind of "fudge factor" since it is hard to calculate. For the small gold particles in the red sol, $M = 4 \times 10^{-4}$ cm^2/s-V, and $\zeta = -0.058$ V.

The negative sign indicates that the gold migrates to the anode, and so has a negative charge. The Helmholtz equation gives $M = 4.15 \times 10^{-4}$, which doesn't prove much except that we know how to use the equation. Since the equation is written in esu, we must divide by $(300)^2$ to convert volt to statvolt.

In order to create a lyophobic sol, we must either reduce a mass to colloidal size, called *dispersion*, or we must build the colloidal particles from molecules, called *condensation*. In either case, a third substance, a *peptizing* agent, may have to be added to stabilize the sol.

This agent can supply ions that will be adsorbed on the particles resulting from dispersion or condensation to give them a stabilizing charge. For clays, the OH^- ion is a peptizing agent, which can be supplied by alkalis. Dispersion can be done mechanically, in a *colloid mill* that grinds the substance into small, equal particles.

Another method is with an electric arc. Metal electrodes are used, at a current of 5-10A and voltage of 30-40V. Bredig made particles of about 40 nm by this method, and it was improved by Svedburg to obtain sols of many metals down to 5 nm particle size. Ultrasonics can also be used to disperse sols. When a sol is created in a nonpolar solvent, the particles may not be charged (and must be stabilized by some other means). They do not then exhibit electrophoresis or other electrical phenomena depending on granule charges.

COLLOIDS AND LIGHT

A homogeneous phase or solution does not disturb the propagation of light, except to change its phase velocity to v = c/n, where c is the speed of light in vacuum and n is the index of refraction. In a gas, density fluctuations that are a natural result of the free movement of the molecules can scatter light. *Scattering* is the emission of light in all directions, which decreases the intensity of the ordered beam.

This Rayleigh scattering is proportional to the inverse fourth power of the wavelength, so blue light is scattered more than red, giving the blue colour of the clear sky. Scattering should be distinguished from *absorption*, which is the conversion of the energy of the light to other forms. Both scattering and absorption cause *attenuation* of the light beam. The transmitted beam and the scattered light may be colored if the scattering or absorption is not constant with wavelength. The blue sky and the orange sunset colors have the same cause, Rayleigh scattering by density fluctuations.

A colloidal system contains particles that affect a light beam by scattering and absorption. If the particles are of a size comparable to the wavelength of light or larger, they scatter or absorb light independently.

The same thing happens if they are separated by distances comparable to or greater than the wavelength of light. The wavelength of visible light is 400-700 nm, with the maximum sensitivity at 555 nm. This is in the middle of the range of colloidal dimensions, so colloids can be expected to have significant effects on a light beam.

One common effect of colloids is *turbidity*, an effect like that of stirring up mud in water. Slight turbidity may not be noticed until a beam of light passes through the colloid. Colloidal systems need not be turbid: a gel may be quite transparent when the particles are small.

Solutions, as homogeneous phases, are not turbid. The turbidity causes scattering so that the path of the light beam can be clearly seen. This is called the *Tyndall effect*, and the observed scattered light is called the *Tyndall cone*. John Tyndall (1820-1893) was Faraday's successor at the Royal Institution. He suggested that the blue sky was caused by scattering by dust particles, but Rayleigh later found the true cause, showing that the sky would be blue even if the air were pure.

The scattered light is polarized perpendicularly to the direction of the beam if the particles are approximately spherical and small. If the particles are nonspherical, or larger than a wavelength, the scattered light will be partially polarized.

The Tyndall effect is a common atmospheric phenomenon. Searchlights produce Tyndall cones in slightly hazy air. Smoke is often blue when seen in scattered light, orange in transmitted light. This was quite clear in the summer of 2002, when forest fires near Denver put smoke in the air, and the sunlight became a strange reddish colour.

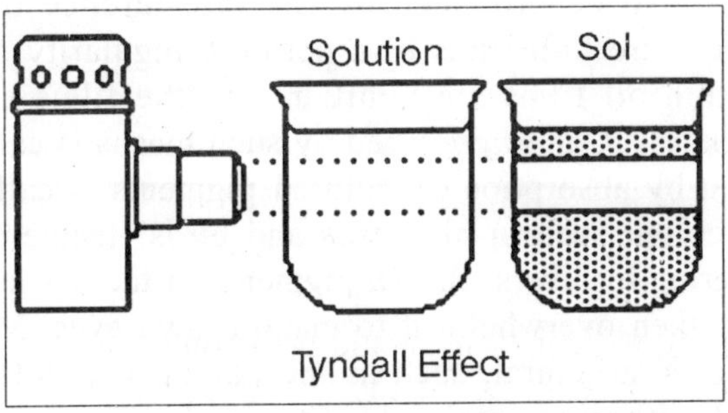

Fig. Tyndall Effect

The crepuscular rays seen at sunset, which seem to radiate from the position of the sun, are parallel Tyndall cones. A laser beam may make a distinct Tyndall cone in dusty air. If you have a piece of polarizing filter, try to determine the polarization of any Tyndall cones you may observe. All the Tyndall cones that you see are evidence of lyophobic sols.

If the sol is composed of transparent particles with an index of refraction considerably different from that of the external phase, and they are present in sufficient concentration, the sol wil be come opaque and white, the limit of turbidity. If the particles have the same index of refraction, the Tyndall effect will be small. This property is used in enamels, which are opaque glasses fused onto a metal substrate, and in pottery glazes. Stannic oxide, for example, gives a white enamel or glaze.

A beam of bright sunlight entering through a window may be marked with moving bright specks that are light scattered from colloidal dust grains that are always present in the air. The grains themselves, which may be submicroscopic, are not seen, only the light they scatter.

This principle is used in the *ultramicroscope* that allows individual, submicroscopic particles to be observed. Only the light from them is detected against a dark background, so that they can be counted and their motion observed; no image of the particles can be formed.

The instrument consists of a bright light source, a slit and optics to focus the slit on the sample, and a microscope. A thin slice of the sample is illuminated by the slit. By turning the slit 90° the depth of the area viewed can be determined. The direction of viewing is at right angles to the illumination. Some ultramicroscopes are coaxial and use a different illumination method. If the number of particles per unit volume is found in this way, by counting using a squared graticule, and the weight of colloid per unit volume is known, then the size of the particles can be determined.

Colloids can produce colour. The red of a gold sol or ruby glass has already been mentioned, where the colour is due to the wavelength dependence of scattering and absorption.

Colour can also be produced by the interference of white light, especially if there are thin films or a periodic regularity in the density. The colors in thin oil films are familiar, and the films are, of course, colloidal in thickness. Colour produced by such means is called *structural*. Colour produced by absorption by colored pigments is called *pigmental*.

The blue colour seen in blue eyes and birds' feathers is structural, caused by scattering by fibers. Brown pigments in the iris modify the blue colour to green, then overwhelm it to make brown eyes. Much colour in the insect world is structural, such as the colors of a butterfly's wings, often combined with pigmental colour to produce a great variety. The greenish colors of crude oil and its products are a result of colloidal suspensions.

AEROSOLS

An aerosol of colloidal solid particles may be called a *smoke*, while if the particles are liquid, it is a *fog*. Sometimes the two are combined, in a suspension of solid particles with an adsorbed liquid film on the surface. The original *smog* was a smoke with a liquid film of sulphuric acid, which made it excessively unpleasant to breathe.

The Great Smog occurred in London on 5 December 1952, killing nearly 4000 people who had respiratory problems, and stimulating clean air legislation. Now it usually lacks the smoke, and is a fog of some unpleasant liquid coming from motorcars instead. Aerosols have the usual characteristics of lyophobic sols: a strong Brownian movement, the scattered blue light of the Tyndall effect, and stabilization with particle charges.

Clouds are aerosols of water particles, supported by the Brownian motion like any sol. The droplets are produced by condensation of water vapour on *condensation nuclei*, which are usually hygroscopic particles of dust, or positive ions. The mist above splashing water is positively charged, with compensating negative charge in the form of unhydrated negative ions. For condensation to occur, the air must be supersaturated for water vapour. The radius of the droplets is large for colloidal particles, and they are often supported by updrafts more than by Brownian motion.

The radius of cirrus cloud droplets may be 2 μm. When they reach a radius of about 0.04 mm, the droplets fall as rain, often coalescing with others. The largest raindrops have a radius of about 3.6 mm, and such large drops are rare.

Raindrops can be blown upwards into freezing air by updrafts, gathering water by coalescence with others, and freezing to a solid ball. This can happen repeatedly, forming large hailstones that eventually fall. The largest hail reported may be the 3" diameter hail that fell in

Bloomington, Ind. in 1917. A farmer was killed by hail near Lubbock, Texas in the 1930's, but this is the only reported fatality.

Saturated air can be expanded adiabatically to cool it and achieve supersaturation. This can be done in the laboratory with the Wilson cloud chamber with an adiabatic expansion against a piston or the equivalent, which does work that cools the air.

The cooling is given by $T_2/T_1 = (V_1/V_2)^{k-1}$, where k is the ratio of the specific heat at constant pressure to the specific heat at constant volume (1.4 for air). In clean air, condensation occurs for a volume ratio greater than 1.25. From 1.25 to 1.34, the condensation falls as fine rain.

For 1.25 to 1.28 the condensation is on negative ions, and the particle radius is 200μm. For 1.28 to 1.34 the condensation is on the positive ions, and the particle radius is 20μm. For larger expansion ratios, the mists can be colored: 1.408 gives green, 1.422 purple, and 1.429 red.

At the lower expansion ratios, condensation is difficult enough that it may occur along the tracks of alpha particles, which ionize strongly, making the tracks visible. The condensation soon settles out, since the droplets are large. The chamber can work with alcohol vapour as well as water vapour.

The vapour pressure p' of a small drop of radius r is greater than the vapour pressure p of a plane surface of the liquid. The vapour pressure p' is given by $\ln (p'/p) = (1/RTd)(2\gamma/r - q^2/8\pi r^4)$, if q is the charge density on the surface. γ is the surface tension of the liquid.

This shows that a charge of either sign will stabilize a small droplet against evaporation. In the absence of a charge, small droplets will evaporate in favour of large drops. Electric charges may explain the stability of clouds, and the fact that rain may not fall from them.

An aerosol of starch grains, with a density of $100 g/m^3$, has a very large surface area, and adsorbs O_2 from the atmosphere. Starch grains in flour are from 5 nm to 200 nm in diameter.

It is no wonder that it is a violet explosive. Coal ground to pass through a 200-300 mesh sieve can be blown directly into a fire as a convenient fuel. It can also make a sol with fuel oil to form a *colloidal fuel* that can be used exactly like fuel oil.

Coal dust is reputed to be a fire hazard, but it is not as dangerous as starch dust. Many "coal dust" explosions may have another cause.

Smudge pots are used to protect agricultural plants from frost. They are small fires producing dense smoke. The heat produced by the fires is probably one of the most important effects.

The smoke may provide condensation nuclei for the dews of evening, increasing the heat by the latent heat of the condensed vapour. The aerosol blanket slows loss of heat by radiation, which can be very important into a clear night sky. Smoke is also used for military purposes, for concealment or signalling. White phosphorus burns to hygroscopic P_2O_5, which forms a dense white fog in humid air. Silicon tetrachloride mixed with ammonium hydroxide gives a dense white smoke of ammonium chloride and metasilicic acid.

This smoke is used in skywriting, since it is easily made from liquids when needed. The navy's smoke for concealment was made by restricting the air supply to the boilers, producing thick black smoke from the funnels. Radar and the end of gunnery has rendered smoke screens useless. Colored smoke may be used for signalling.

EMULSIONS

An *emulsion* is a colloidal system in which both phases are liquid. If the liquids were miscible, they would form a solution, so emulsions are lyophobic colloids. The typical example is water and oil.

The internal phase is determined by which component has the higher surface tension. This component will form spherical bubbles immersed in the other, which will be a continuous phase. The granules of an emulsion may be large, even microscopic. An emulsifying agent is usually required to form a stable emulsion.

The emulsifying agent, or *protective colloid*, is *surface-active*, meaning that it reduces the surface tension of the liquid, and so tends to concentrate in boundary films. In the case of water and oil, sodium oleate, a *soap* reduces the water surface tension and raises that of the oil, so that the emulsion will be oil droplets in water, and quite stable. There are many other *detergents*, but sodium oleate will serve as a good example.

Oleic acid is a fatty acid with the formula $CH_3(CH_2)_7CH = CH(CH_2)_7$ COOH, a monounsaturated carboxylic acid with a long hydrocarbon chain. Fats are glyceryl esters of this acid. Glycerol has three OH groups, each of which can take the H off the end of the oleic acid and stick the rest to the glycerol framework, making a *triglyceride*. Boiling the fat with NaOH makes three molecules of sodium oleate, a liquid soap.

The hydrocarbon chain nestles up to the oil, the sodium to the water, and peptizes the oil. The emulsion can be made by simply shaking oil, water and soap together, but this will not make droplets of a uniform size. The larger droplets may "cream off" in this case, and float to the surface of the emulsion. The emulsion may be *homogenized* by blowing it through small orifices under a pressure of 4000 to 5000 psi. The small, uniform drops that result will make a stable, long-lasting emulsion.

Milk is an emulsion of oil ("butterfat") in a watery sol of the hydrophilic protein casein in which the external phase is a solution of lactose and various salts. Milk will "cream" unless homogenized, since the fat globules range from 100 nm to 22 μm in diameter. Cream can contain from 29% to 56% fat, in packed globules stabilized by casein. Human milk contains albumin in addition to casein. Mayonnaise is an emulsion of oil in water, using egg yolks as an emulsifying agent.

Hollandaise sauce is another emulsion, of butter in lemon juice, again using egg yolks as an emulsifying agent. Butter itself is an emulsion, this time of water droplets in oil. The cosmetic "vanishing cream" is an oil-in-water emulsion. If this is reversed, "nourishing cream" is a water-in-oil emulsion. Crude oil as it comes from the well is often emulsified with water. In this case, heating is sufficient to "break" the emulsion, and separate the oil from the salt water.

Many insecticides are oil-in-water emulsions for spraying. The oil wets the oily leaf surfaces and sticks, while the water carrying the poison evaporates. Lubricating grease is a water-in-oil emulsion.

The emulsifying agent, calcium oleate, is soluble in oil, not in water, and so makes the water the internal phase. The idea here is to make the lubricant stiff, so it will not drip off.

Whether a component of an emulsion is the internal or external phase is determined by the relative surface tensions, not be the amounts of the components. Equal-sized globules can close-pack like spheres to occupy 74% of the volume.

The internal phase can be even higher in concentration if the globules are not all the same size, so the smaller can huddle in the voids left by the larger. Because of the presence of an emulsifying agent, emulsions are very stable, and "breaking" them when required can be difficult.

To find out whether an emulsion is oil-in-water or water-in-oil, the effect of adding a small amount of either oil or water to a sample of the emulsion on a microscope slide is observed. If you add the external phase, it will mix easily and quickly, but the internal phase will not mix and remain a drop. This is called the *dilution test.*

FOAMS

A foam may be an internal phase of gas in an external phase of liquid or solid. In a liquid foam, a colloidal adsorptive agent forms a film that bounds the gas bubble. Bubbles blown with soap solution are related to foams, but are quite large and have an independent existence. Smaller bubbles in a mass form the more usual foam.

People washing dishes or clothes are reassured by a thick layer of soap foam on the water, showing that there is still detergent left to emulsify additional oil. Such foams are mainly air, with very little liquid.

The colloidal dimension in a foam is the thickness of the film, not the size of the bubble. The bubble is lighter than its surroundings, and will rise to the top, where it joins the foam. In beer, the foam is stabilized by albumin and by the hop resins added to the beer. When carbon dioxide is released, it uses the adsorptive agent it finds to make the foam. This "head" on Guinness stout is famous and creamy—and carefully engineered. Meringue is a dried foam using egg albumin. Marshmallows use sugared gelatin for the same purpose.

Ore flotation depends on the property of the adsorptive agent to wet the valuable metallic sulphides or other ore, but not the silicates, which are preferentially wetted by water. Note again the central role played by surface chemistry. A froth is made with water and the adsorptive agent, and mixed with the ore to be beneficiated.

The froth floats to the surface, where it is skimmed off, together with the enriched ore. In a few special cases, the valuable mineral is wetted by the water, while the gangue sticks to the oil. We use the buoyancy of the bubbles to effect the separation.

A fire-fighting foam is made from mixing water, aluminium sulphate and sodium bicarbonate with an adsorptive agent.

The carbon dioxide that is released makes a dry foam, while the other ingredients form a kind of gel. This foam can be used on all kinds of fires, including burning oils.

Examples of solid foams are pumice, meerschaum, and Ivory soap. The white soap has colloidal air beaten into it so that it will float. Meerschaum, "sea foam" in German, is a light-colored metamorphic rock associated with serpentine, and is a magnesium silicate, also known as *sepiolite* from its resemblance to light cuttlefish (sepia) bone.

It is soft, smooth, light and translucent, used mainly for carving smoking pipes. It is fibrous and porous, with gaseous inclusions, apparently a dried gel. It is so porous that it floats on water, in spite of its mass density of 2.0 g/cc. Pumice, an extrusive igneous rock, is a solidifed foam of volcanic glass, usually obsidian. It also floats on water, and makes a good gentle abrasive. Diatomaceous earth consists of the microscopic shells of diatoms, very common marine plants, made of opaline silica and very porous.

It adsorbs nitroglycerine to make dynamite, rendering it much less sensitive and much safer to handle. Bread is, of course, a dried foam as well. The protein *gluten* makes the film that surrounds the CO_2 bubbles produced by the enzyme zymase secreted by the yeast.

Zymase acts on the sugars (hexose) produced from starch by other enzymes, and also makes ethyl alcohol at the same time as the carbon dioxide. The alcohol is often the desired product! Hydrophilic gluten and water make a good gel in "strong" flours that include as much CO_2 as possible and hold the starch granules. Rye flour has little gluten, and will not make light bread by itself. Foams aid digestion by providing as large a surface area as possible.

GELS

So far we have looked at lyophobic colloids, which will have nothing to do with the other phase. A lyophilic colloid, by contrast, actively seeks out the other phase and adsorbs it strongly. In most cases, the other phase is water, so we are dealing with hydrophilic colloids. Aluminum hydroxide and orthosilicic acid are inorganic examples.

Most organic colloids are hydrophilic. Gelatin is of animal origin, derived from protein, made by boiling bones and horny parts. Gums are of vegetable origin. Gums are branching polysaccharides, soluble in water. They include the *mucilages* carrageenan (from Irish moss, a seaweed) and agar-agar (also from seaweed), which are sulphate esters.

Gum arabic is from the acacia tree, and gum mastic is from the *Pistacia lentiscus*. Gums are secreted by trees to protect and seal wounds. When you find any of these in a list of food ingredients, they are for the purpose of making a gel. Ice cream may contain guar gum, cellulose gum, locust bean gum and carrageenan. These gums are not digestible.

A *resin* is similar to a gum in purpose, but is insoluble in water. Resins are notably secreted by evergreens, and are terpene derivatives, soluble in turpentine and similar solvents. Rosin is the residue when turpentine is distilled. Amber and copal are other natural resins, while alkyd and phenolic resins are artificial, used in plastics and paint.

Gelatin or gums dissolve in hot water. As the clear colloidal system cools, its viscosity increases steadily. The viscosity of the external phase is not affected in hydrophobic sols, by distinction. At some point, the system *gels*, forming a wobbly but definitely solid body.

This is really an extraordinary thing to happen. If you heat the gel, it will melt and form a viscous liquid. On cooling, it will gel again. If you dry it out, it will shrink and look horrible.

On adding water, it will plump up again into a wiggly gel. These colloids are called *reversible* or *elastic*. The gels formed from inorganic hydroxides will not reform a gel once they have dried out, and the dry form will be brittle. The pore spaces will still be there, however, and will absorb moisture and other substances. Silica gel is a widely used substance. Though it can be renewed by heating and live to absorb again, it will never again be wobbly and gelatinous.

Gels are used as culture media for microorganisms. Gelatin was originally used, but it melts at 37°C, and so cannot be used to study microorganisms at body temperature. Agar-agar makes a gel that can stand higher temperatures, so it is used in preference as a culture medium, poured into the familiar Petri dish. It was first used by Robert Koch. The colloidal phase in a gelatin is fibrillar, composed of fibers of colloidal cross-section. When a gel sets, these fibers form a tangled mass like a pile of brush, that holds the system together.

Droplets in a gel are lens-shaped, showing the packing. There has been quite a bit of controversy over the structure of gels, but the fibrillar structure seems correct. The original idea of a cellular structure is untenable. The fibers adsorb large quantities of water, and there are also droplets of water, so that the gel is mainly water, given a doubtful rigidity by the stacks of fibers.

Pectin is a gelatin-like protein substance found in the rind of citrus fruit, in apples and generally in fruits. If a slightly acid solution of pectin is made 65% or 70% sucrose, it will gel. This is the reaction used to make jams. The verb *pectize* is used to describe the creation of a gel, as peptize is to create a sol.

Gels have some curious properties. As they age, *syneresis* may occur, which is the loss of liquid. This is a result of the closer agglomeration of the colloid. The reduced active area requires less water, so the excess water is eliminated.

It does not mean that the gel is deteriorating. Another property is *thixotropy*. On agitation, the gel becomes fluid, but reverts to the gel when left alone.

Fresh gelatin gels are quite thixotropic, and advantage can be taken of this to add ingredients to a jelly. The "thixo-" comes from the future of the Greek verb for "touch," *thixomai*, and a thixotropic substance would be one whose state changes with touching.

COLLOIDS AND MEMBRANES

Let's review the phenomenon of *osmosis*. In the diagram, we have put some solution into a container closed off by a *semipermeable membrane*, and put the container into a vessel containing the pure solvent. Consider the dots as representing the dissolved substance.

This applies to a sol, as well as to a solution, but there will be more "solut" particles in the solution. The membrane can be cellophane, but ordinary wrapping cellophane cannot be used, since it has been lacquered to close the pores.

The pores must be small enough to prevent the solute from passing through. The solvent is found to cross the membrane and enter the solution, making it more dilute. The level of the solution rises to some height h in equilibrium.

If d is the density, then $\pi = dgh$ is the *osmotic pressure*. This is the pressure required on the solution side to make the rate of movement of the solvent the same in both directions across the membrane. If the solution is dilute, van't

Hoff's equation gives the osmotic pressure: $\pi V = nRT$, where the volume V of the solution contains n moles of solute.

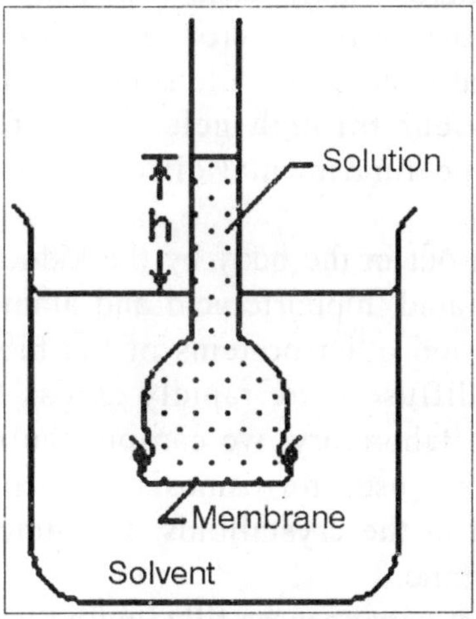

Fig. Osmosis

The osmotic pressure of a solution containing 1 mole of solute per 1 kg of solvent, called a 1 molal solution, is quite large, about 25 atmospheres (as we can estimate from van't Hoff's equation). Not only is this hard on cellophane membranes, but would require a liquid column 825 feet high, which is inconvenient.

Nevertheless, osmotic pressures are high enough to push fluids to the tops of high trees (this is actually a complex subject about which there is controversy). Strong semipermeable membranes can be made by precipitating insoluble salts in the pores of unglazed porcelain.

Sols will exert quite modest osmotic pressures, since there are many fewer particles per unit volume. In fact, the pressures are so low that they cannot be used in most cases to find the number of particles per unit volume. However, the molecular weight of haemoglobin was determined in this way. Cell walls are semipermeable membranes.

If an erythrocyte (red blood cell) is put into pure water, it will swell and burst because it contains a solution of electrolytes in a semipermeable membrane. If put into strong salt solution, it will become dehydrated and shrivel. Solutions of the same osmotic pressure are called *isotonic*. Isotonic solutions are used to prevent damage to biological systems.

Erythrocytes are about 8.6 μm in diameter and 2.6 μm thick, and slightly concave on the faces. They are a bit large for colloidal particles, though the blood plasma is a colloid that can gel under certain circumstances. Like platelets, which are definitely colloidal, they are not living cells like the leucocytes that accompany them.

The separation of colloids and crystalloids is called *dialysis*, and can be carried out with a suitable semipermeable membrane, just as in osmosis. Chemists actually distingush dialysis, in which the components diffuse at different rates across a membrane, from *ultrafiltration*, in which larger particles are mechanically stopped while smaller ones are allowed to pass. Ultrafiltration can occur through gels, while dialysis uses dialytic membranes. There is no fundamental difference between the processes, however.

Dialysis is carried out in the body by the kidneys, which separate the crystalloids urea, uric acid, hippuric acid and ammonia compounds from the colloidal albumin and other proteins of the blood.

The crystalloids diffuse more rapidly across the interface than the larger particles. In the laboratory, we can put the sample to be dialyzed into a container like that used for osmosis, and change the solvent as it becomes concentrated in the crystalloids. Parchment paper can be used for the dialytic membrane.

A bag of parchment paper can be filled with the sample to be dialyzed, and it can be suspended in a bath of moving warm water for rapid dialysis. Membranes can also be prepared from other animal membranes, or collodion (nitrocellulose dissolved in alcohol and ether), or artificial sheet polymers. Electrodialysis can also be used, taking advantage of ion migration in an electric field.

Dialysis has many industrial applications. Sugar is extracted from sugar beets by using the cell walls as dialytic membranes, washing cut beets in warm water. Dialysis is used in the artificial fibre industry to separate alkali from the colloidal fibre material, and in the pharmaceutical industry for the purification of colloidal medicines. Dialysis is used in artificial kidney machines to simulate the action of the kidneys.

COLLOIDS AND ADSORBENTS

As has been pointed out, absorption is a volume effect, while adsorption is a surface effect. Colloids, having large surface areas for a given volume, are excellent at adsorption.

A cubic centimeter of charcoal can have a surface area of 1000 m^2, so a little charcoal can do a lot of adsorbing. Adsorption may be specific. For example, Ni, Pt and Pd in colloidal form adsorb H_2 up to 1000 or 3000 times their volume.

Palladium is the best at this. At red heat, palladium metal will absorb hydrogen readily, and release it at even higher temperature. Black palladium powder can be made by reducing $PdCl_2$ in solution. Because of the adsorption, these metals, in colloidal form, make excellent catalysts for hydrogenation.

Usually, adsorption is not very specific, and a wide variety of substances can be adsorbed on a certain medium. Adsorption is temperature-sensitive, being much more effective at low temperatures, and evolving the adsorbed substances at higher temperatures.

Carbon at room temperature does not adsorb oxygen and nitrogen, but does so at liquid-nitrogen temperatures. Adsorption is generally accompanied by a negative change in enthalpy, so it is *exothermic*. The temperature dependence is a consequence of LeChatelier's Principle.

The three most commonly used adsorbents are carbon as charcoal, alumina and silica. When specially prepared as adsorbents, they are called *activated*. Activation is usually a matter mainly of heating, and perhaps some chemical cleaning. A saturated adsorbent may be re-activated by heating, say at 175°C for 6-8 hours in air.

Charcoal may be prepared from wood, bone, blood and sugar. The charcoal from different sources has different impurities and therefore somewhat different characteristics. Charcoal is a *nonpolar* adsorbent that is good for organic vapors and nonpolar substances in general. Alumina and Silica are *polar* adsorbents and are best for polar substances, like water.

Charcoal is used in gas masks, since it readily adsorbs toxic organic vapors rather indiscriminately.

It does not adsorb carbon monoxide or ammonia well, so specal adsorbents for these must be included. Carbon monoxide is usually oxidized to the dioxide, and ammonia by silica gel. Therefore, a complete gas mask canister contains a mechanical filter, charcoal, silica gel, and an oxidant for CO. Charcoal can be used to produce ultra-high vacuum by cooling it to liquid nitrogen temperatures.

If charcoal in a stout glass tube in a U-shape is saturated with chlorine or sulphur dioxide, and then sealed off, the gas is liquefied when the charcoal end of the tube is heated, and the other end is cooled. Natural gas may be filtered through charcoal to remove higher alkanes. Pentane and hexane are more strongly adsorbed than the lighter methane and ethane. Charcoal is, in general, used when organic substances are to be removed from a gas.

Alumina gel and silica gel are produced by drying the gel produced by precipitating the hydroxides in water. Alumina gel is good for water vapour, carbon dioxide and alcohol. It makes an excellent laboratory dessicant. A small amount of colorimetric indicator is added that is blue when the alumina is dry, but turns pink when it has adsorbed water and is alkaline.

The dessicant can then be regenerated by heating in an oven. Alumina gel is generally used for drying organic liquids, which it also decolorizes, and for removing acids from oils. It also removes oil vapour from compressed air. Alumina gel is a good example of the nonspecifity of adsorption.

Silica gel is the most generally useful adsorbent. It is especially good at adsorbing benzene, for example from coke-oven gas. Silica gel is used to dry carbon dioxide, hydrogen and oxygen before they are liquefied or solidified. It is often found in small bags in sealed packages that must be protected from humidity.

These bags are easily re-activated by heating. Silica gel can be used to dry natural gas for pipeline transport, to prevent the formation of ice clathrates at low temperatures. These clathrates can form above the freezing point and easily clog pipelines.

Other colloidal adsorbents found naturally are *diatomite*, also known as diatomaceous earth or kieselguhr, which is composed of diatom shells made from opaline silica, and *bentonite*, a strongly hydrophilic colloidal clay consisting mainly of montmorillonite.

Bentonite can clarify and deodorize petroleum, purify and soften water, make drilling mud, plug leaks as a grout, improve cleaning powders, and destroy building foundations by swelling. Diatomite has perhaps even more uses than bentonite. Both are valuable colloidal minerals.

Adsorbents are imporant in dyeing fabrics. Often the fabric will not adsorb the dye directly, since the dye may be polar and the fibre nonpolar. However, gels like amphoteric metal hydroxides, especially aluminium hydroxide, may cling to the fibers while strongly adsorbing the dyes. Such intermediates are called *mordants*, which are usually colloids.

The dye and the mordant together, without the fibre, is called a *lake* (from the same word that gave "lacquer"). Purple of Cassius is a famous lake, formed on stannic hydroxide gel by colloidal gold.

HARMONIC MOTION

Why do sinusoidal variations with time seem to crop up everywhere in Physics? By *sinusoidal* I mean any variation that is a linear combination A sin $2\pi ft$ + B cos $2\pi ft$, where A and B are constants, and f (sometimes ν)is the *frequency* in per second, or hertz (Hz). The *angular frequency* $\omega = 2\pi f$, in radians per second, is often convenient.

The unit Hz should not be used with angular frequency, to avoid confusion. Our sinusoidal variation can also be written C cos $(\omega t + \varepsilon)$, where $C = (A^2 + B^2)^{1/2}$ is the *amplitude* and $\varepsilon = \tan^{-1}(A/B)$ is the *phase constant*. The quantity $(\omega t + \varepsilon)$ is the *phase*. A variation of this type is called *harmonic*, or, more precisely, *simple harmonic*, since it involves only one frequency f.

The Figure at the right shows a graphic representation of simple harmonic motion. The vector of length A rotates anticlockwise from its initial position at time 0 as shown by the dotted vector. The projection of the terminus of the rotating vector on the x-axis represents a simple harmonic motion (as does the projection on the y-axis).

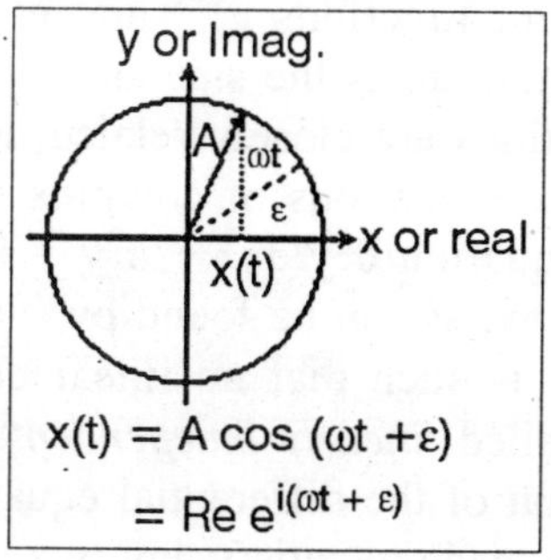

Fig. Simple Harmonic Motion

A vector (x,y) can also be expressed as a complex number in the form $x + iy = Ae^{i\omega t} = A(\cos \omega t + i \sin \omega t)$. It is very convenient to express a harmonic variation in complex exponential form, $Ae^{i(\omega t + \varepsilon)}$, because then all the rules of manipulation of exponents are available.

The complex constant $Ae^{i\varepsilon}$ is called a *phasor*, and is very useful for the specification of a harmonic variation.

Engineers use j instead of i for the imaginary unit to avoid confusion with the symbol for electrical current. Physicists often take the time variation as $e^{-i\omega t}$. In any case, two parameters are necessary for the complete specification of a simple harmonic motion.

With these definitions out of the way, we can return to the question. The small-amplitude natural vibrations of bodies are observed to be simple harmonic, or else a sum of several simple harmonic variations.

Ohm's Law (of acoustics) asserts that the fundamental aural perception is that of a simple harmonic variation. In fact, this is the origin of the word. Fourier's Theorem shows that an arbitrary periodic time variation can be expressed as a sum of simple harmonic variations, usually an infinite number.

These are only a few examples of the ubiquity of harmonic motion. They all are consequences of the mathematical behaviour of simple harmonic motion, often combined with the principle of superposition.

The two essential mathematical properties of simple harmonic motion are: (1)the sum of any number of such motions is also a harmonic motion of the same frequency, with at most a difference of amplitude and phase constant, and (2) the derivative (or integral) of a harmonic motion is also a harmonic motion of the same frequency, again with at most a difference of amplitude and phase constant.

The exponential function e^{ax} also has these properties, but increases without bound for large positive or negative x (depending on the sign of a), so it is unsuitable for the description of a steady motion. The harmonic vibration, on the other hand, is periodic with period $T = 1/f$. Also, the two linearly independent functions e^{+ax} and e^{-ax} cannot be turned into one another by differentiation, as the sine and cosine can. The exponential and trigonometric functions are closely related, as we know, and the sine and cosine are linear combinations of complex exponentials.

The differential equation $d^2y/dx^2 + a^2y = 0$ has solutions e^{-iax} and e^{+iax}, or sin ax and cos ax, as can be found by substitution in the equation. Each pair of solutions is such that no linear combination of them is a constant, so they are called *linearly independent.*

The general solution of the differential equation is $y(x) = A \sin ax + B \cos ax$, and it can be made to satisfy any two algebraic conditions, such as having specified values at two different points, or for having a specified value and derivative at a certain point.

These conditions are variously called *initial* or *boundary* conditions. For example, $B = y(0)$ and $A = y'(0)/a$, if the initial position and slope are specified at the origin. This differential equation occurs over and over, so it is good to remember its solutions and their properties.

For example, consider a mass M hanging from a fixed point O by a flexible, inextensible, weightless cord of length L. The motion can be described in terms of the distance x that M moves to the right or left of its equilibrium position. When it is displaced by x, the weight Mg acts directly downward, but the force of the cord on M is angled a bit.

The net horizontal force on M is then $-Mg(x/L)$, if we assume that the displacement x is always small compared to L (say, less than 0.1L). By Newton's Second Law, $Md^2x/dt^2 = -Mg(x/L)$, or $d^2x/dt^2 + (g/L)x = 0$. This is just the equation we were talking about, so M moves harmonically with angular frequency $\omega = (g/L)^{1/2}$.

The frequency depends on the strength of gravity, and on the length of the cord, but remarkably not on the mass M, nor on the amplitude of vibration. The *isochronous* and *harmonic* nature of the motion depend on the smallness of the motion. For large swings, the pendulum no longer moves with simple harmonic motion! Finite motion is a different and more difficult problem (but one easily soluble).

The way to solve more difficult problems of this sort was shown by J. L. Lagrange. One begins by selecting any convenient parameters, called *generalized coordinates*, that define the state of the system. If only one such parameter q is necessary, we have a system with one *degree of freedom.* For the pendulum, q could be the angle θ that the line OM makes with the vertical. The velocity of M can then be expressed in terms of the rate of change of θ, as $v = Ld\theta/dt$.

The kinetic energy T of M is then $M(Ld\theta/dt)^2/2$. Lagrange showed that the partial derivative of T with respect to the rate of change of the coordinate is a generalized momentum. In this case, $p = ML^2(d\theta/dt)$, which is the angular momentum about O, but the result also holds in general. Lagrange now expresses the potential energy V in terms of the coordinate, which in this case is MgL cos θ.

The negative of the derivative of the potential energy with respect to the coordinate is the generalized force. In this case, it is -MgL sin θ, which is the torque about O. Newton's Second Law then gives $MLd^2\theta/dt^2 + Mg \sin\theta = 0$, or $d^2\theta/dt^2 + (g/L)\sin\theta = 0$.

This is not our previous equation of motion, because the sine appears instead of just the angle. We won't solve it, since we just wanted to show how to use generalized coordinates to get equations of motion. The momenta and forces that are associated with a generalized coordinate are appropriate to the type of coordinate. Since we chose an angle, we get angular momenta and torques in place of linear momenta and forces.

In our example, T was independent of q, and V was independent of q' = dq/dt. Lagrange also showed how to handle more complex cases by defining the Lagrangian function L = T – V, and then writing the equation of motion as d(δL/δq')/dt = δL/δq (the δ's represent partial differentiation).

This is *Lagrange's equation*, and further details are to be found in books on advanced mechanics.

There is one such equation for each generalized coordinate, in systems with more than one degree of freedom. It is a very powerful way to find equations of motion. With very special choices of the generalized coordinates, each of Lagrange's equations may contain only a single generalized coordinate, and so the system is easily solved. These special generalized coordinates are called *normal coordinates*.

Thus far, the pendulum does not move unless it is given an initial disturbance, say by being held away from the equilibrium position and released, or by being given a knock when hanging still.

Our equations imply that the pendulum will continue moving with the same amplitude from then on, but we know from experience that the amplitude slowly decreases, and the pendulum eventually comes to rest, due to air resistance and friction in the support.

To make the pendulum move continuously, it must be driven or *forced*. Suppose we do this by moving the support from side to side a distance y = D cos ω't. We suppose $D \ll L$, so the added force on M is +Mg(y/L). Now, the equation of motion is $d^2x/dt^2 + (g/L)x = (g/L)D \cos \omega't$, after removing the common factor M. We require any solution that will give the new right-hand term, called a *particular solution*, to which we can add the solution we already know that describes the natural motion and contains two constants that we can choose so that boundary conditions are satisfied./P>

Let us try x = C cos ω't. When it is substituted in the equation, we get $-C\omega'^2 + C\omega^2 = D\omega^2$, omitting the common factor cos ω't. We have a solution, provided that $C = D\omega^2/(\omega^2 - \omega'^2)$.

This motion is at the frequency of the disturbance, and proportional to the magnitude of the disturbance. If the forcing frequency ω' is less than the natural frequency, the forced motion is in phase with the force, and as ω' approaches zero, it is also of the same amplitude.

If ω' is greater than ω, the forced motion is 180° out of phase with the force, and its amplitude becomes less and less as the forcing frequency is raised. When ω' and ω are about equal, the amplitude of the forced motion becomes very large, which is called *resonance*.

This is all borne out very nicely by the pendulum. In the natural motion, the pendulum swings about the point of support, and the effective length of the cord is L. Forced at a low frequency, the pendulum swings about a higher support, and the effective length L' is greater than L. It is simply the length of a pendulum that would swing at the forcing frequency. At a high forcing freqency, the pendulum swings about a lower support with effective length L", again corresponding to a pendulum whose natural frequency is the forcing frequency.

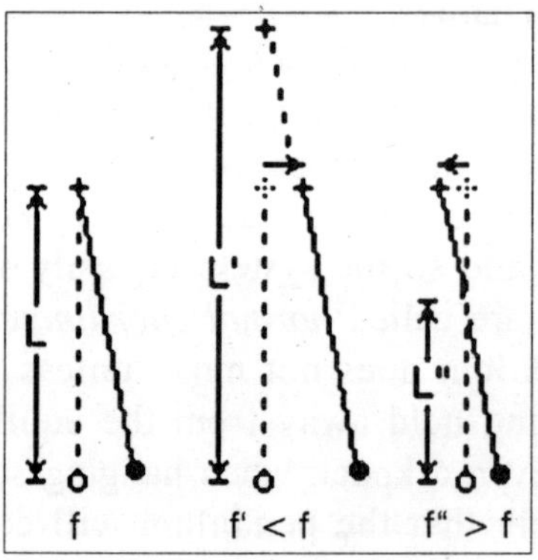

Fig. Forced Osillation

The phase relations are very clearly shown. At zero forcing frequency, the pendulum hangs vertical as the support slowly moves from side to side. At a very high forcing frequency, M remains practically fixed, as the support moves rapidly from side to side. As another example, consider a vehicle of mass M supported by springs of spring constant K between body and wheels.

Its natural frequency will be $(K/M)^{1/2}$. If the springs compress a distance d when the body is put on them, the natural frequency will be (g/

$d)^{1/2}$. When the wheels experience a short sharp vertical shock, they will move upwards while the body remains largely undisturbed.

The mass M will experience an upward impulse, so the body will begin to move upward in a harmonic motion at its natural frequency that will die out quickly if sufficient frictional resistance is provided. If the vertical disturbance is a rapid oscillation, the body will hardly be disturbed. On the other hand, disturbances near and below the natural frequency will give rise to considerable motion, especially in the case of resonance.

It is possible to find the response of a system to an arbitrary force f(t), for example by the method of variation of parameters. The solution is then expressed as an integral of f(t) times sine and cosine functions. A force $f(t) = (\mu/\pi\tau)/(t^2 + \tau^2)$ can represent the application of a force for a short time, where μ is the impulse, or integral of force times time, and τ is a time parameter.

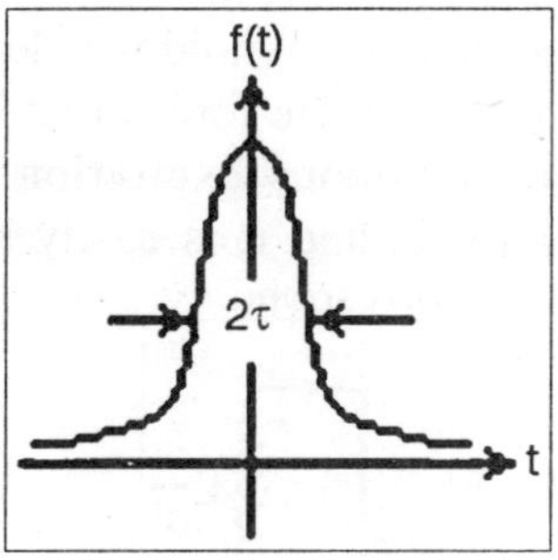

Fig. An Impulsive Force

The force is peaked, with a width at half-maximum of 2τ. The motion of the oscillator, assumed to be intially at rest in its equilibrium position, is then $x(t) = (\mu/\omega)e^{-\omega\tau}\sin \omega t$.

This shows the smoothing effect clearly, when τ is comparable with the natural period of oscillation.

Suppose we apply D cos ω't to a pendulum initially at rest, and then ask what happens when ω' = ω. The conditions x = dx/dt = 0 at t = 0 then give the solution $x = D[\omega^2/(\omega^2 - \omega'^2)](\cos \omega' t - \cos \omega t)$.We have chosen the arbitrary constants A and B in the free motion to satisfy the boundary conditions. The difference of the cosines is $2 \sin[(\omega' - \omega)/2]\sin[(\omega' + \omega)/2]$. As ω' approaches ω, the first sine can be replaced by its argument, which cancels the difference in frequencies in one factor of the denominator.

Finally, when ω' = ω, the limit is $x = (D/2)\, \omega t \sin \omega t$. The amplitude of oscillation increases linearly with time. This can only go on for a short time, however, for our approximations become invalid with large amplitudes. A pendulum 1 m long has a period quite close to 2 s. If we drive the pendulum by shaking the support by the small amplitude of 0.1

mm at 0.5 Hz, it will take 10.5 minutes to reach an amplitude of 10 cm, which is about the limit at which our approximations are valid. Since losses increase with amplitude, the amplitude will eventually be limited at a value for which the energy contributed by the exciting force balances the losses.

All this analysis applies not only to mechanical systems, but also to electrical circuits. Here, the definitions of inductance L and capacitance C are $v = -L(di/dt)$ and $q = Cv$, or $i = C(dv/dt)$. In the RLC circuit shown below, the positive direction of the current, and the polarity of the voltages are shown. In the present case, $R = 0$, so

$$dv/dt = -L(d^2i/dt^2) = i/C, \text{ or } d^2i/dt^2 + (1/LC)i = 0,$$

which is the familiar equation of motion of a vibrating system.

The natural angular frequency is $\omega = (LC)^{-1/2}$. One analogy between mechanical and electrical systems is i: dx/dt, v: $-F$, L: M, C: $1/K$. Another is i: F, v: dx/dt, L: $1/K$, C: M. In either case, one can imagine electrical and mechanical systems that will behave identically. All our results for forced vibrations and impulsive excitation still hold for electric circuits.

So far we have not considered the forces that bring a vibrating system to rest in the absence of continuous excitation. In an electric circuit, resistance provides such a force, and it is easily handled mathematically.

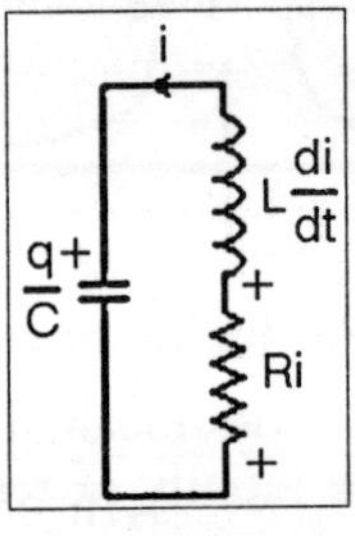

Fig. RLC Circuit

If a capacitor C, an inductance L and a resistance R are connected in series, the equation of motion is $d^2i/dt^2 + (R/L)di/dt + (1/LC)i = 0$, found by summing the voltages around the circuit and equating the sum to zero, then differentiating with respect to t.

If a solution e^{at} is assumed, and substituted in the equation, we find that it is a solution provided the equation $a^2 + (R/L)a + (1/LC) = 0$ is satisfied. This means that $a = -(R/2L) + [(R/L)^2 - 1/LC]^{1/2}$, or $a = -(R/2L) + j\omega[1 - 1/Q^2]^{1/2}$, where $\omega = (1/LC)^{1/2}$ and $Q = (L/C)^{1/2}/R$.

If $Q >> 1$, the resonant frequency is little changed, but there is an exponential decay $e^{-Rt/2L}$, or an energy decay time constant (for the square of the current) $\tau = L/R = Q/\omega$. Q can also be expressed as $\omega L/R$, a more familiar form. Every inductance L has some R associated with it (which includes magnetic losses as well as resistance of the wire).

As Q becomes smaller, the decay becomes more rapid, and the frequency of vibration decreases. Even at $Q = 2$, the frequency is still 87% of its value without damping.

However, at $Q = 1$, the vibration ceases to be oscillatory, and the motion last only a fraction of the natural period. This case is called *critical damping*. In the *RLC* circuit, it occurs when $R = (L/C)^{1/2}$. If R is proportional to L, halving L will increase Q by the square root of 2.

The presence of a little dissipation makes forced motion easier to understand. Let's introduce some dissipation, and write the equation of motion as $q'' + 2rq' + \omega^2 q = De^{j\omega' t}$ (primes on the q's indicate time derivatives). The forcing term is expressed as an exponential, to make it easier to handle phase shifts, since they will now not simply be π or $-\pi$.

Assuming a solution $Ce^{j\omega' t}$, we find $C = D/[(\omega^2 - \omega'^2) + 2jr\omega']$. The first thing to do is to express this amplitude in polar form, so we can see the magnitude and phase directly.

The result is $C = De^{-j\varepsilon}/[(\omega^2 - \omega'^2)^2 + 4(r\omega')^2]^{1/2}$. The phase constant $\varepsilon = \tan^{-1}[2r\omega'/(\omega^2 - \omega'^2)]$. We recover our earlier results when $r = 0$, so we are on the right track.

Exactly at resonance, $\omega' = \omega$, $C = e^{-j\varepsilon}D/2r\omega$ with $\varepsilon = \pi/2$. The amplitude is finite, and the phase shift is $\pi/2$. This means the external force is in phase with the velocity, so it does the maximum work on the oscillator. The phase shift increases continuously from 0 to π as ω' goes from 0 to infinity, with most of the change close to resonance.

Near resonance, the expressions can be thrown into a more symmetrical approximate form. We write $\omega' - \omega = \Delta\omega$, $\omega' + \omega = 2\omega$, and elsewhere replace ω' by ω. The result is: $C = e^{-j\varepsilon}(Dr/\omega)/(2r^2 + \Delta\omega^2)$, with $\varepsilon = \tan^{-1}(r/\Delta\omega)$. The symmetrical response curve we have met before; it is called Lorentzian, and its Fouier transform is the exponential.

The amplitude of oscillation decreases exponentially in the time domain, which is a consequence of the Lorenzian response in the frequency domain.

LORE OF WAVES

Suppose that a function of space and time $\psi(x, y, z, t)$ satisfies the partial differential equation

$$\frac{\partial^2\psi}{\partial x^2} + \frac{\partial^2\psi}{\partial y^2} + \frac{\partial^2\psi}{\partial z^2} = \frac{1}{v^2}\frac{\partial^2\psi}{\partial t^2}$$

where v is a constant with the dimensions of a velocity. The requirement that v be a constant is a very strict one, implying that the medium of propagation is homogeneous and isotropic (same in all directions). Then direct substitution shows that an arbitrary function $\psi = f(n \cdot r \pm vt)$ satisfies the equation, where n is a unit vector pointing in any direction and r is a position vector from an arbitrary origin. These solutions are waves of arbitrary shape whose surfaces of equal phase are planes, moving in any direction at speed v. Moreover, any linear combination of solutions

$$\psi = \sum c_i \psi_i (n_i \cdot r - vt)$$

is also a solution, since the equation is linear in ψ. The solutions are said to be *superimposed*. From the obvious properties of these solutions, they are called *wave functions*, and the partial differential equation is called the *simple wave equation*. Do not be misled into thinking that *all* waves behave in this manner, or even that *many* do.

In fact, most physical waves, such as those of sound in air or waves on the surface of water, or an electromagnetic wave moving along a transmission line, or light propagating down a fibre, are not precisely described by an equation of this type.

The great virtue of the wave equation is that it well describes the behaviour of a wide variety of waves whenever the *phase velocity* v depends very little on the shape of the wave. This is quite accurately true for electromagnetic waves, including light, propagating in vacuum, and very close to true for sound waves of low intensity propagating in a homogeneous isotropic medium, for example, which explains the great utility of the simple wave equation.

The expression on the left of the simple wave equation is called the Laplacian, and expresses in some way the curvature or strain or departure from equilibrium of a medium whose displacement in some sense is ψ. If ψ gives the shape of an elastic membrane, then in equilibrium its Laplacian is zero. If ψ is the electrostatic potential in a region, then its Laplacian is zero.

The expression on the right is in some way an acceleration, or mass times acceleration, so that the wave equation resembles Newton's second equation of motion. Any true physical interpretation depends on the type of wave under consideration, but it will involve the balancing of inertial and strain forces whose constants will appear in the expression for the phase velocity v. For mechanical waves, these will be the density and some elastic constant.

One unacquainted with the methods of science may well ask why recognising that a wave is described by some mathematical equation is so important. Isn't it just one of many properties of a wave, and not a very informative one at that? By no means. In the language of geometry, it is a *postulate*, something not proved but assumed as the basis for argument. Euclid's postulates defined Euclidean space, in which all his later results lived.

By a slight change to one postulate, we get geometry on the surface of a sphere, which differs in some respects (two parallel lines meet at two points instead of none, for example).

The verification that space is actually Euclidean, or that a wave satisfies the wave equation, follows from experience, not proof. Starting from the postulate of the wave equation, a great number of mathematical results can be obtained by logical proof, and we know all are valid provided the postulate is true. By recognizing that a wave satisfies the wave equation, we immediately acquire a gigantic kit of tools and properties established by mathematical analysis over the past several centuries.

Many properties of waves satisfying the wave equation will be familiar from different branches of Physics. Huyghens' Principle is one, that a later wavefront is the envelope of wavelets from a present wavefront. Fermat's Principle, that the optical path length is a minimum, is another. We have written the wave equation in rectangular coordinates, which gives the plane-wave solutions naturally. These solutions have the property that the amplitude of the wave does not change as the wave moves. We could also use other coordinate systems.

Spherical coordinates give waves expanding from a centre, or contracting on a centre, whose amplitude varies inversely as the distance from the centre. Cylindrical coordinates give waves expanding from a line or contracting on a line, whose amplitude varies inversely as the square root of the distance from the line. In both cases there can be different angular dependences, and the various solutions can be superimposed to satisfy a variety of conditions.

The wave equation can also be generalized to the case of a velocity that depends on position or propagation direction, $v = v(x,y,z,n_x,n_y,n_z)$. This introduces very great mathematical complexity (making the equation insoluble in terms of elementary functions), but has the expected results. Huyghens' and Fermat's principles still apply in many cases, as Huyghens' in double refraction (non-isotropic medium) and Fermat's in inhomogeneous media.

A further, and extremely useful generalization is to consider wave functions that are periodic functions of time, $\psi = g(x,y,z)e^{i\omega t}$, where $\omega = 2\pi f$ is the *angular frequency* in radians per second, and f is the frequency in Hz. Now we can do the time derivative and reduce the equation to $\nabla^2 g = -\omega^2 g/v^2 = -k^2 g$. Here, ∇^2 stands for the space derivatives (the Laplacian) and $k = \omega/v = 2\pi f/v = 2\pi/\lambda$, where λ is the *wavelength*, defined by the equation $v = \lambda f$.

The *amplitude* g then satisfies the equation $(\nabla^2 + k^2)g = 0$, called the Helmholtz wave equation. The Helmholtz equation is much easier to solve than the wave equation with the time.

For a plane wave, a solution to this equation is $g = e^{-ik\mathbf{n}\bullet\mathbf{r}}$ and we recover the same dependence on space and time as in a solution to the simple wave equation, but now the function is no longer arbitrary, but must be complex exponential, or *sinusoidal*.

Such functions are also called *harmonic*. This is really not a great restriction, because Fourier's Theorem allows us to expand an arbitrary function as a linear combination of harmonic functions. If v is independent of frequency, we recover our previous results completely.

However, a very great flexibility has been introduced. Let us suppose that the phase velocity v is a function of the frequency. Now, at any given time, let us Fourier analyse a certain wave in terms of its sinusoidal or harmonic components. If the phase velocity were a constant, this superposition would describe the same wave at all later times.

If the phase velocity depends on the frequency, the different frequencies move at different speeds, and their relation to one another changes. This can greatly change the shape of the wave as it moves along. In optical media, the *index of refraction* is the ratio of the speed of light in vacuum to the speed of light in the medium, $n = c/v$. It varies with the frequency of light, usually becoming larger as the wavelength decreases.

This causes refracted rays of different frequencies to follow different angles. White light is analysed into its frequency components, which are spread out into a *spectrum*.

For this reason, the phenomenon is called *dispersion*. Waves on the surface of deep water are also strongly dispersive, the phase velocity being proportional to the square root of the wavelength. The ocean performs a Fourier analysis of the disturbance due to a storm, the long waves arriving first, the short waves later, at great distances. Another example is the propagation of electromagnetic waves on a transmission line, such as a coaxial cable or optical fibre.

For any dispersive medium, the dependence of the phase velocity on the frequency is most conveniently given by a *dispersion relation* $\omega = \omega(k)$, which can be plotted with w and k along rectangular axes. The phase velocity is $v = \omega/k$. The slope of the curve, $u = d\omega/dk$, gives the speed of movement of a wave pulse consisting of a small range of frequencies, such as a brief burst of monochromatic light, and is called the *group velocity*.

There are waves for which the speed of propagation depends on the amplitude of the wave.

One example is the propagation of strong sound waves in air. Whatever the initial shape of the wave, small variations tend to move forward when the pressure is higher than normal, and backward when the pressure is less than normal. The front and back of the wave become steep and eventually become shock waves, with a shape like the letter N.

The speed of waves on water depends on their height when the water is shallow, so they steepen when moving onshore and break, all their energy dissipated in throwing turbulent water forward (and destroying any rock or structures in their way).

Such waves obviously do not satisfy the principle of superposition, and cannot be described by the simple wave equation. They are called *nonlinear* waves, because their wave equation is not linear in the amplitude, and they may have very peculiar properties. In some cases, they exhibit *solitons*, which are pulse wave solutions that do not spread out indefinitely, but retain their shape and interact with one another in complicated ways.

A *monochromatic* wave is one whose time dependence is described by $e^{i\omega t} = \sin \omega t + i \cos \omega t$. The use of complex numbers here may seem strange, but it is much easier than using sines and cosines individually, because the exponentials multiply and divide using the laws of exponents. We used it to derive Helmholtz's equation simply for convenience.

We could have used the sine or cosine equally well, or even the sine with a phase constant δ, $\sin(\omega t + \delta)$. But there is another reason to choose the exponential. Suppose we have to find $\sin \omega t + 2 \cos(\omega t + 45°)$.

The sine by $e^{i\omega t}$, and the cosine by $2e^{i135°} e^{i\omega t}$, and their sum by $(1 + 2e^{i135°}) e^{i\omega t}$. This is $(2.4142 - i1.4142) e^{i\omega t} = 2.7979e^{-i30.36°} e^{i\omega t}$. The real part of this, or $2.7979 \sin(\omega t - 30.36°)$. The rectangular-polar conversion facility of your pocket calculator is convenient for this calculation.

If you go ahead and find this trigonometrically, you will deeply appreciate the wonderful effects of complicating things with complex numbers. All we have to do in this case is really add the two vectors (1,0) and (1.4142 - i1.4142), and forget about the factor $e^{i\omega t}$, which goes through with no change or effect.

The vectors we have just used are called *phasors* or *amplitudes*, and expressing them as complex numbers is a way to include both magnitude and phase in the same quantity.

We use bold-face letters to stand for them, as $\mathbf{A} = Ae^{i\delta} = A_r + iA_i$, where the equivalent polar and rectangular forms are shown. Scientific pocket calculators have special functions to convert from one form to the other quickly. They are used everywhere in alternating current calculations, and in quantum mechanics. The factor $e^{i\omega t}$ is imagined as a rotation with angular velocity ω in the 2-dimensional complex phasor space, and the actual physical quantites are the projections on one of the axes or the other as this rotation takes place. Our calculations, however, never include it.

Phasors are added and subtracted, and multiplied or divided by complex numbers, but are never multiplied or divided themselves, since $e^{i\omega t} \times e^{i\omega t} = e^{i2\omega t}$, and this introduces things at double frequencies. There is one very common case, however, in which we want to multiply two phasors, and that is in the calculation of the *intensity* of the wave, because this is proportional to the square of the amplitude of the wave.

To see what's involved, consider a wave A sin $(\omega t + \delta)$. Its square is $A^2\sin^2(\omega t + \delta) = (A^2/2)[1 + \cos(2\omega t + 2\delta)]$, which consists of a constant term $A^2/2$, and a double-frequency term.

In the long run, the double-frequency term is as often positive as negative, so the time average intensity is just $A^2/2$. How can we get this from the phasor $A = Ae^{i\delta}$? Well, AA*/2 will do the job nicely, where A* is the complex conjugate of A (change sign of imaginary part or negate the angle). So we simply multiply the phasor by its complex conjugate, and divide by 2.

There is more, however. In electrical circuits, the power is the product of the voltage and the current, p = vi. Suppose $v = V \sin \omega t$ and $i = I \sin(\omega t + \delta)$. Then $p = VI \sin \omega t \sin(\omega t + \delta) = VI [\sin^{2\omega} t \cos \delta + \sin \omega t \cos \omega t \sin \delta)$. Double-frequency terms again appear, and all average to zero. What is left is $P = \langle p \rangle = (VI/2) \cos \delta$.

In AC circuits, small letters stand for instantaneous values, and capital letters for amplitudes. Now, how can we get this from the phasors? The phasors are $V e^{i\omega t}$ and $I e^{i(\omega t + \delta)}$, and if we do what we did before, we get $(VI/2)e^{-i\delta.}$ The real part of this is $(VI/2) \cos \delta$, which is what we wanted. Our formula for intensity is now P = Re(AB*)/2 = Re(A*B)/2.

You can see why the second formula gives the same result, so it does not matter which phasor is conjugated. This formula is often required.

Incidentally, the reason that j is used instead of i for $\sqrt{-1}$ in electrical engineering is the possibility of confusion with the symbol i for current. Also in engineering, the magnitude of a phasor usually stands for $V/\sqrt{2}$ instead of for V itself, so the power is simply P = Re(VI*). That is, phasors are *rms* (root-mean-square) values, which are commonly used in specifying voltages and currents. That is, 240V means an amplitude, or maximum value, of 339V.

Consider the wave $Ae^{i(-kx + \omega t)}$. This sinusoidal wave has amplitude A, and moves to the right (in the direction of the x-axis) with phase velocity x/t = ω/k. This is the speed of any point of constant phase on the wave. The wave $Ae^{i(kx+\omega t)}$ moves to the left at the same speed. The corresponding phasors are $A_+ = Ae^{-ikx}$ and $A_- = Ae^{ikx}$.

A travelling wave is associated with a phase that varies linearly with distance, and an amplitude that does not. If these two waves are superimposed, we get $A = A(e^{ikx} + e^{-ikx}) = 2A \cos kx$. The phase of this wave does not change at all, but the amplitude is a sinusoidal function of distance! This is called a *standing wave*. The points at which the amplitude is greatest are called the *loops*, and the smallest *nodes*.

Now let us superimpose two waves travelling in the same direction, one with wave number $k - \Delta k/2$ and frequency $\omega - \Delta \omega/2$, the other with wave number $k + \Delta k/2$ and frequency $\omega + \Delta \omega/2$. Let both waves have the same amplitude, unity. Their superposition gives $e^{i(-kx + \omega t)}[e^{i(\Delta kx/2 - \Delta \omega t/2)} + e^{i(-\Delta kx/2 + \Delta \omega t/2)}] = 2 \cos(\Delta kx/2 - \Delta \omega t/2)e^{i(-kx + \omega t)}$. This is a wave with the average k and ω of the two waves travelling to the right, but with an amplitude that varies cosinusoidally between +2 and -2, and is itself a wave travelling with the velocity Δ k/Δ ω, which we recognize as the group velocity (when the difference Δ ω is small).

The waves are said to be *beating* at the angular frequency Δ ω/2 at a fixed position x. When the waves are at exactly the same frequency, the beats disappear, and the situation is called *zero beat*. This is a powerful method of tuning two frequencies to be the same when the ear is used as the detector. In radiotelegraphy, a variable beat frequency oscillator's output is mixed with the received signal and adjusted to give a beat frequency that is in the audio range and comfortable to the ear.

Our analysis of waves so far has said nothing about the sources of the waves, but simply describes waves that have somehow come into being. The wave equation without a source term is called the *homogeneous wave equation*. The *inhomogeneous wave equation* contains an additional term that represents the radiation of waves from a source.

For mechanical waves it may be the movement of a physical boundary, or the creation or annihilation of fluid at some point, or varying applied forces. The source of electromagnetic waves is an accelerated electric charge. When a source radiates, it sends energy into the surrounding medium, and so must exert a reaction on the source, the *radiation reaction*, that accounts for the energy transfer.

These questions are much more difficult than those concerned merely with the propagation of waves that are already there, but at the same time extremely valuable.

We have not had to say anything about the nature of the wave function, beyond the fact that knowing it allows all properties of the wave to be determined. A wave dependent on a single parameter is called a *scalar wave*. There are more complicated waves in which more than one quantity must be known. Each of these quantities may separately satisfy the wave equation, but there are also relations between the several quantities.

A sound wave is determined when the difference of the pressure from the equilibrium pressure in the medium, the *overpressure*, is known. Displacements of the medium are also involved, but these are always in the direction of propagation, normal to the wave front, and determined by the overpressure. The sound wave is a *longitudinal* wave.

An electromagnetic wave involves vector electric and magnetic fields. If the electric field is known, the magnetic field can be found, so in general three quantities are required, and we have a *vector* wave. There is a condition that the electric field must be in the wave front, or *transverse*, so there are really only two independent wave functions.

An electromagnetic wave is a *transverse* wave. The two components are the two polarizations of the wave, which may be specified in a number of ways. For electromagnetic waves in which only one polarization state exists, or the two states are mixed equally, as in white light, and for phenomena that are nearly independent of polarization, such as diffraction, scalar waves give an adequate description.

Let us again look at the superposition of two waves. If the two waves are not of the same frequency, the resultant amplitude will vary rapidly at any point, the waves being now in phase, now out of phase. If they are at nearly the same frequency, the amplitude variation will have the nature of beats. With waves of the same frequency, the resultant amplitude will be constant at any point. Representing the waves by phasors, the resultant amplitude is easily written down using the Law of Cosines for the vector addition:

$$a^2 = a_1^2 + a_2^2 - 2a_1a_2 \cos d$$

The intensities are proportional to the squares of the amplitudes, so this expression can be used at once.

It says that the resultant intensity of two superimposed waves is equal to the sum of the intensities of the two individual waves, plus an *interference term* that can be positive or negative, and is proportional to the cosine of the phase difference between the two waves.

If the waves are of equal amplitude, the intensity varies from four times the intensity of one wave alone, to zero. The phase difference is equal to $k(s_1 - s_2)$, where s_1 and s_2 are the optical path lengths from the two sources to the point under consideration.

As one moves about, then, there should be great variations in resultant intensity, called *interference fringes*. The existence of interference fringes is conclusive evidence that waves are being observed.

In ordinary life we know that such fringes are not observed with light, though obvious in the surface waves on water. Their absence made a particle model of light quite plausible. It was not until the delicate diffraction phenomena were explained by Thomas Young as the result of interference that it became clear that light was, in fact, a wave.

When we say 'wave' here, we mean that amplitudes and phases are involved, which is the case in both classical wave theory and quantum mechanics (which, indeed, is also called *wave mechanics*). It is important to explain why interference phenomena are not commonly observed, and why special arrangements are required to see them at all.

The first reason is the extremely high frequency of light, in the area of 10^{14} Hz. This means that we can only observe average intensities, not the amplitudes, and that the patterns must be constant in time. Secondly, light is emitted by charges moving on an atomic scale, in individual random bursts that last no longer than about 10^{-8} s, so that the corresponding wave is no longer than a metre or so in the best case. The wave bursts from two separate sources can never have constant phase relationships. We have only a chaos of overlapping, constantly changing, interference fringes that average out to uniformity, making the resultant intensity equal to the sum of the individual intensities. Finally, most light has a broad spectrum of frequencies.

Indeed, this is the requirement for white light. Even if the interference from each frequently separately were stable, the fringes would be of different scales and would overlap. The only exception would be near where the path difference for the two waves was zero, which would be the same for any colour. Considering the nature of ordinary light, it is no wonder that interference is not observed.

The laser is a different case. Here, the atomic radiators are made to radiate in step, and to stay in step for a reasonable length of time, to a very good approximation. It has been possible to observe interference between light coming from separate lasers, including fringes and beats. Lasers are much more like the light sources one imagines when studying optics. The light from other kinds of sources, even when made nearly monochromatic by a spectroscope or filter, is at best only narrow-band noise. Two light beams that can make interference fringes when superimposed are called *coherent*.

Coherence is limited both in time and space. For example, consider a wave train from some source that oscillates smoothly and predictably, except for arbitrary sudden changes of phase an average time t apart. Two points on the wave will be coherent at any fixed time if they are less than a distance ct apart (on average), and coherent at one point for time intervals of t or less. These numbers are the *coherence length* and the *coherence time*.

The search for interference in common light is not hopeless. A first step is to reduce the bandwidth of the light by passing it through a filter or spectrometer, or by using a nearly monochromatic souce like a sodium lamp or a discharge tube. This makes possible a coherence length of at least a few wavelengths. The second step is to divide each individual wave in the chaotic mixture into two coherent parts, send these parts over different paths, and recombine them later. Now the fringes for each wavelet will register, and may be visible. It is often necessary to restrict the size of the source by using a pinhole or slit. Before lasers, this was the only way to observe interference, and even with lasers it is still a very good idea, since lasers make adjustment and observation very much easier.

The parts can be divided by taking different areas of the wavefront (*wavefront-splitting*) or else by partial reflection (*amplitude-splitting*). Each setup is analysed the same way, by evaluating the phase difference and using the simple equation above. The most usual way of producing interference is to make two virtual images of the same source, and letting the waves from the two images interfere. This can be done by refraction (using a prism with two faces slightly inclined to each other), or reflection (using two mirrors inclined slightly to each other).

Both of these methods carry the name of Fresnel (biprism and mirrors, respectively). An interesting variation of the latter method is to use only one mirror, and let the virtual image interfere with the actual source (Lloyd's mirror). This cannot be done with the prism, since one wave will pass through the glass, the other through air, and the path difference will be excessive (more than the coherence length).

These methods are all wavefront-splitting, and the source is usually a fine slit. The slit must be parallel to the fringes, and helps to get a much higher intensity than would be obtained from a pinhole. Suitable mirrors are optical flats at grazing incidence. The Fresnel biprism and mirrors have a bright fringe at the centre of the pattern (zero path difference), but the Lloyd's Mirror has a dark fringe, showing that there is a phase change of half a wavelength on reflection from rare to dense. Both Fresnel's Biprism and Lloyd's Mirror are easy to set up and straightforward to analyse. The Fresnel Mirrors require a little more work.

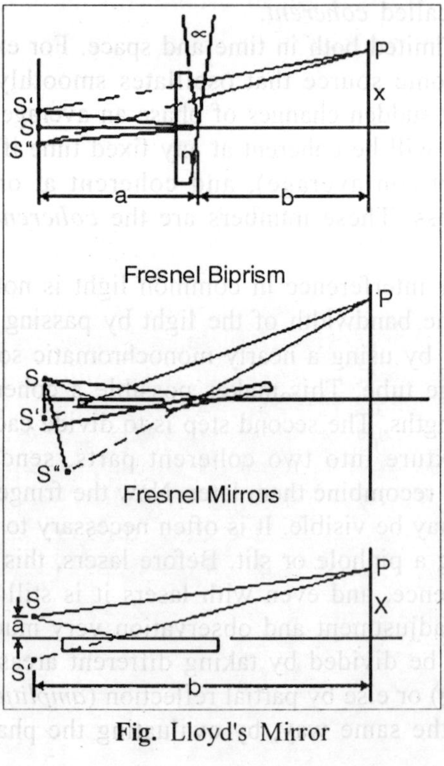

Fig. Lloyd's Mirror

The diagram shows these arrangments, and gives formulas for the fringe spacing Δx. For the biprism, if a = 100 mm, b = 900 mm, n = 1.5 and α = 1°, the fringe spacing is 0.57 mm at 500 nm wavelength. For the Lloyd's mirror, if a = 0.5 mm and b = 1000 mm, the fringe spacing is 0.5 mm for 500 nm wavelength. The fringe spacing is measured with a micrometer eyepiece. This is not a precise way to measure wavelengths, but is a very instructive and satisfying experiment.

Every physics text presents another arrangement called Young's Experiment. Young did not actually carry it out in this way, because the resulting intensities are so feeble, but it can be explained very clearly. A slit in front of a source makes a cylindrical wavefront, which falls on two closely-spaced slits in a second screen.

The waves diffracted through these slits fall on a screen, where they are superimposed and interfere. The double passage of slits and the dependence on diffraction from narrow slits render the intensity very feeble. The experiment can in principle be done with pinholes, but conditions are even worse. Young used a wire obstacle instead of slits.

The light diffracted at top and bottom of the wire appears to come from line sources, just as if it came from two slits, but much brighter. In this way he first measured the wavelength of light, showing that it went from about 400 nm to 700 nm in going from blue to red. The fringe spacing is given by essentially the same formula as for Lloyd's Mirror.

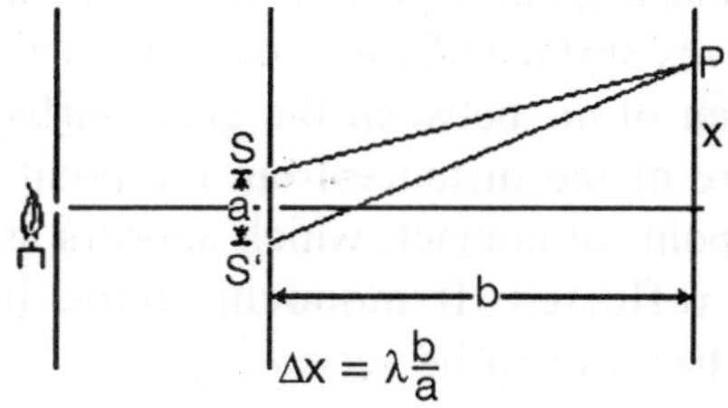

Fig. 'Young's Experiment'

Note that all the fringes mentioned above can actually be projected on a screen. Another kind of interference arises from a thin transparent film, for example a layer of oxidized oil on the surface of water. Fresh oil collects in droplets, but old oil dropped by motorcars will spread over the surface, gradually spreading and becoming thinner.

When we look at any part of the film, we see the sky reflected in both the top and bottom interfaces of the oil layer, giving two coherent waves of roughly equal intensity but with a path length difference depending on the thickness of the oil. The light that passes through is absorbed by the bottom of the pool, if we are lucky, so it does not drown out the delicate reflected light.

The two coherent waves are here produced by amplitude splitting, and are focused by our eyes on the retina, where they are superimposed and interfere. Such fringes cannot be projected on a screen.

They are said to be localized in the film (but actually are produced on the retina). The colours we see are the result of the removal of the complementary colours by interference.

Their pattern reveals the inequality in thickness of the film. In an ideal case, the film is thickest where the original drop fell and thinnest at the edges of the spot.

The colours change with time as the film thins. The thinnest films disappear, because only the top reflection involves a half-wavelength phase change (the refractive index of the oil is greater than that of both the air and the water), and all wavelengths are removed.

The film becomes 'black.' These brightly-coloured spots are really the most obvious evidence of interference. Perhaps they did not become at all common until the motor age. Remarkably, they do not demand a localized monochromatic source.

At the thin edge of the spot, the blue end of the spectrum will first not be destroyed by interference, and a greenish colour will be seen. When green is removed, the remainder will be purplish, perhaps magenta. Removal of red will leave a bluish green or cyan. A still thicker film might remove both blue and red, leaving a green. These colours have been extensively studied, and Minnaert gives a list of them.

If you set the convex surface of a lens of long focal length on an optical flat, there is a thin film of air between the glass surfaces whose thickness increases as the square of the distance from the point of contact. Colours can be seen near the point of contact, which appears as a small black disc, when white light is reflected. If monochromatic light is used, many concentric bull's-eye fringes can be seen.

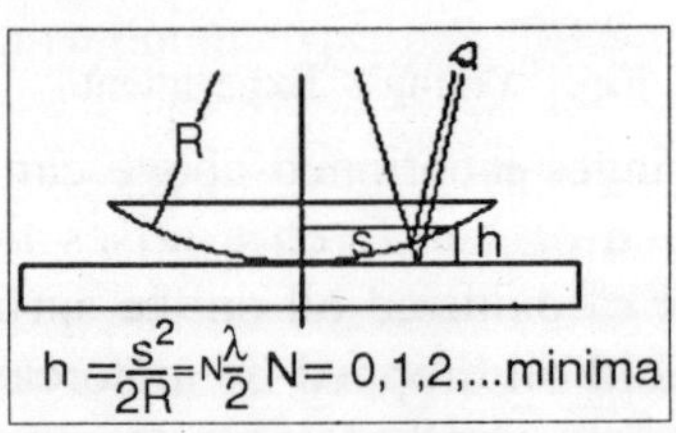

Fig. Newton's Rings

These are Newton's Rings, described by him in the Opticks, and whose explanation caused him considerable difficulty. In fact, he did not get it right, since his model of light was inadequate.

He had to assume that the light particles were alternately reflectible and refractable as they travelled, lending a kind of 'wavelength,' but this was grasping at straws, and he knew it. The blackness of the central disc again shows the relative phase shift of half a wavelength on reflection. Like the colours of thin films on water, this manifestation of interference is easily seen, but was never noticed as anything unusual.

The same principle is used in the Michelson interferometer, in which the two beams are completely separated and can be subjected to any desired treatments.

The thin film here is the space between two virtual images of the source. This space can be varied at will, and made wedge-shaped or parallel, so a variety of fringes can be produced.

When the images are parallel, the fringes are circular, expanding from or contracting on the centre when the spacing is changed. When the film is angular, the fringes are straight lines (approximately) that move right or left as the spacing is changed.

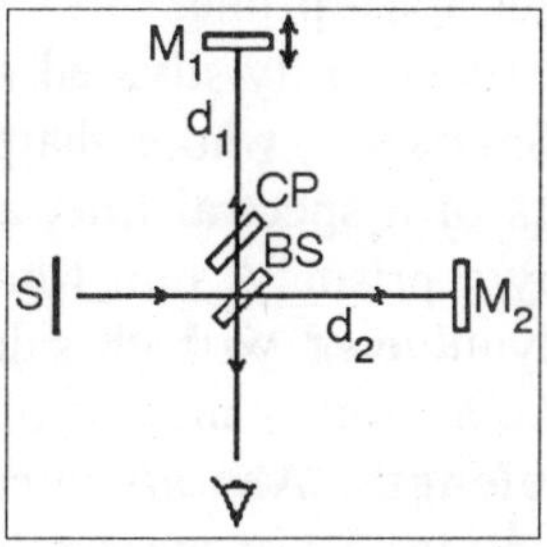

Fig. Michelson Interferometer

The colours that are produced in white light are the same as those made by oil films, but now the thickness of the film can be changed at will. The Michelson interferometer is suitable for accurate measurement of distance or wavelength.

It was originally intended for detecting ether drift (since one arm could be in the direction of motion through the ether, and the other normal to it), and proved that there was no ether drift, consistent with relativity. Adjustment of the interferometer is straightforward, but not easy. In the diagram, BS is the beam splitter, and CP is the compensating plate, exactly the same as the beam splitter except not silvered. It is used so that both paths pass through the same thickness of glass.

When only two waves interfere, the intensity in the fringes varies sinusoidally, and the dark and light intervals seem equally wide. If the sides of the thin film are made more reflective than what arises simply from the difference of refractive indexes, then many more beams are produced, with equal phase delays.

Now if N beams are in phase, the resultant intensity is proportional to N^2, which can be very big.

A very small phase shift between successive beams (imagine the phasors rotated by a small amount) can cause the vector sum to be zero, the phasors forming a polygon of N sides, so we have darkness very close to the bright fringe. It is easy to see that the fringes have now become quite sharp, the sharper the more reflective the surfaces are made, separated by areas of weak intensity.

Sketch vector diagrams of the phasors for various phase delays to see this clearly. It is much easier to measure the location of these sharp fringes, so they are preferred for accurate measurement. (The only way to achieve accuracy with a Michelson interferometer is to count hundreds or thousands of fringes, which can be tedious unless it is done by machine.)

There are good examples of multiple-beam interference. The *diffraction grating* consists of a large number of parallel scratches, each one of which makes a wavelet that can interfere with the others. When all the phasors line up, we get a sharp line.

Two parallel optical flats, partly-silvered on the inner sides, make a *Fabry-Perot* ıtalon (interferometer), whose sharp fringes can even separate the hyperfine components of a spectral line, after a preliminary sorting out of the wavelengths by a prism. A thin film whose thickness is to be investigated can be lightly silvered on both sides.

Then the fringes make a contour map of its thickness, giving answers to a fraction of a wavelength. An *interference filter* consists of a semitransparent metal coating, a transparent layer around a wavelength thick, and another semitransparent metal coating on a glass support. This is like a Fabry-Perot, with sharp pass bands widely spaced. All but the one wanted are eliminated with wide-band absorption filters. The result is a filter that passes only a very small wavelength interval, producing nearly monochromatic light.

Suppose we coat glass with a thin transparent film about 125 nm thick that has an index less than that of the glass.

The light reflected normally from front and back of the film will be a half-wavelength out of phase, and so will interfere destructively to reduce the total light scattered. Note that there is no extra half-wavelength in this case, since both reflections are rare to dense.

The cancellation will be exact only for light of wavelength 500 nm (green), but will be pretty good over the visual spectrum. If we can find a material whose index is the square root of the index of the glass, then the two reflected waves will be equal in amplitude, and the cancellation will be total: the glass will not reflect visible light, but transmit it all.

MgF_2 has an index of 1.38, which is a bit too large (1.22 is what is needed), but the film is hard and durable, and gives a considerable reduction. A lens with this coating appears a faint magenta in reflected light. Double and triple layers, using materials of higher indexes than glass as well as the low-index ones, can give more exact results. Antireflection coatings do not merely reduce light loss; reflected light in a multielement system can wind up where it should not, producing glare.

Finely and regularly layered transparent media can produce interesting effects due to interference of the many beams produced. Reflection and transmission can be strongly wavelength-dependent, producing bright colours. High reflectance is also possible, and multilayer coatings are used to make excellent mirrors, much better than metal ones.

Bird feathers, butterfly wings, and beetles may acquire their colours in this way. The play of colours in opals and feldspars, and the bronzy reflectivity of biotite mica have a similar source. White-light holograms also make use of layered media.

The waves of wave mechanics do not obey the simple wave equation, but a similar one, the Schrodinger equation, which has the first derivative with respect to time on the right-hand side instead of the second. Nevertheless, the wave function is an amplitude with phase, its average square has a physical meaning (probability density) and is subject to interference and diffraction. In fact, waves are only one aspect of quantum mechanics. The nature of the interaction of light with matter is deeply modified, and the well-known quanta, photons, are introduced.

These are definitely not Newton's light particles, since they have many strange properties not shared by material bodies. On the other hand, things like electrons are found to behave like waves in that they can be diffracted, and interfere, even making fringes like light. The lore of waves is an essential part of the universe.

Chapter 8

Sound Waves

Sound is a vibratory motion of the air. In our ears, it gives the sensation of hearing. It is excited by vibrating bodies and transmitted to our ears without sensible motions of the air.

These things have been known since ancient times, but now we can enquire more deeply into what is happening. How fast and how far do the particles of the air move in transmitting the wave? What are the pressure, density and temperature variations in the wave? How much energy is transmitted by the wave? The answers to this question depend on a dynamical theory of sound, which was first worked out by Newton.

PLANE WAVES

To make the problem as simple as possible, without losing any of the important details, let us consider a *plane wave* of sound. By a plane wave, we mean one in which quantities vary only with the distance along a certain direction, and of course with the time.

We also restrict ourselves to sound in a fluid medium, which cannot support shear stresses. The wave motion must be longitudinal, along the direction we have chosen.

Let distance along this direction referred to an arbitrary origin be x, so that we can label each particle of the medium by its position x when at rest (we use the word particle in an abstract sense, as if the gas were a continuum without molecular structure, which is a satisfactory approximation in many cases). We further restrict ourselves to propagation in a gas, with a density ρ_0 and a pressure p_0 when at rest.

The wave can be completely described in terms of the displacement $\xi(x,t)$ of the particles originally at any coordinate x at any time t. The time derivative of the displacement is the particle velocity, and the space derivative of the displacement is the *dilatation* Δ, equal to $v - v_0/v_0$, the relative change in unit mass $v = 1/\rho$.

When this is very small compared to unity (as in a normal sound wave), the dilatation is the negative of the *condensation* $s = \Delta\rho/\rho$, the relative change in density. Nothing condenses; this is merely a convenient and conventional term. A *condensation* is part of a sound wave where the density is higher than normal, while in a *rarefaction* the density is less than normal. Higher density means higher pressure, and lower density lower pressure.

The force on a thin sheet of gas of thickness dx is the difference in the pressures on the two sides of the sheet.

This must equal the mass of the sheet times its acceleration, by Newton's Second Law. Since we know the difference in the condensation on the two sides of the sheet, which will be the second derivative of the displacement with respect to x, times the thickness dx, we will know the pressure difference if we know how condensation is related to pressure. In an ideal gas, the pressure is proportional to the density when the temperature is held constant, so in this case the fractional change in pressure will be equal to the fractional change in density, or the condensation.

If we let p denote the change in pressure, then $p = p_0 s$. The pressure variations in a sound wave are so rapid, however, that there is no time to reach isothermal conditions.

What occurs is an *adiabatic* process in which the pressure difference is greater due to the heating or cooling of the gas as it is compressed or expanded. In an adiabatic process, $p = \gamma p_0 s$ instead. γ is the ratio of the specific heat at constant volume to the specific heat at constant pressure. This was a problem that baffled Newton, when the calculated and measured speeds of sound did not agree, and was only settled later by Laplace.

PROPERTIES OF SOUND WAVES

Now we can write down Newton's Law, and it is seen to be the simple wave equation, whose properties and solutions are so well known. Just obtaining the wave equation says a great deal about sound waves, and lets us apply techniques developed for optics, such as Snell's Law and ray tracing. Under the conditions we have assumed, sound waves travel without dispersion at the phase velocity $c = \text{sqrt}(\gamma p_0/\rho_0)$, and obey the principle of superposition. The conditions for the validity of our analysis are that $s << 1$, $p << p_0$, $u << c$, that the gas obeys the ideal gas law, and that γ is constant. For air at 0 °C, p_0 = 1 014 000 dyne/cm^2, ρ_0 = 0.00129 gm/c^{3}m and $\gamma = 1.41$.

This gives c = 333 m/s or 1092 ft/s, agreeing well with measurements. Using the ideal gas law, we also find that $c = \text{sqrt}(\gamma RT/M)$, where R is the universal gas constant, and M is the molecular weight of the gas. In case the gas cannot be considered ideal, our analysis is still completely valid so long as we use the correct adiabatic compressibility of the gas instead of γp_0. In this case, c may not vary exactly as the square root of the absolute temperature.

A precise value of c = 331.46 m/s for dry air containing 0.03% CO_2 at 0 °C is given in tables, but the actual velocity is affected by vibrational relaxation in O_2, the presence of H_2O, and other small factors, so the precision is somewhat illusory.

At 0 °C the speed of sound in distilled water is 1482 m/s, in sea water of 3.5% salinity 1522 m/s, in mercury 1450 m/s, in steel rails 5050 m/s, and in bulk soft iron 5957 m/s, which should give some idea of the range of phase velocities. One source says the phase velocity in water *decreases* by 2.4 m/s per °C, though in most liquids it increases with temperature, as in a gas. Another source gives the phase velocity in water as $c = 1403 + 4.9t - 0.05t^2 + 0.16p$ m/s, where t is the Celsius temperature, and p is the pressure in bars (10^6 dy/cm^2. This gives an increase with temperature.

HARMONIC (SINUSOIDAL) WAVES

A harmonic plane wave travelling towards positive x has a displacement $\xi = A \exp[i(\omega t - kx)]$. If it is travelling in the other direction, simply change the sign of k. The phase velocity $c = \omega/k$, $\omega = 2\pi f$ and $k = 2\pi/\lambda$. Now the time derivative is the same as multiplication by $i\omega$ and the space derivative is the same as multiplication by -ik. Now all the quantities in the wave are algebraically related, as shown at the right.

Note that the overpressure and condensation are in phase, and lead the displacement by 90° for a wave travelling towards +x, but lag by 90° for a wave in the opposite direction. The particle velocity always leads the displacement by 90°. The power in the wave is the product of the overpressure p and the particle velocity u.

If these are expressed as phasors, then the average power is half the product of one phasor times the complex conjugate of the other, just as in alternating current circuits.

In fact, the ratio of the pressure (potential) to the particle velocity (current) is a quantity $r = \gamma p_0/c$ analogous to a wave resistance, and is called the *specific acoustic impedance*, or *wave impedance*. For air, r = 42.6 dyne-sec/cm^3, and for water, 14 800.0 dyne-sec/cm^3. The specific acoustic impedance is also the product of the density and the phase velocity. Its units can also be expressed as g/s-cm^2 in cgs, or kg/s-m^2 MKS. There has been some effort to name this unit the *rayl* in honour of Rayleigh, but this has apparently not met with much enthusiasm.

Harmonic Wave

$$\xi = Ae^{i(\omega t - kx)}$$

$$u = i\omega\xi$$

$$s = ik\xi$$

$$p = i\gamma p_0 k\xi$$

$$p = \frac{\gamma p_0}{c} u = ru$$

$$p = \frac{1}{2}\rho\bar{u} - \frac{1}{2}r|u|^2 - \frac{|p|^2}{2r}$$

To get an idea of the magnitudes involved, suppose a loudspeaker cone of 15 cm diameter is radiating 100 W of acoustic power at 1000 Hz. This will be a rather strong sound wave.

Now $P = 5.6 \times 10^6$ erg/s-cm^2, so u = 1629 cm/s is the peak particle velocity. The condensation amplitude is s = u/c = 0.05, approximately, which is indeed small compared to unity. The overpressure p = ru = 6940 dy/cm^2, and so $p/p_0 = 0.007$, which is satisfactorily small.

This is a very conservative estimate, since a normal overpressure is closer to 1 dy/cm^2. Note that we have not yet used the frequency, which we need only to find the particle displacement. At 1000 Hz, the wavelength is 33.3 cm, and the wave number k = 0.1887 cm^{-1}, so the particle displacement amplitude is.05/ 0.1887 = 0.265 cm. The ratio $\xi/\lambda = 0.008$, so the particle displacement is a very small fraction of a wavelength.

The sound pressure level (SPL) in dB is conventionally 20 log (p/2 × 10^{-5} N/m^2), where p is the rms value of the overpressure (peak value divided by the square root of 2). The reference pressure of 20 μPa corresponds to an intensity of 10^{-12} W/m^2, near the threshold of hearing.

VELOCITY POTENTIAL

The description of the motion by the particle displacement ξ as a function of time and the initial position of the particle is very convenient in one-dimensional problems. It is called the *Lagrangian picture* of fluid motion.

An alternative description that is much easier to work with for three-dimensional problems and for deeper theoretical analysis is the *Eulerian picture* in which the values at a fixed point as a function of time are considered. In this picture, the basic description is in terms of the vector velocity v(x,y,z,t), which is a vector field. The pressure and condensation will be scalar fields, ρ(x,y,z,t) and s(x,y,z,t).

The fundamental equations are Newton's law and the equation of continuity, which express the conservation of momentum and mass, respectively. They are most easily expressed in vector calculus notation.

Because of browser restrictions, we must use grad, div and curl instead of the del operator. Newton's law is ρDv/Dt = -grad p – ρg, where the last term, the gravitational force, stands for any volume force on the fluid, and D/Dt = √/∂t + v·grad is the rate of change operator moving with the fluid. We will omit the gravitational force, and any other constant force, which will have no effect on acoustic disturbances.

The equation of continuity is $\partial\rho/d\rho + \text{div}(\rho v) = 0$. These are the basic equations of hydrodynamics, and are derived in any fluid mechanics text.

In sound waves, p and ρ vary little from their steady values (by less than one part in a thousand), and v is much less than the phase velocity of sound. Under these conditions, the equations become very closely linear, and things are very much simpler.

For acoustics, the equation of motion is $\partial v/\partial t = -(\text{grad } p)/\rho$, and the equation of continuity is $\partial\rho/\partial t = -\rho \text{ div}(v)$ or $\partial s/\partial t = -\text{div}(v)$. In these equations, ρ is to be taken as a constant, its equilibrium value, and p is the pressure minus the equilibrium pressure.

We shall consider the medium to be an ideal fluid of zero viscosity. This has the important consequence that microscopic rotational motion cannot arise if it was initially absent (as is the case when a sound wave passes through the medium). Kelvin stated this in the form $(D/Dt)\int v\cdot ds$, which says that the circulation around any closed curve remains constant (at zero).

This is equivalent to the differential statement curl v = 0, which may be familiar from electrostatics. Just as in electrostatics, this means that we can define a function φ such that v = grad φ (there is no minus sign here, as in electrostatics, as a matter of convention). The function φ(x,y,z,t) is called the *vector potential*, and allows us to represent the three components of the vector velocity in terms of a single scalar function.

If we write the fundamental equations in terms of φ instead of v, we get $\text{grad}(\partial\varphi/\partial t) = -\text{grad}(p/\rho)$ and $\partial s/\partial t = -\text{div grad } \varphi$. The first equation is equivalent to $\partial\varphi/\partial t = -(c^2/\rho)s$, where we have used the compressibility relation $p = \rho c^2 s$. Now we can eliminate either φ or s between these two equations.

Let's eliminate s. The time derivative of the second equation is $\partial^2\varphi/\partial t^2 = -c^2\partial s/\partial t$. Using the first equation, we get $\partial^2\varphi/\partial t^2 = c^2 \text{div grad } \varphi$. Recalling that div grad is the sum of the second partials, we recognize this as the three-dimensional wave equation.

Let us suppose that φ is known, perhaps as the solution of the wave equation. The other variables are easily expressed in terms of it. First, v = grad φ, from the definition of φ.

Then, $p = -\rho(\partial\varphi/\partial t)$, and $s = -p/\rho c^2 = -c^2(\partial\varphi/\partial t)$. The space derivatives of φ give the velocity, while the time derivative gives the pressure and condensation. The boundary conditions on φ are that φ = 0 at an open end (p = 0), and that $\partial\varphi/\partial n = 0$ at a closed end (or rigid boundary). Distance normal to the boundary is represented by n ("normal").

For a harmonic plane wave travelling in the direction k, where |k| = k

= ω/c, time derivatives are equivalent to multiplication by $j\omega$, and grad is equivalent to multiplication by $-jk$. Then, $v = -jk\varphi$, $p = -j\omega\rho\varphi$ and $s = -j\omega c^2\varphi$. The wave impedance $p/v = \rho c$ is real, so p and v are in phase. The particle velocity v is in the direction of propagation (rather than opposite to it) for a positive condensation.

The set-up for the Eulerian picture is more elaborate than for the Langrangian picture, but once we get to the wave equation, everything falls out nicely, and we can treat more complicated problems.

The particle velocities will be opposite but equal in the condensations and rarefactions of a propagating wave. If the density were constant, as much air would travel forward in a condensation as would return in the rarefaction. This is the normal case often emphasized in wave motion, where the wave and its energy travel, but the medium does not.

However, the density is $\rho(1 + s)$, which is greater in the condensation and less in the rarefaction. Therefore, more air will move forward than returns, though the effect is proportional to sv, the product of small quantities. Strong waves will be associated with mass motion in the direction of propagation. A similar effect occurs for the wave velocity, as a result of the neglected terms in the substantial derivative, making strong waves travel faster than weak ones. These are typical finite-amplitude effects that are interesting to study, but of little practical interest in most cases.

ABSORPTION

One should distinguish between *attenuation*, which is a dimunition in intensity for any cause whatever, and *absorption*, in which energy in the sound wave is transformed into some other form, usually heat.

No simple mechanism for the absorption of sound energy by a perfect gas immediately suggests itself, and, in fact, sound does propagate with remarkably little absorption. It is attenuated mainly by spreading, scattering, and absorption by surfaces.

The finite viscosity, heat conduction, and molecular mean free path of the medium do give rise to an absorption that increases at higher frequencies. One interesting effect is that the temperature changes in the adiabatic processes may equilibrate with molecular internal degrees of freedom at low frequencies, while at higher frequencies equilibration does not occur.

The result is an effective change in γ, which not only changes the phase velocity, but also introduces an absorption because of the phase lag that is effective in the band of frequencies where the phase velocity is changing. The main other contribution to absorption is called viscothermal because it involves these transport properties of the gas, and increases as the square of the frequency. It is also called *classical* absorption. The distance at which the amplitude of a sound wave is diminished by a factor of 1/e by viscous absorption is $(3c/8\pi^2\nu)\lambda^2$, where ν is the kinematic viscosity, 0.132 cm^2/s for air. At 1000 Hz, this distance is more than 10 km. The classical intensity absorption coefficient for air is given in tables as $\alpha = 1.61 \times 10^{-10} f^2$ dB/m, where $\ln(I_0/I_d) = 2\alpha d$. At low pressures, absorption occurs when the wavelength becomes comparable to the molecular mean free path (about 66nm in air at STP).

Most of these effects are considerable only at frequencies well above the audible range or at very low pressures. If the mean free path is taken as inversely proportional to the pressure, then the mean free path becomes 1/10 of the wavelength for 1000 Hz at a pressure of about 1.3 μHg, a low vacuum.

The effect of humidity on sound propagation is small, but rather complicated. At 68°F, the phase velocity increases from about 1127 ft/s for dry air to about 1131 ft/s at 100% humidity, mainly due to the decrease in density.

The absorption at 100 Hz is 1.67 dB/km for dry air, 0.38 dB/km at 50% humidity, and 0.22 dB/km at 100% humidity. At 2000 Hz, the figures are 4.14, 7.14 and 6.29 dB/km. These figures are much larger than the classical absorption calculated from the equation in the preceding paragraph, and are probably due to vibrational and rotational relaxation in water vapour, oxygen, and carbon dioxide.

Although absorption in gases is well accounted for by viscothermal and relaxation effects (observed absorption only slightly higher than predicted), absorption in some liquids, such as water or alcohols, is much higher than would be expected on these grounds. The exess absorption can be explained as due to a structural relaxation, a change in the molecular arrangement, during the passage of the wave.

Fine wires and threads offer little resistance to the passage of sound. Tyndall found that a piece of felt half an inch thick stopped sound less well than a wet pocket handkerchief.

In the latter case, the water closed the pores in the cloth so that it acted like a solid sheet, while the felt did not. This is the reason a hedge is a poor sound barrier, but a tight fence is a much more satisfactory one. Rain and fogs similarly have little effect on sound. In fact, the calmness often found in fogs may actually improve the transmission of sound. However, the presence of moisture catalyzes the absorption by vibrational relaxation in oxygen.

SOUND OUTDOORS

In spite of small absorption, sound cannot be heard for any great distance outdoors. One reason for this is that the temperature of the atmosphere decreases rapidly with altitude (roughly 6 °C for each 1000 m). The lower phase velocity at altitude means that an initially vertical wave front will be tilted backwards, so the rays of sound are bent upwards, creating a shadow at ground level.

A strong temperature inversion, or a wind blowing from the source of sound towards the observer, will have the opposite effect, and sound may be heard at a considerable distance. What is important in the case of the wind is that the wind speed generally increases with altitude, a *wind shear*, not simply a uniform wind. We will consider the effect of the wind in detail below. The unpredictability of the range of foghorns is well known, and can be largely ascribed to such effects.

Fog and rain have little effect, since the scale of the disturbances is much smaller than a wavelength. Raising the source of sound has the effect of increasing its range; bells in church towers take advantage of this.

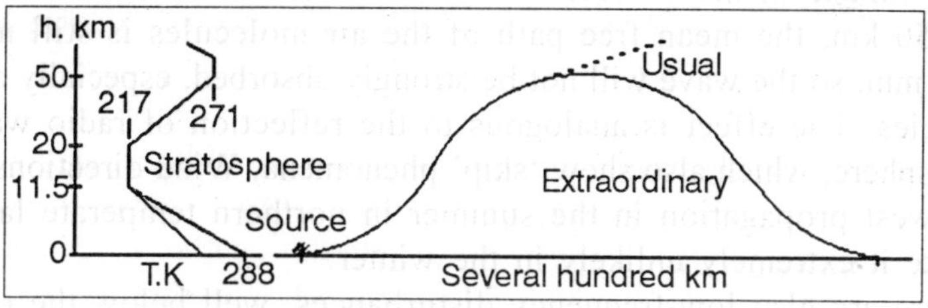

Fig. Long-Distance Propagation of Sound

A strange effect was long noticed in the audibility of very large explosions and similar noises. The sound is observed in a region surrounding the source, perhaps extending 50 km or more.

Then there is a zone of silence, but at several hundred kilometres, the sound, or at least its lower-frequency components, is sometimes again heard, and with extra delay.

Guns on the continent were heard in England, for example. In the 1930's, this *anomalous propagation* was finally recognized as the effect of temperatures comparable to those on the ground, far above the stratosphere.

The sound was reflected by these hot layers and again bent downward towards the earth. Sound from sources like jet aircraft or rockets can also be trapped as in a waveguide in the stratosphere, with higher temperatures both above and below, and can be detected thousands of miles from its source by a receiver in the stratosphere.

A plane wave propagating in a thermally stratified atmosphere can be represented by a ray normal to the wave front whose inclination to the horizontal changes so that the velocity with which the line of intersection of the wave front with a horizontal plane moves, called the trace velocity, is constant. If θ is the inclination, and c(h) the phase velocity as a function of altitude, then $c(h) \sec \theta = c(0) = 340$ m/s, if the wave started horizontally at the surface.

As c(h) decreases, θ increases, and the wave climbs. At about 11.5 km, the stratosphere is reached, where c = 295 m/s. The inclination of the ray remains at about 30° through the stratosphere. At 20 km, the temperature begins to climb again (due to absorption of solar radiation by ozone), and reaches a maximum at about 49 km, where c = 330 m/s.

The ray bends over, and at the maximum is inclined only 14° with the horizontal. Should the temperature up here be a little hotter than normal, and the temperature on the ground a little colder than normal, or a strong wind shear in the direction of propagation exist on high, or else a strong inversion exist on the ground, the ray will become horizontal, and then follow the mirror of its previous path down to the ground again, about 200km from the source.

At 50 km, the mean free path of the air molecules is still no more than 0.1 mm, so the wave will not be strongly absorbed, especially at lower frequencies. The effect is analogous to the reflection of radio waves by the ionosphere, which also show 'skip' phenomena. Wind directions favour east to west propagation in the summer in northern temperate latitudes, and make it extremely unlikely in the winter.

There are also low-frequency disturbances, well below the range of hearing, that can be detected with a microbarograph or hot-wire microphone, caused by various phenomena around the earth such as volcanoes, earthquakes, ocean waves and so forth.

These include the strongly dispersive acoustic gravity waves, trapped in the stratosphere, which are more hydrodynamic phenomena than acoustic ones, that can propagate right around the earth. These disturbances have periods of up to minutes.

A flight of stairs may reflect an acoustic impulse in the form of a musical tone. For reinforcement, the depth of the tread of the stairs should be a half wavelength. Since this depth is about 22 cm, the tone should have a frequency of 769 Hz, about a g" (second g above middle C). Rayleigh remarked that a flight of stairs on his estate of Terling returned a handclap as a sound resembling the chirp of a bird.

The best situation for observing this effect would be a circular flight of stairs, with the observer standing at the centre of curvature so the sound will be reflected back.

So that the delay between the clap and the return will be enough that the return will be easily detected, say about 0.1 s, the distance to the steps and back should be about 33 m, or 100 ft, so the radius of the steps should be about 50 ft. The result from an ordinary flight of stairs in a house, probably because one must be too close for the echo to be distinguished.

The Doppler Effect

A purely geometrical effect conceived by Christian Doppler in 1842 is the apparent increase in frequency of sound when the source and observer are approaching, and the decrease of frequency when they are receding. He apparently thought of it with respect to the light from double stars, as a red or blue shift. This was ignored in acoustics until railways finally gave speeds high enough relative to the speed of sound to make the effect noticeable. Buijs Ballot put musicians on an open carriage pushed by a locomotive at various speeds in 1845, and measured the change in frequency. Scott Russell, the naval architect, also investigated the effect.

Mach studied it with a rotating whistle in 1861. Sound propagates through the medium with a speed independent of the motions of source and observer, of course. The motions of both source and observer relative to the medium is what is important, not simply their motions relative to one another, as in the case of light.

There is also *aberration*, or an apparent change in direction, when the source and observer are not moving directly toward each other or apart. This is easily noticed with low-flying jet aircraft, where the sound appears to come from behind the aircraft. The wavefront emitted by a source at S just reaching the observer O, at which time the source is actually at S'. If a wind is blowing, but the source and observer are stationary, then there is no change in frequency of the observed sound.

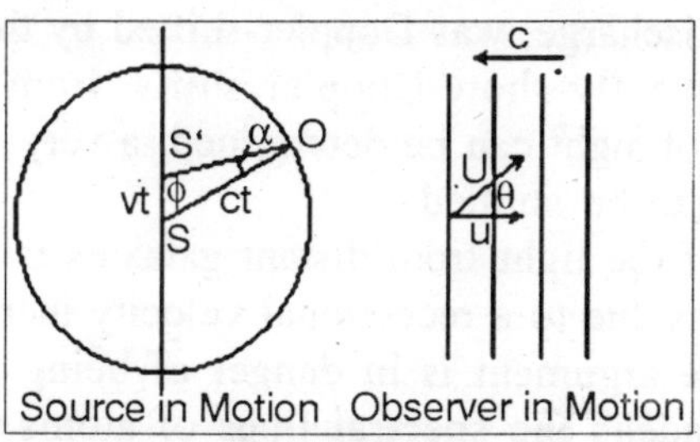

$$f' = \frac{c + U\cos\theta}{c - V\cos\phi} f$$

$$\sin\alpha = \frac{V\sin\phi}{\sqrt{V^2 + c^2 - 2cV\cos\phi}}$$

Fig. The Doppler Effect

A ratio of frequencies of the twelfth root of two (1.0595), or a semitone, results from a moving source approaching a stationary observer at about 42 mph, which was easily available to Doppler.

A full tone difference is attained at 82 mph, and a minor third at 125 mph. A major third is not reached until 150 mph. The speed of sound is taken as 750 mph in these calculations. If the observer were moving at twice the speed of sound away from an orchestra, the music would be heard in the correct tune and time, but backwards, as Rayleigh remarked.

Christian Johann Doppler (probably Doppler; 1803-1853) probably had such double stars in mind when he wrote in 1842 that perhaps the difference in colors were due to the motion of the stars, since the frequency of the light from an approaching star should be increased, and from a receding star decreased.

The speeds of stars are far too small for this to have any observable effect, but C. H. D. Buijs-Ballot (Buys Ballot) deduced that the effect should be much larger with sound, because of its smaller velocity. Accordingly, he experimented with musicians on a locomotive and at lineside in 1845, on the railway between Utrecht and Maarsen, and determined that Doppler's conjecture was indeed fact. The sound was produced by a trumpet on the locomotive, and three at lineside, with 14 observers, distributed on the train and ground. The phenomenon of the frequency shift with a moving source or moving observer has since then been known as the *Doppler Effect*, Though Buijs-Ballot should be remembered.

The optical Doppler effect was not observed until 1905, when J. Stark found that the light emitted by canal rays, which are fast ions from the cathode of a glow discharge, was Doppler-shifted by the expected amount. Stellar spectra frequently show Doppler shifts, from which the relative velocity in the line of sight can be determined, a very useful quantity, and double star orbits can be studied.

The red shift of the light from distant galaxies is also famous, and is usually interpreted as due to a recessional velocity increasing linearly with distance, though the argument is in danger of being a circular one. The Doppler effect broadens the spectral lines of atoms as a result of their random thermal velocities. It also has an important application in surveying by means of satellite radio signals.

Classical Doppler Effect

The Doppler effect in the case of waves in a material medium will be called the "classical Doppler effect ∂ here. We shall assume that all velocities are much less than the speed of light, so that relativistic considerations do not enter. The most important characteristic is that there is a preferred reference frame K, in which the medium is at rest. Referred to this system, the waves will be supposed to have a phase velocity c and an angular frequency $\omega_o = 2\pi f_o$. The wavelength λ will then be given by $\lambda = c/f$. The magnitude of the wave vector $k = 2\pi/\lambda = \omega/c$. Harmonic waves moving in the direction of the unit vector n can be described by a wave function $a(t,x) = A \exp[i\omega(t - n\cdot x/c)]$. Here, x is the usual radius vector. The argument of the exponential is $i\varphi(t,x)$, where φ is the *phase*, a function of time and position.

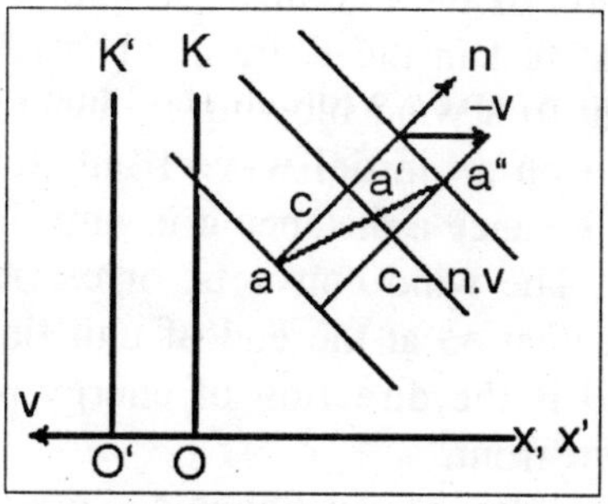

Fig. Wave front Convection

At any time t, the phase is constant over *wave fronts* $n\cdot x$ = constant = d,. where d is the distance of the wavefront from the origin, the length of a line from the origin drawn normal to the wave front. Hence, $\varphi = \omega t - 2\pi d/\lambda$. For a constant phase of zero, $d = (\omega\lambda/2\pi)t = ct$, so we verify that any wave front moves forward with velocity c.

If a reference frame K' moves with constant velocity v relative to K, then $x' = x - vt$, and, of course, $t' = t$.

This is a *Galilean transformation*, which preserves Newton's Laws, but not the laws of Electromagnetism. If we now consider a point P fixed in K, it must experience the same phase history whether described in K or the moving frame K', so that $\omega(t - n\cdot x/c) = \omega'(t - n'\cdot x'/c')$. In K', ω, n and c may have new values that are constant over the wavefront, which must still be plane, since we have made a linear transformation. Therefore, we may equate the coefficients of t and each component of x. First, substitute

$$x = x' + vt,$$

to find

$$\omega\,(t - n\cdot v/c - n\cdot x'/c) = \omega'\,(t - n'\cdot x'/c').$$

Equating the coefficients of t, we find that $\omega' = \omega\,(1 - n\cdot v/c)$. Then we have left $(\omega/c)n\cdot x' = (\omega'/c')n'\cdot x'$.

This condition is satisfied by $\omega/c = \omega'/c'$ and $n = n'$. From the first equality we find $c' = c - n\cdot v$. This means that the wave front is carried along by the component of the velocity normal to it. The component of the velocity in the wave front must then correspond to a convection of energy in that direction. We note that the direction of the wave front does not change in system K'. These deductions are confirmed by the behaviour of material waves.

If the source is moving with velocity V, then the frequency observed in the K system is $\omega = \omega_0/(1 - n\cdot V/c)$.

If an observer is moving with velocity v, then the observed frequency is $\omega' = \omega\,(1 - n\cdot v/c) = \omega_0[(1 - n\cdot v/c)/(1 - n\cdot V/c)]$. Both the velocities here are absolute velocities; that is, relative to the system in which the medium is at rest.

A graphical analysis of the classical Doppler effect is shown at the right. The medium is at rest in the K frame. The K' frame, moving with velocity v, is equivalent to a wind blowing in the -v direction.

Consider a point a on an initial wave front. If there were no wind, this wave front would advance a distance c in unit time, and the energy at a would then be at a'. The wind convects, or carries, the wave front so that the energy from a is at a∂ at the end of unit time.

The direction a-a∂ is the direction of energy propagation, which is not normal to the wave front.

The wave front then moves a distance $c - n\cdot v$ in unit time, which is the new phase velocity c'. Since the new frequency f' is defined by $f'/c' = f/c$, we have $f' = f(c'/c) = f(1 - n\cdot v/c)$.

If we want v to be the wind velocity instead of the velocity of K', simply change the sign of v in all the equations. The diagram really contains all that is in the algebraic derivation, and makes the meaning clearer.

If the source velocity is small compared to c, we can expand the denominator and find that $\omega' = \omega_0[1 - n\cdot(v - V)/c]$, approximately. In this approximation, the frequency shift depends only on the relative velocity of observer and source, and not on the absolute velocities. The first-order correction factor to the second term is $1 + n\cdot V/c$.

The speed of sound at 15°C is about 340 m/s. Let us find the speed of a moving source so that the difference in pitch when the source is approaching and receding is one full tone, or a frequency ratio of 8/9. In this case, $8/9 = (1 + v/c)/(1 - v/c)$, which gives $v/c = 1/17$, or $v = 20$ m/s. This is 72 kmph or 45 mph, which is easily attained.

Relativistic Doppler Effect

Relativity must be taken into account for the optical Doppler effect, since light, with its velocity of c, is an essentially relativistic object. Also, light has no medium, and it is impossible to find absolute velocities as in the case of a material wave. It is, fortunately, even easier to find the relativistic Doppler effect than the classical one.

The phase is $\omega(t - n\cdot x/c) = (\omega/c)ct - (\omega n/c)\cdot x$, and it will be a Lorentz invariant just as it is a classical invariant. The quantities (ct,x) are known to form a contravariant Lorentz 4-vector x^α. The phase is the inner product of this 4-vector with a set of four quantities $(\omega/c, -\omega n/c)$, and the result is a Lorentz invariant. Therefore, these four quantities must transform as a covariant 4-vector k_α, so that $\varphi = k_\alpha x^\alpha$ will be an invariant.

Since we know that $k^\alpha = (\omega/c, \omega n/c)$ is a contravariant 4-vector, we know that it will transform accordingly. If n makes an angle θ with the direction of the velocity v, we find at once that $\omega' = \gamma(\omega - \beta\omega\cos\theta)$, or $\omega' = \gamma\omega(1 - \beta c$

$$\frac{\partial^2 \xi}{\partial t^2} = \frac{c^2 \dfrac{\partial^2 \xi}{\partial x^2}}{\left(1 + \dfrac{\partial \xi}{\partial x}\right)^{\gamma+1}}$$

Very Strong Sound Waves

Our analysis is most likely to be found inadequate in practice with very strong sound waves.

A more accurate wave equation, in which the condensation is not neglected compared to unity, is shown at the left. Note that (1 + s) now appears in the denominator on the right-hand side. Since this equation is nonlinear, its solutions will not obey the principle of superposition, and the shape of a wave will change as it moves.

Analysis of this equation shows that the phase velocity is slightly greater where the condensation is greater. Essentially, the particle velocity is added to the phase velocity. The more condensed portions of the wave move forward relatively to the less condensed, and the front of the wave becomes steeper. Such behaviour is familiar from water waves, which are also nonlinear in shallow water (as well as being generally dispersive). Nonlinearity also results in the production of sum and difference tones, and other new frequencies, when waves are superimposed.

Should the front of a wave become so steep that it becomes a discontinuity, or *shock wave*, viscosity and heat conduction become essential to the analysis. By the action of spreading and dissipative processes, the shock wave can again become a continuous sound wave. In nearly all cases except those dealing with explosions and shock waves, nonlinear effects in the propagation of sound waves are completely negligible.

Reflection and Refraction

When a plane sound wave encounters a plane material interface, it is reflected and refracted according to the laws familiar from optics. The amounts reflected and transmitted depend on the relative acoustic impedance of the two media. We can easily solve the problem for normal incidence by using the results we have already found.

Quantities in the incident wave are unprimed, those in the reflected wave singly primed, and those in the transmitted wave are doubly primed.

The waves are described by their displacements ξ. The volume elasticity κ is used for generality, instead of γp for an ideal gas. At the boundary, the two media must remain in contact as the wave passes, and the pressures must be equal on the two sides of the interface.

These two boundary conditions can be expressed as shown in terms of the displacements.

The pressure condition is put in terms of the displacement by assuming a harmonic wave, a very convenient, but not essential, way to proceed here. Everything can be expressed in terms of the acoustic impedances r and r∂ of the two media. Finally, the simultaneous equations are easily solved for ξ' and ξ∂ in terms of ξ, and the results appear in the box.

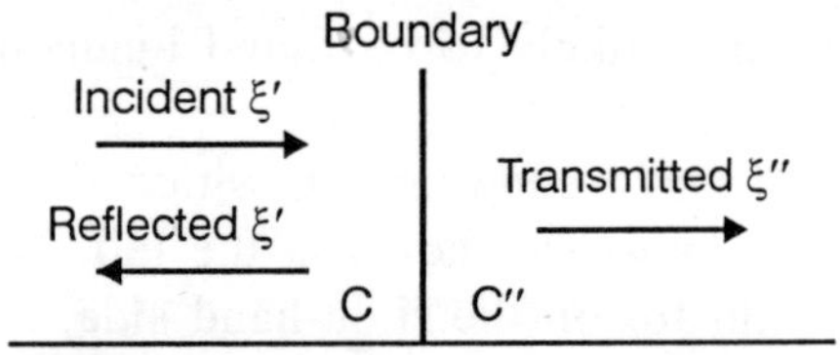

Fig. Reflection at an Interface

1. $\xi + \xi' - \xi''$ equal displacements
2. $p + p' + p''$ equal pressures

$$k\kappa(\xi - \xi') = k''\,\kappa''\,\xi'' \quad p = \;\; p = \kappa s = \gamma p_0 s$$

$$\xi - \xi' = \frac{r''}{r}\xi'' \qquad\qquad s = \pm\, ik\xi$$

$$\boxed{\xi' = \frac{r - r''}{r + r''} \qquad \xi'' = \frac{2r}{r + r''}} \qquad k\kappa = \frac{\omega}{C}\kappa = \omega r \qquad c^2 = \frac{\kappa}{\rho_0}$$

A dense, rigid medium has a very large acoustic impedance. In this case, ξ' = –ξ, and ξ∂ = 0, so the wave is completely reflected. If the second medium has a much smaller acoustic impedance, the reflected wave is in phase instead. Although actual materials cannot be perfect reflectors, the impedance match between air and a solid or liquid is generally so poor that very little energy goes into sound waves in the denser medium. The behaviour at a solid is quite complicated, because three modes of waves exist in a solid, not just the longitudinal compression wave.

Waves from the air do not penetrate into water, and sound waves in water do not penetrate into the air. The two media are acoustically separate. This may be of interest to anglers, who can feel free to talk while they fish. They should beware of casting moving shadows on the water, however.

Reflection and refraction in three dimensions is only a little more complicated. Many texts complicate the matter, but it is really easy to understand if you look at it in terms of fundamentals, and not just as a mathematical exercise.

The nature of the interface between the media deserves some consideration first. We'll consider two fluid media with a plane interface of a thickness much less than a wavelength. Ideally, we should think of a massless, flexible film separating the two media. When the two media are gaseous, diffusion at least will blur the transition, so such a film would be necessary. The interface between a gas and a liquid, or between two immisicble liquids, would generally suit our condition quite naturally.

The wave quantities will be assumed to vary harmonically in time as $e^{j\omega t}$. We shall further restrict the analysis to plane waves, which vary harmonically in space according to $e^{-jk \cdot r}$.

The wave vector k points in the direction of propagation, and is of magnitude k = ω/c. The *wavefronts*, or surfaces of constant phase, are the planes k·r = d + ct. Pressure and particle velocity are functions varying like this, and if p and v are their peak magnitudes (phasors), then v = p/ρc, where ρc is the *wave impedance* that we defined above. Its units are g/s-cm^2 in the cgs system. Its value for air at STP is 43 g/s-cm^2. For a positive p, v is in the direction of propagation.

Now let's suppose an incident plane wave of amplitude A strikes the interface. The intersection of two planes is a straight line, which we shall call the *trace* of the wavefront on the interface. As the wavefront moves forward with velocity c, the trace will move parallel to itself with the *trace velocity* v_t.

In a plane normal to the trace, both the interface and the wavefront will appear as straight lines, and the wavefront will make an angle θ with the interface, called the *angle of incidence*. A normal to the wavefront, or *wave normal*, will make the same angle with a normal to the interface. It is now easy to see that the trace velocity is c/sin θ, always greater than or equal to the phase velocity, and infinite for normal incidence.

The wavefront does not penetrate the interface, but is replaced on the other side by a *refracted* wavefront of amplitude C, moving with phase velocity c'. This wavefront must maintain a constant phase relationship with the incident wavefront, and so they have the same trace and, therefore, the same trace velocity.

Then, c/sin θ = c'/sin θ', which gives the angle of refraction θ' in terms of θ, c and c'.

This is the *law of refraction*, or Snell's Law. From our construction, we see that the incident and refracted wave normals, and the normal to the interface, all lie in the same plane, the *plane of incidence*. It was exceedingly difficult to discover this general law empirically, and it was only recognized properly as late as the 17th century. However, we see how easily and naturally it follows from a wave model, and could be extended from acoustics to optics by analogy.

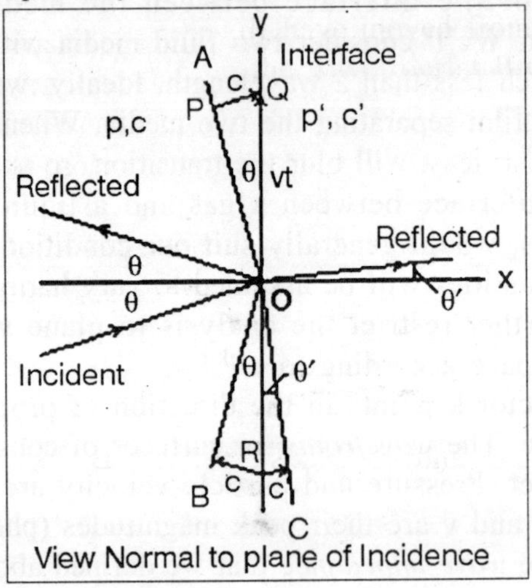

Fig. Geometry of Refraction and Reflection

It is found that there is generally a third wave of amplitude B, reflected back into the first medium.

Just as in the case of the refracted wave, the wavefront must also have a constant phase relation to the incident wave, and so its trace must be the same as the traces of the incident and refracted waves, and move with the same trace velocity.

This implies that c/sin θ = c/sin θ∂, where θ∂ is the angle of reflection, so that θ∂ = θ, the familiar *law of reflection*. Of course, the reflected wave normal also lies in the plane of incidence.

The laws of refraction and reflection depend only on the constancy of trace velocity, and so are valid within rather wide limits, and may hold well even when our theoretical relations between the amplitudes do not (as for a thick, but laterally uniform, interface).

The figure shows the wavefronts at the instant of time when the trace is at the origin O. The x-coordinate is normal to the interface, the y-coordinate in the interface, and the z-coordinate normal to both. The z-axis is the common trace of the three wavefronts.

The incident wave is given by $Ae^{-jk(x \cos \theta + y \sin \theta)}$, the reflected wave by $Be^{jk(x \cos \theta - y \sin \theta)}$, and the refracted wave by $Ce^{-jk'(x \cos \theta' + y \sin \theta')}$. The wave vector k' for $x > 0$ is given by $k' = \omega/c'$. Note that ω is the same for all the waves involved. On the interface, $x = 0$, so the waves are $Ae^{-jyk \sin \theta}$, $Be^{-jyk \sin \theta}$ and $Ce^{-jk' \sin \theta'}$. We have already found that $k \sin \theta = k' \sin \theta'$, so the exponential factors giving the y-dependence are all the same.

We come to the important conclusion, then, that if the boundary conditions are satisified at one point on the interface, then they are satisfied at all points on the interface. After a unit time interval, the trace of the wavefront will have moved from O to Q with velocity v_t, while a point originally on the wavefront at P will have moved with velocity c to Q.

Therefore, $c = v_t \cos \theta$, as asserted above. At a unit time interval earlier, the traces were at point R, from which points on the reflected and refracted wavefronts have moved as shown.

For an ideal plane interface, the relations between the amplitudes A, B and C are easily found. At the interface, the pressure due to A and B on one side must be equal to the pressure due to C on the other.

If this were not true, there would be a finite force on an element of the interface of zero mass, producing an infinite acceleration, which is insupportable.

At the point $x = 0$ and $y = 0$ this gives $A + B = C$, and the equality of pressures is guaranteed to hold everywhere on the interface and for any time. Under the action of the wave, the interface must move normally to the surface, in the x-direction. We assume the medium has zero viscosity, so the fluid can slide at will in the y and z directions.

We require, therefore, that the normal velocities on each side of the interface be equal, so a gap is not left.

The incident pressure wave A creates a velocity $v = A/\rho c$ along the wave normal. The x-component of this velocity is $A(\cos \theta/\rho c)$. The reflected wave contributes an x-component $-B(\cos \theta/\rho c)$.

The sum of these is the net x-velocity on the left of the interface. On the right-hand side, we have just $C(\cos \theta'/\rho c')$. Therefore, the boundary condition of equal velocity x-components is $(A - B)(\cos \theta/\rho c) = C(\cos \theta'/\rho c')$. If we compare this with the similar relations for normal incidence that we obtained at the beginning of this section, we see that they are exactly the same, if we replace ρc by $\rho c/\cos \theta$, an equivalent "angular∂ wave impedance that varies from ρc for normal incidence to infinity for grazing incidence.

The final results are $B = [(\rho'c'/\cos \theta' - \rho c/\cos \theta)/(\rho'c'/\cos \theta' + \rho c/\cos \theta)]\, A$, and $C = [(2\rho'c'/\cos \theta')/(\rho'c'/\cos \theta' + \rho c/\cos \theta)]\, A$, which are very similar to the expressions for normal incidence, if the angular wave impedances are used.

In considering energy relations, it must be remembered that when a finite beam of width a is refracted, the refracted beam is of width b =

a(cos θ'/cos θ). Reflected and incident beams are, of course, of the same width. If ρ'c' is less than ρc, θ' will reach 90° for some angle of incidence θ_c less than 90°. Then we will have |B| = |A| for this and larger angles of incidence, and all the incident energy will be reflected.

There will, in general, be a phase difference between A and B. C will not be zero, but it will be the amplitude of a boundary, or *evanescent* wave that carries no energy normal to the interface, but is exponentially attenuated in the second medium. It does carry energy along the boundary, which can have some interesting effects for finite beams. This phenomenon is called *total internal reflection*, and is studied in optics. For air and water, the phase velocities of 330 m/s and 1500 m/s, respectively, give a relative index of refraction of 4.55, which means that the angle of incidence for total reflection is only 13°.

As the angle of incidence approaches 90°, the angular wave impedance approaches infinity, so that B = -A, approximately. There is again total reflection, but in this case the phase of the reflected wave is reversed, as in a dense-to-rare reflection, and C = 0.

If ρ'c'/cos θ' = ρc/cos θ there will be no reflected wave. Using (1/c')sin θ' = (1/c)sin θ, we can eliminate cos θ' from this relation, and find sin θ = (c/c'){[1 – (rho;'c'/ρc)^2]/[1 – (ρ'/ρ)^2]}$^{1/2}$.

A physical angle θ will exist if the quantity inside the curly brackets is greater than zero. If this is so, then waves incident at this angle will be totally transmitted. This means, of course, that the normal velocity of the incident wave equals the normal velocity of the refracted wave, so that no reflected wave is necessary.

The reflection of sound from solids is a rather complex study, because of the variety of surfaces and the existence of transverse waves in solids. Bulk longitudinal waves in solids are a little faster than longitudinal waves in fluids, because of the additional shear forces.

Near normal incidence, shear waves will not be excited, and our equations can be used if the proper parameters are inserted. Solid surfaces may be porous, and offer considerable dissipative resistance, as in sound-absorbing wall coverings.

A solid surface may be characterized by its *specific surface impedance* z, the ratio of pressure to normal velocity. Applying the boundary conditions for an incident wave A and a reflected wave B, we have A + B = C and (A – B)(cos θ/ρc) = C/z.

Then, B = [(z – ρc/cos θ)/(z + ρc/cos θ)] A. In general, z can be complex, z = r + jx, with r and x functions of frequency. Let β = z/ρc. The energy reflection coefficient α = |B/A|^2 = [(β – 1)/(β + 1)]^2, or β = (1 = "α)/(1 + "α). Assuming that z is real, an absorption coefficient of 0.5 corresponds to β = 0.172 or z = 7.4 g/s-cm^2.

A z of 43 g/s-cm^2 would mean $\alpha = 1$, or perfect absorption. Surface impedances can be measured by experiments analogous to electromagnetic transmission line experments, using the Smith chart and the standing-wave ratio. It is easy to see that for oblique incidence, the reflectivity becomes greater, and the absorption less. At glancing incidence, the surface absorption vanishes.

Sound is absorbed by viscous friction at surfaces, so a sound absorber must present a large amount of surface.

A permeable material is required, such as glass wool or cotton wool, and the incident sound must directly contact a permeable surface. Foamed plastics are quite unsuitable as sound absorbers, though their lightness may encourage transmission.

An absorbing surface of this type is equivalent to the absorbing resistance in parallel with an acoustic inductance, which short-circuits the resistance at low frequencies. Above 1 kHz, a porous surface may have $\alpha = 0.9$, but the absorption drops at low frequencies.

What is usually called "styrofoam" usually isn't. Styrofoam is extruded polystyrene (XPS) made by Dow Chemical, and is blue. It has no porosity and is an excellent thermal insulator for moderate temperatures. "Styrofoam" cups and such are made from expanded polystyrene (EPS), which consists of small spheres fused together, which are quite evident on close inspection. A sample I tested had a density of 14 kg/m^3 and a Young's modulus of about 500 psi or 3.5 GPa.

If Poisson's ratio was about zero, then the velocity of longitudinal waves would be about 500 m/s, and the specific acoustic impedance 700 g/s-cm^2. A Poisson's ratio of 1/3 would raise the phase velocity only to about 600 m/s. Neglecting the effect of any porosity, the absorption coefficient for normal incidence would be 0.22 on the basis of transmission alone. Intensity reflected from the back surface should be taken into consideration.

A *membrane absorber* is a volume between two membranes with some kind of absorber, perhaps at the circumference. An example is a double-glazed window, but the idea here is more to reduce transmission than to absorb incident sound.

This device acts like an absorbing resistance in parallel with an acoustic capacitance (and with an acoustic inductance in series, representing the mass of the membrane), absorbing best at low frequencies, but with α as high as 0.5 below about 300 Hz.

A final type of absorber is the Helmholtz resonator with an absorbing plug in its neck. This is a resonant absorber, which is effective only over a narrow frequency range, and which is applicable only in special circumstances.

The reverberation time of a room of volume Vm^3, which is the time for a sound to drop 60 dB from an initial value, is roughly given by Sabine's formula, t = 0.16V/A seconds, where A is the absorbing area in m^2 times the value of α, summed over all absorbing area in the room.

It is clear that large rooms require some help in absorption if the reverberation time is not to increase above a certain maximum, which depends on the size and use of the room. Small rooms should have reverberation times of around 1 second, which is the boundary between "dead" and "live" rooms.

Most acoustical wave phenomena have analogies in optics, and vice-versa, except, of course, for polarization phenomena. The existence of polarization in light retarded the acceptance of the wave theory of light, and put the analogies with sound in doubt.

The analogies in interference and diffraction are particulary interesting. A single sheet of cloth reflects very little sound, since the cloth is porous. However, if parallel sheets are a half-wavelength apart, the weak reflections will all be in phase, and the reflection will become strong. The action is similar to a multiple-layer reflective coating in optics.

WIND AND WAVES

When considering the Doppler effect, we assumed that the waves always moved with the phase velocity c in the air, independently of the movement of the source.

When a wind blows, the waves are carried with the air, *convected.* It is a common observation that sounds can be heard better downwind of the source than upwind, or that 'the wind carries the sound.' Indeed it does, but this cannot be the explanation for this observation, since wind speeds are always much less than the speed of sound, usually less than a twentieth of the sound speed of 750 mph. In fact, a constant wind will have very little effect on the propagation of sound, not even affecting the frequency heard in different directions.

However, when a wind blows, it is retarded at the surface—a sort of boundary layer effect—and increases in speed aloft.

This is a *wind shear,* that can be expressed by a gradient dU/dy, where U is the wind speed and y is the height. A wavefront propagating with the wind will have its top inclined forward, so it will tend to return to the surface, while a wavefront propagating against the wind will be deflected upwards. This is a much better explanation of the fact that sounds can be heard better downwind than upwind.

Let us consider the case of a horizontally stratified atmosphere with temperature and wind varying with altitude, and a plane wavefront propagating with its normal making an angle φ with the horizontal. Note that this is a different convention than we used above in considering the effect of temperature, where we used the convention familiar from optics. The wind U convects the wavefront, so that the rays are no longer normal to the wavefronts, but deviate slightly in the direction of the wind. Let the inclination of the ray with the horizontal be ψ.

The Figure at the right shows how φ and ψ are related. We can use the fact that U is much less than c to simplify the formulas. We assume nothing regarding the magnitude of φ or ψ.

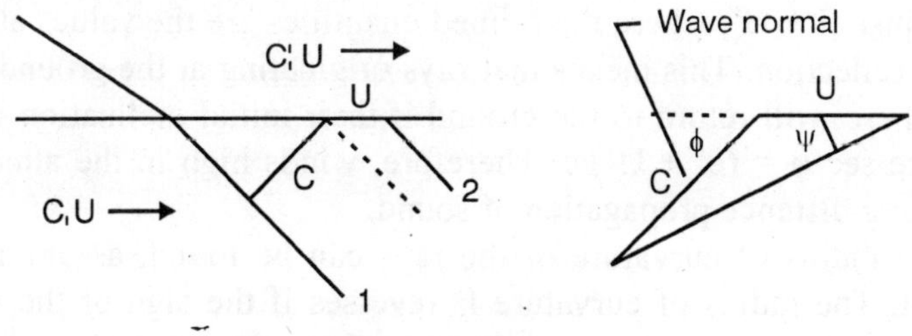

Fig. Radius of Curvature

Trace velocity = c sec f + U = const.

$$\sin(\phi - \psi) \doteq \phi - \psi = \frac{U}{c}\sin\phi$$

$$\sec\left(\psi + \frac{U}{c}\sin\psi\right) + \frac{U}{c} = \text{const.}$$

$$\sec\psi + \frac{U}{c}\sec^2\psi = \text{const.} (U \ll c)$$

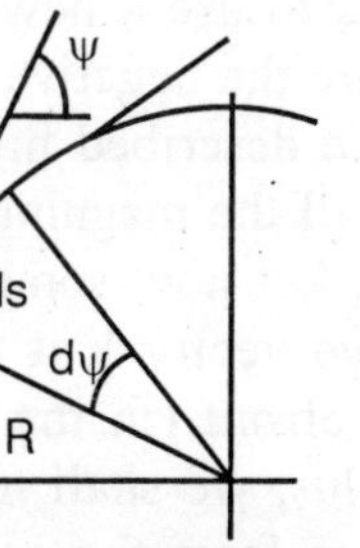

Fig. Effect of the Wind

$$ds = R\,d\psi \qquad R = \frac{1 + \frac{2U}{c}\sec\psi}{\frac{1}{c}\frac{dU}{dy}}$$

$$\frac{d\psi}{ds} = \frac{1}{R} \qquad (c) = \text{const.})$$

Our analysis is based on the constancy of the trace velocity on horizontal planes, as in the case of temperature, which the present analysis includes. This can be expressed in terms of the ray inclination, making it possible to trace the paths of rays, which are the directions of energy propagation.

The result is quite simple, a quadratic equation for sec ψ, that reduces to our previous result if U = 0. We notice (with Rayleigh) that sec φ = 1 corresponds to the boundary for total reflection, and the constant in this case is just c' + U', where the primed quantities are the values at the height of total reflection. This means that rays originating at the ground, assuming U = 0 there, will return to the ground if their initial inclination is less than φ, where sec φ = (c' + U')/c. Therefore, winds high in the atmosphere do affect long-distance propagation of sound.

The radius of curvature of the rays can be found, assuming that c is constant. The radius of curvature R reverses if the sign of the wind shear reverses, and, of course, when the direction of propagation changes. We have been cavalier with the signs here, but they can be worked out easily if you are bothered.

RADIATION, SCATTERING AND DIFFRACTION

A large number of interesting problems involve waves that are not plane, for example radiation from vibrating bodies, scattering by obstacles, and diffraction. The method of basing the analysis on the displacement becomes very inconvenient, since it is a vector, so that all three components have to be considered.

The way out of this is to use a new function, the *vector potential* φ, whose spatial derivatives are the negative components of the velocity. Now the wave can once more be described in terms of a single quantity. If we also use harmonic waves, all the magnitudes can be related to one another as in the plane-wave case, but now some of them are vectors.

In particular, the wave vector k is the wave number k times a unit vector normal to the wavefront (in the direction of the ray). Since the analysis uses vector calculus, we shall not give it here, but merely quote some interesting deductions from the analysis.

Consider a small volume of any shape much less than a wavelength in size. When the wave comes by, it refuses to compress or expand as the air it replaces would do, and moreover does not move back and forth with the particle velocity. The same effect would be produced by a double source (analogous to an electric dipole) aligned along the direction of the displacement, which would radiate sound in all directions.

The result is that some of the energy in the wave is *scattered* in all directions. As in optics, the scattered amplitude is proportional to the inverse square of the wavelength, so the scattered intensity is inversely proportional to the fourth power.

This strong wavelength dependence means that shorter wavelengths are preferentially scattered by small obstacles, so the lower-frequency components in the transmitted wave are relatively stronger. A complex musical note reflected by a wood may return an octave higher, since the octave is scattered sixteen times as strongly as the base note.

We are not normally aware of phenomena such as shadows and diffraction with sound as we are with light. However, there is really not much difference in the two cases, only a vast difference between the wavelength and the sizes of obstacles.

The usual obstacles to sound are comparable in size to the wavelength, so diffraction is very strong and sound readily bends around obstacles. Comparable obstacles to light are so small as to be invisible to the naked eye.

Sound Indoors

The principal application of acoustics has always been to structures erected for large meetings and public performances, and the assignment to make the communication between source of sound and the audience as satisfactory as possible. Vitruvius discusses the problem at length, including the effects of the echo of the first part of a word on the intelligibility of its ending, which is rather important in Latin.

He faced the difficult problems of an open amphitheatre, stone surfaces, different audience sizes, and the strength of single voices. A wall behind the actors reflected sound into the audience, while the presence of the audience decreased the reverberation time to a comfortable level. Brass resonating chambers were set in the remoter parts of a theatre to reinforce frequencies that were attenuated.

Reverberation time is a very important parameter. It is very difficult to speak to a considerable audience in the open, or in a *dead* chamber where the walls are highly absorbing and the reverberation time short. Anechoic chambers are required for acoustic measurements, but are highly unsatisfactory for speech. A long reverberation time is equally unpleasant, as earlier sounds intefere with later ones.

Rayleigh noticed that the Baptistery in Pisa had a reverberation time of about 12 seconds. The notes of a common chord, sounded one after the other, are heard together for an extended time. The ear appreciates a comfortable intermediate value, such as is commonly reached in the furnished rooms of houses. Before the furniture has been brought in, the empty rooms are characteristically *live*.

Many performance halls have been found unsatisfactory on acoustic grounds, from unwanted resonances or focussing, improper reverberation time, and badly directed sound. Few are now built without extensive acoustic design, which can be effective if well-done.

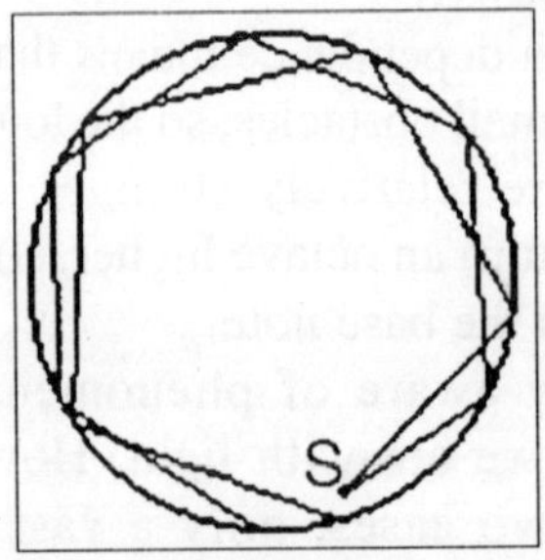

Fig. Whispering Gallery

When a sound wave strikes a solid surface, any velocity components along the surface are resisted by viscosity, and fluid motions in any pores or cracks in the surface are also subject to the same dissipative forces. A boundary layer of a thickness about h = sqrt($8\pi^2\nu/f$) governs these phenomena. Here, f is the frequency, and ν is the kinematic viscosity, as above. The phase velocity of waves propagating down a tube of radius d (much greater than h) is reduced by $hc/4\pi d$, and the linear absorption coefficient is $\alpha = fh/4\pi dc$. The asorption is much greater than for waves propagating in the open. When d becomes smaller than h, the motion is dominated by viscosity, and inertia has little influence.

Waves are rapidly stifled in such pores. Wall hangings, carpets and acoustic ceiling tiles are, for this reason, quite effective in reducing echoes. Only in this way can sound die out in an enclosed space.

The sound absorption coefficient of a surface is $\alpha = 1 - r$, where r is the energy reflection coefficient. This can be evaluated for diffuse incident sound, so that it is applicable to buildings. α at frequencies of 125 Hz, 1000 Hz, and 4000 Hz for several surfaces are: acoustic tile, 0.10, 0.75, 0.50; brick, 0.02, 0.04, 0.07; oak panelling on 1" battens, 0.20, 0.05, 0.05; and snow, 0.15, 0.75, 0.85.

In the gallery around the base of the dome in St Paul's cathedral in London, it has long been noticed that the voices of people on the other side of the dome are easily heard near the wall.

This is an example of a *whispering gallery*, of which others are known. Some have regarded the abnormal audibility as due to the focusing effect of the spherical surface. Indeed, sound issuing from one focus of an ellipsoid will be directed towards the other focus, and sound from one end of the diameter of a circle may be intensified at the other end of the diameter, which lies on a caustic surface.

Sir George Airy thought this the explanation of St Paul's whispering gallery. Lord Rayleigh, on the other hand, has given a plausible explanation that is more generally applicable. The sound emitted in a solid angle along a tangent to the curved wall is restricted to lie within a certain spherical shell, so its intensity will decrease with spreading only as $1/r$, not as $1/r^2$. Therefore, sound tends to cling to a concave wall in general.

APPLICATIONS OF ACOUSTICS

This paper has been limited to the fundamental properties of sound waves, and some of the lore directly related to this subject. The theory can be immediately extended to other than plane waves, and the important subjects of radiation and scattering can be taken up.

This requires a recasting more along the lines of fluid mechanics, since using the displacement as a fundamental quantity becomes quite inconvenient in more general investigations, where potentials are more convenient. The science of sound also usually includes the treatment of the vibrations of solid bodies, which is again quite mathematical and incidentally, rather successful. Waves in solids is a distinct field with its own applications, such as seismometry, and which also plays an important role in ultrasonics.

Ultrasonics is an applied branch of acoustics concerned with the production and use of frequencies above about 40 kHz, which are used for testing, measurement, ranging, communication and materials processing. Engineering acoustics is concerned with the production and measurement of sound, especially random sound or *noise*, and the acoustical properties of materials.

Architectural acoustics studies the acoustic properties of buildings, as mentioned above. The auditory sense, or physiological acoustics, is another branch of study, whose most famous and successful investigator was Hermann von Helmholtz (1821-1894).

Chapter 9

Conservation of Energy

The conservation of energy is undoubtedly the single most important idea in physics. Strangely enough, although the basic idea of energy conservation was familiar to scientists from the time of Newton onwards, this crucial concept only moved to centre-stage in physics in about 1850 (i.e., when scientists first realized that heat was a form of energy).

According to the ideas of modern physics, energy is the substance from which all things in the Universe are made up. Energy can take many different forms:e.g., potential energy, kinetic energy, electrical energy, thermal energy, chemi- cal energy, nuclear energy, etc. In fact, everything that we observe in the world around us represents one of the multitudinous manifestations of energy.

Now, there exist processes in the Universe which transform energy from one form into another: e.g., mechanical processes (which are the focus of this course), thermal processes, electrical processes, nuclear processes, etc. However, all of these pro- cesses leave the total amount of energy in the Universe invariant.

In other words, whenever, and however, energy is transformed from one form into another, it is always conserved.

For a closed system (i.e., a system which does not exchange energy with the rest of the Universe), the above law of universal energy conservation implies that the total energy of the system in question must remain constant in time.

ENERGY CONSERVATION DURING FREE-FALL

Consider a mass m which is falling vertically under the influence of gravity. We already know how to analyse the motion of such a mass. Let us employ this knowledge to search for an expression for the conserved energy during this pro-

cess. The physics of free-fall under gravity is summarized by the three equations. Let us examine the last of these equations:

$v^2 = v_0^2 - 2gs$. Suppose that the mass falls from height h_1 to h_2, its initial velocity is v_1, and its final velocity is V_2. It follows that the net vertical displacement of the mass is $s = h_2 - h_1$. Moreover, $v_0 = v_1$ and $v = v_2$. Hence, the previous expression can be rearranged to give

$$\frac{1}{2}mv_1^2 + mgh_1 = \frac{1}{2}mv_2^2 + mgh_2.$$

The above equation clearly represents a conservation law, of some description, since the left-hand side only contains quantities evaluated at the initial height, whereas the right-hand side only contains quantities evaluated at the final height. In order to clarify the meaning of equation, let us define the kinetic energy of the mass,

$$K = \frac{1}{2}mv^2,$$

and the gravitational potential energy of the mass,

$$U = mgh.$$

Note that kinetic energy represents energy the mass possesses by virtue of its motion. Likewise, potential energy represents energy the mass possesses by virtue of its position. It follows that equation can be written

$$E = K + U = \text{constant}.$$

Here, E is the total energy of the mass: i.e., the sum of its kinetic and potential energies. It is clear that E is a conserved quantity: i.e., although the kinetic and potential energies of the mass vary as it falls, its total energy remains the same.

Incidentally, the expressions and for kinetic and gravitational potential energy, respectively, are quite general, and do not just apply to free-fall under gravity.

The mks unit of energy is called the joule (symbol J). In fact, 1 joule is equivalent to 1 kilogram meter-squared per second-squared, or 1 newtonmeter. Note that all forms of energy are measured in the same units (otherwise the idea of energy conservation would make no sense).

One of the most important lessons which students learn during their studies is that there are generally many different paths to the same result in physics. Now,

we have already analyzed free-fall under gravity using Newton's laws of motion. However, it is illuminating to re-examine this problem from the point of view of energy conservation. Suppose that a mass m is dropped from rest and falls a distance *h*. What is the final velocity *v* of the mass? Well, according to equation, if energy is conserved then

$$\Delta K = -\Delta U:$$

i.e., any increase in the kinetic energy of the mass must be offset by a corresponding decrease in its potential energy. Now, the change in potential energy of the mass is simply $\Delta U = mg\, s = -mg\, h$, where $s = -h$ is its net vertical displacement. The change in kinetic energy is simply ΔK

$= (1/2)\, mv^2$, where v is the final velocity. This follows because the initial kinetic energy of the mass is zero (since it is initially at rest). Hence, the above expression yields

$$\frac{1}{2}mv^2 = mgh,$$

or

$$v = \sqrt{2gh}.$$

Suppose that the same mass is thrown upwards with initial velocity v. What is the maximum height h to which it rises? Well, it is clear from equation that as the mass rises its potential energy increases. It, therefore, follows from energy conservation that its kinetic energy must decrease with height.

Note, however, from equation that kinetic energy can never be negative (since it is the product of the two positive definite quantities, m and $v^2/2$). Hence, once the mass has risen to a height h which is such that its kinetic energy is reduced to zero it can rise no further, and must, presumably, start to fall.

The change in potential energy of the mass in moving from its initial height to its maximum height is m g h. The corresponding change in kinetic energy is $-\,(1/2)mv^2$; since $(1/2)mv^2$ is the initial kinetic energy, and the final kinetic energy is zero. It follows from equation that $-\,(1/2)mv^2 = -\,mg\,h$, which can be rearranged to give

$$h = \frac{v^2}{2g}.$$

It should be noted that the idea of energy conservation—although extremely useful—is not a replacement for Newton's laws of motion. For instance, in the previous example, there is no way in which we can deduce how long it takes the mass to rise to its maximum height from energy conservation alone—this information can only come from the direct application of Newton's laws.

WORK

We have seen that when a mass free-falls under the influence of gravity some of its kinetic energy is transformed into potential energy, or vice versa. Let us now investigate, in detail, how this transformation is effected. The mass falls because it is subject to a downwards gravitational force of magnitude m g. It stands to reason, therefore, that the transformation of kinetic into potential energy is a direct consequence of the action of this force.

This is, perhaps, an appropriate point at which to note that the concept of gravitational potential energy—although extremely useful—is, strictly speaking, fictitious. To be more exact, the potential energy of a body is not an intrinsic property of that body (unlike its kinetic energy). In fact, the gravitational potential energy of a given body is stored in the gravitational field which surrounds it.

Thus, when the body rises, and its potential energy consequently increases by an amount ΔU; in reality, it is the energy of the gravitational field surrounding the body which increases by this amount. Of course, the increase in energy of the gravitational field is offset by a corresponding decrease in the body's kinetic energy. Thus, when we speak of a body's kinetic energy being transformed into potential energy, we are really talking about a flow of energy from the body to the surrounding gravitational field. This energy flow is mediated by the gravitational force exerted by the field on the body in question.

Incidentally, according to Einstein's general theory of relativity (1917), the gravitational field of a mass consists of the local distortion that mass induces in the fabric of space-time.

Fortunately, however, we do not need to understand general relativity in order to talk about gravitational fields or gravitational po-tential energy. All we need to know is that a gravitational field stores energy without loss: i.e., if a given mass rises a certain distance, and, thereby, gives up a certain amount of energy to the surrounding gravitational field, then that field

will return this energy to the mass—without loss—if the mass falls by the same distance. In physics, we term such a field a conservative field.

Suppose that a mass m falls a distance *h*. During this process, the energy of the gravitational field decreases by a certain amount (i.e., the fictitious potential energy of the mass decreases by a certain amount), and the body's kinetic energy increases by a corresponding amount. This transfer of energy, from the field to the mass, is, presumably, mediated by the gravitational force -m g (the minus sign indicates that the force is directed downwards) acting on the mass. In fact, given that $U = mg\ h$, it follows from equation that

$$\Delta K = f\Delta h.$$

In other words, the amount of energy transferred to the mass (i.e., the increase in the mass's kinetic energy) is equal to the product of the force acting on the mass and the distance moved by the mass in the direction of that force.

In physics, we generally refer to the amount of energy transferred to a body, when a force acts upon it, as the amount of work W performed by that force on the body in question. It follows from equation that when a gravitational force f acts on a body, causing it to displace a distance x in the direction of that force, then the net work done on the body is

$$W = fx.$$

It turns out that this equation is quite general, and does not just apply to gravitational forces.

If W is positive then energy is transferred to the body, and its intrinsic energy consequently increases by an amount W.

This situation occurs whenever a body moves in the same direction as the force acting upon it. Likewise, if W is negative then energy is transferred from the body, and its intrinsic energy consequently decreases by an amount $|W|$.

This situation occurs whenever a body moves in the opposite direction to the force acting upon it. Since an amount of work is equivalent to a transfer of energy, the mks unit of work is the same as the mks unit of energy: namely, the joule.

We have made two assumptions which are not universally valid. Firstly, we have assumed that the motion of the body upon which the force acts is both 1-dimensional and parallel to the line of action of the force. Secondly, we have assumed that the force does not vary with position. Let us attempt to relax these two assumptions, so as to obtain an expression for the work W done by a general force f.

Let us start by relaxing the first assumption. Suppose, for the sake of argument, that we have a mass m which moves under gravity in 2-dimensions.

Let us adopt the coordinate system, with z representing vertical distance, and x representing horizontal distance. The vector acceleration of the mass is simply $a = (0, -g)$. Here, we are neglecting the redundant y-component, for the sake of simplicity. The physics of motion under gravity in more than 1-dimension is summarized by the three equations -. Let us examine the last of these equations:

$$v^2 = v_0^2 + 2a{\cdot}s.$$

Here, v_0 is the speed at $t = 0$, v is the speed at $t = t$, and $s = (\Delta x, \Delta z)$ is the net displacement of the mass during this time interval. Recalling the definition of a scalar product [*i.e.*, $a{\cdot}b = (a_x\, b_x + a_y\, b_y + a_z\, b_z)$], the above equation can be rearranged to give

$$\frac{1}{2}mv^2 - \frac{1}{2}mv_0^2 = -mg\Delta z.$$

Since the right-hand side of the above expression is manifestly the increase in the kinetic energy of the mass between times t = 0 and t = t, the left-hand side must equal the decrease in the mass's potential energy during the same time interval. Hence, we arrive at the following expression for the gravitational potential energy of the mass:

$$U = mgz.$$

Of course, this expression is entirely equivalent to our previous expression for gravitational potential energy equation. The above expression merely makes manifest a point which should have been obvious anyway: namely, that the gravitational potential energy of a mass only depends on its height above the ground, and is quite independent of its horizontal displacement.

Let us now try to relate the flow of energy between the gravitational field and the mass to the action of the gravitational force, $f = (0, -mg)$. Equation

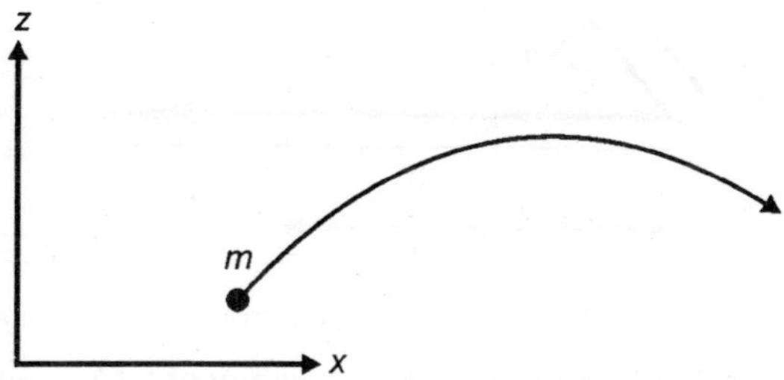

Fig. Coordinate System for 2-dimensional Motion Under Gravity

can be rewritten

$$\Delta K = W = f \cdot s.$$

In other words, the work W done by the force f is equal to the scalar product of f and the vector displacement s of the body upon which the force acts. It turns out that this result is quite general, and does not just apply to gravitational forces. Figure is a visualization of the definition. The work W performed by a force f when the object upon which it acts is subject to a displacement s is

$$W = |f|\,|s| \cos\theta.$$

where θ is the angle subtended between the directions of f and s. In other words, the work performed is the product of the magnitude of the force, $|f|$, and the displacement of the object in the direction of that force, $|s| \cos\theta$. It follows that any component of the displacement in a direction perpendicular to the force generates zero work.

Moreover, if the displacement is entirely perpendicular to the direction of the force (*i.e.*, if $\theta = 90^\circ$) then no work is performed, irrespective of the nature of the force.

As before, if the displacement has a component in the same direction as the force (i.e., if $\theta < 90^\circ$) then positive work is performed Likewise, if the displacement has a component in the opposite direction to the force (*i.e.*, if $\theta > 90^\circ$) then negative work is performed.

Suppose, now, that an object is subject to a force f which varies with position. What is the total work done by the force when the object moves along some general trajectory in space between points A and B (say)? Well, one way in which we could approach this problem would be to approximate the trajectory as a series of N straight-line segments. Suppose

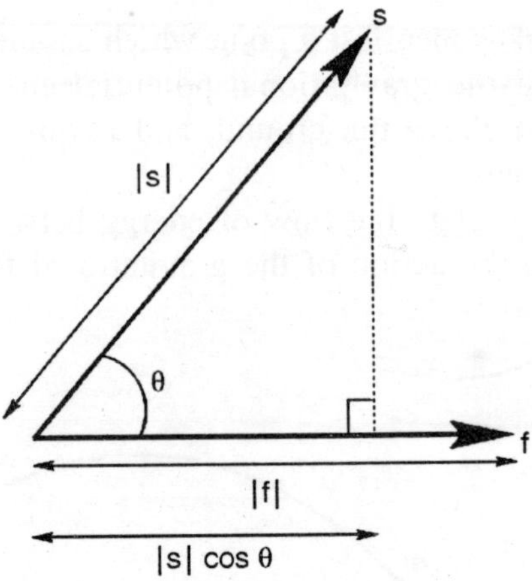

Fig. Definition of Work

That the vector displacement of the ith segment is Δr_i. Suppose, further, that N is sufficiently large that the force f does not vary much along each segment. In fact, let the average force along the ith segment be f_i. We shall assume that formula —which is valid for constant forces and straight-line displacements—holds good for each segment. It follows that the net work done on the body, as it moves from point A to point B, is approximately

$$W \simeq \sum_{i=1}^{N} f_i \cdot \Delta r_i.$$

We can always improve the level of our approximation by increasing the number Nof the straight-line segments which we use to approximate the body's trajectory between points A and B. In fact, if we take the limit N→ ∞ then the above expression becomes exact:

$$W = \lim_{N \to \infty} \sum_{i=1}^{N} f_i \cdot \Delta r_i = \int_A^B f(r) \cdot dr.$$

and the mathematical construct R_{BA} $f(r)$ _dr is termed a line-integral.

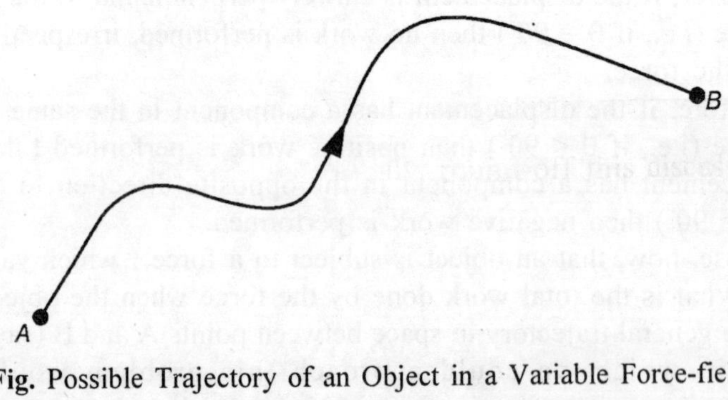

Fig. Possible Trajectory of an Object in a Variable Force-field

The meaning of equation becomes a lot clearer if we restrict our attention to 1-dimensional motion. Suppose, therefore, that an object moves in 1-dimension, with displacement x, and is subject to a varying force $f(x)$ (directed along the x-axis). What is the work done by this force when the object moves from x_A

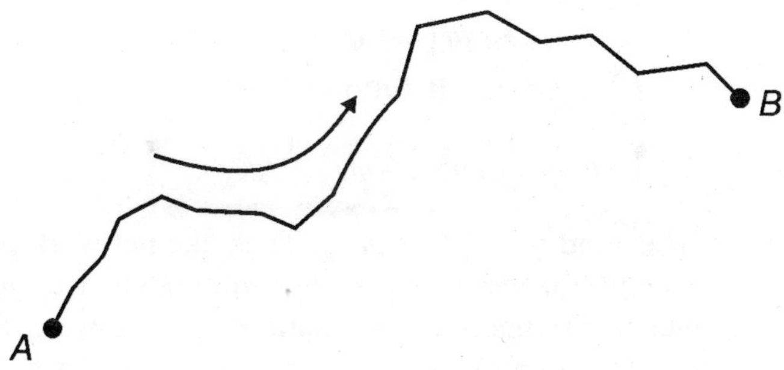

Fig. Approximation to the Previous Trajectory using Straight-line Segments

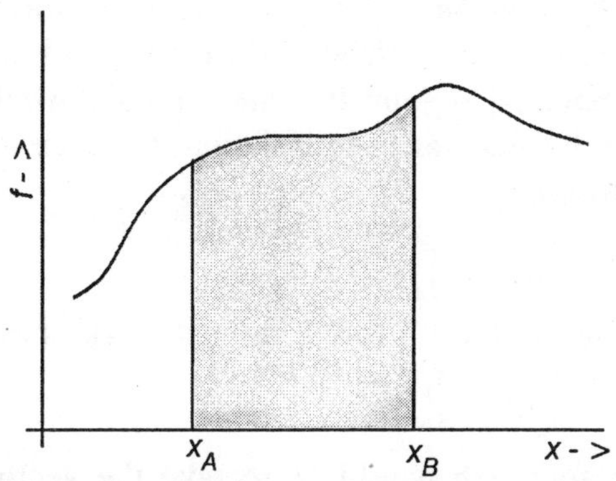

Fig. Work Performed by a 1-dimensional Force

to x_B? Well, a straight-forward application of equation [with $f = (f, 0, 0)$ and $dr = (dx, 0, 0)$] yields

$$W = \int_{x_A}^{x_B} f(x)dx.$$

In other words, the net work done by the force as the object moves from displacement x_A to x_B is simply the area under the $f(x)$ curve between these two points.

Let us, finally, round-off this discussion by re-deriving the so-called workenergy theorem, equation, in 1-dimension, allowing for a non-constant force. According to Newton's second law of motion,

$$f = m\frac{d^2x}{dt^2}.$$

Combining Equations and, we get

$$W = \int_{x_A}^{x_B} m\frac{d^2x}{dt^2}dx = \int_{t_A}^{t_B} m\frac{d^2x}{dt^2}\frac{dx}{dt}dt$$

$$= \int_{t_A}^{t_B} \frac{d}{dt}\left[\frac{m}{2}\left(\frac{dx}{dt}\right)^2\right]dt,$$

where $x(t_A) = x_A$ and $x(t_B) = x_B$. It follows that

$$W = \frac{1}{2}mv_A^2 - \frac{1}{2}mv_A^2 = \Delta K,$$

where $v_A = (dx = dt)t_A$ and $v_B = (dx = dt)t_B$. Thus, the net work performed on a body by a non-uniform force, as it moves from point A to point B, is equal to the net increase in that body's kinetic energy between these two points. This result is completely general, and does not just apply to 1-dimensional motion.

Suppose, finally, that an object is subject to more than one force. How do we calculate the net work W performed by all these forces as the object moves from point A to point B? One approach would be to calculate the work done by each force, taken in isolation, and then to sum the results. In other words, defining

$$W_i = \int_A^B f_i(r)\cdot dr$$

as the work done by the ith force, the net work is given by

$$W = \sum_i W_i.$$

An alternative approach would be to take the vector sum of all the forces to find the resultant force,

$$f = \sum_i f_i,$$

and then to calculate the work done by the resultant force:

$$W = \int_A^B f(r)\cdot dr.$$

It should, hopefully, be clear that these two approaches are entirely equivalent.

CONSERVATIVE AND NON-CONSERVATIVE FORCE- FIELDS

Suppose that a non-uniform force- field $f(r)$ acts upon an object which moves along a curved trajectory, labeled path 1, from point A to point B. The work W_1 performed by the force- field on the object can be written as a line-integral along this trajectory:

$$W_1 = \int_{A\to B:\text{path }1} f\cdot dr.$$

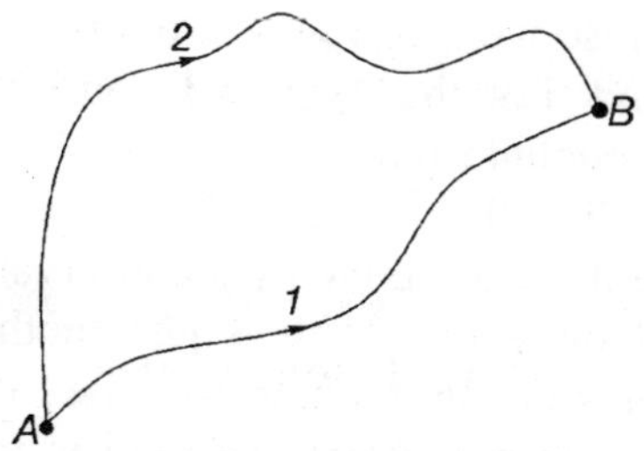

Fig. Two Alternative Paths Between Points A and B

Suppose that the same object moves along a different trajectory, labeled path 2, between the same two points. In this case, the work W_2 performed by the force- field is

$$W_2 = \int_{A \to B:\text{path } 2} f \cdot dr.$$

Basically, there are two possibilities. Firstly, the line-integrals and might depend on the end points, A and B, but not on the path taken between them, in which case $W_1 = W_2$. Secondly, the line-integrals and might depend both on the end points, A and B, and the path taken between them, in which case $W_1 \neq W_2$ (in general).

The first possibility corresponds to what physicists term a conservative force- field, whereas the second possibility corresponds to a non-conservative force- field.

What is the physical distinction between a conservative and a non-conservative force- field? Well, the easiest way of answering this question is to slightly modify the problem discussed above. Suppose, now, that the object moves from point A to point B along path 1, and then from point B back to point A along path 2.

What is the total work done on the object by the force- field as it executes this closed circuit? Incidentally, one fact which should be clear from the definition of a line-integral is that if we simply reverse the path of a given integral then the value of that integral picks up a minus sign: in other words,

$$\int_A^B f \cdot dr = -\int_B^A f \cdot dr,$$

where it is understood that both the above integrals are taken in opposite directions along the same path. Recall that conventional 1-dimensional integrals obey an analogous rule: i.e., if we swap the limits of integration then the integral picks up a minus sign. It follows that the total work done on the object as it executes the circuit is simply

$$\Delta W = W_1 - W_2,$$

where W_1 and W_2 are defined in Eqs. and, respectively. There is a minus sign in front of W_2 because we are moving from point B to point A, instead of the other way around. For the case of a conservative field, we have $W_1 = W_2$. Hence, we conclude that

$$\Delta W = 0.$$

In other words, the net work done by a conservative field on an object taken around a closed loop is zero. This is just another way of saying that a conservative field stores energy without loss: i.e., if an object gives up a certain amount of energy to a conservative field in traveling from point A to point B, then the field returns this energy to the object—without loss—when it travels back to point B. For the case of a non-conservative field, $W_1 \neq W_2$. Hence, we conclude that

$$\Delta W \neq 0.$$

In other words, the net work done by a non-conservative field on an object taken around a closed loop is non-zero. In practice, the net work is invariably negative. This is just another way of saying that a non-conservative field dissipates energy:

i.e., if an object gives up a certain amount of energy to a non-conservative field in traveling from point A to point B, then the field only returns part, or, perhaps, none, of this energy to the object when it travels back to point B. The remainder is usually dissipated as heat.

What are typical examples of conservative and non-conservative fields? Well, a gravitational field is probably the most well-known example of a conservative field. A typical example of a non-conservative field might consist of

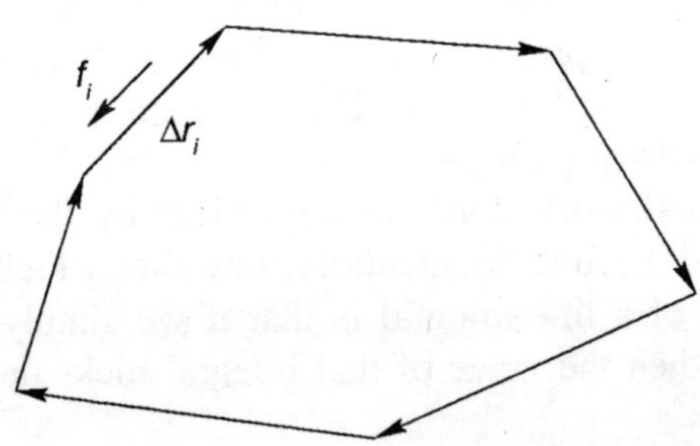

Fig. Closed circuit over a rough horizontal surface

an object moving over a rough horizontal surface. Suppose, for the sake of simplicity, that the object executes a closed circuit on the surface which is made up entirely of straight-line segments.

Let Δr_i represent the vector displacement of the ith leg of this circuit. Suppose that the frictional force acting on the object as it executes this leg is f_i. One thing that we know about a frictional force is that it is always directed in the opposite direction to the instantaneous direction of motion of the object upon which it acts. Hence, $f_i \propto -\Delta r_i$. It follows that $f_i \cdot \Delta r_i =$

$-|f_i|\,|\Delta r_i|$. Thus, the net work performed by the frictional force on the object, as it executes the circuit, is given by

$$\Delta W = \sum_i f_i \cdot \Delta r_i = -\sum_i |f_i||\Delta r_i| < 0.$$

The fact that the net work is negative indicates that the frictional force continually drains energy from the object as it moves over the surface. This energy is actu-ally dissipated as heat (we all know that if we rub two rough surfaces together, sufficiently vigorously, then they will eventually heat up: this is how mankind first made fire) and is, therefore, lost to the system. Hence, friction is an example of a non-conservative force, because it dissipates energy rather than storing it.

POTENTIAL ENERGY

Consider a body moving in a conservative force- field $f(r)$. Let us arbitrarily pick some point O in this field. We can define a function $U(r)$ which possesses a unique value at every point in the field. The value of this function associated with some general point R is simply

$$U(R) = -\int_O^R f \cdot dr.$$

In other words, $U(R)$ is just the energy transferred to the field (i.e., minus the work done by the field) when the body moves from point O to point R. Of course, the value of U at point O is zero: i.e., $U(O) = 0$. Note that the above definition uniquely specifies $U(R)$, since the work done when a body moves between two points in a conservative force- field is independent of the path taken between these points.

Furthermore, the above definition would make no sense in a non-conservative field, since the work done when a body moves between two points in such a field is dependent on the chosen path: hence, $U(R)$ would have an infinite number of different values corresponding to the infinite number of different paths the body could take between points O and R. According to the work-energy theorem,

$$\Delta K = \int_O^R f \cdot dr.$$

In other words, the net change in the kinetic energy of the body, as it moves from point O to point R, is equal to the work done on the body by the force- field during this process. However, comparing with equation, we can see that

$$\Delta K = U(O) - U(R) = -\Delta U.$$

In other words, the increase in the kinetic energy of the body, as it moves from point O to point R, is equal to the decrease in the function U evaluated between these same two points. Another way of putting this is

$$E = K + U = \text{constant}:$$

i.e., the sum of the kinetic energy and the function U remains constant as the body moves around in the forcefield. It should be clear, by now, that the function U represents some form of potential energy.

The above discussion leads to the following important conclusions. Firstly, it should be possible to associate a potential energy (i.e., an energy a body pos-sesses by virtue of its position) with any conservative force-field. Secondly, any forcefield for which we can define a potential energy must necessarily be conservative.

For instance, the existence of gravitational potential energy is proof that gravitational fields are conservative. Thirdly, the concept of potential energy is meaningless in a non-conservative force- field (since the potential energy at a given point cannot be uniquely defined). Fourthly, potential energy is only defined to within an arbitrary additive constant. In other words, the point in space at which we set the potential energy to zero can be chosen at will. This im-plies that only differences in potential energies between different points in space have any physical significance.

For instance, we have seen that the definition of gravitational potential energy is $U = \text{m } gz$, where z represents height above the ground. However, we could just as well write $U = mg\,(z - z_0)$, where z_0 is the height of some arbitrarily chosen reference point (e.g., the top of Mount Ever-est, or 0the bottom of the Dead Sea). Fifthly, the difference in potential energy between two points represents the net energy transferred to the associated force-field when a body moves between these two points. In other words, potential energy is not, strictly speaking, a property of the body—instead, it is a property of the forcefield within which the body moves.

HOOKE'S LAW

Consider a mass m which slides over a horizontal frictionless surface. Suppose that the mass is attached to a light horizontal spring whose other end is anchored to an immovable object.

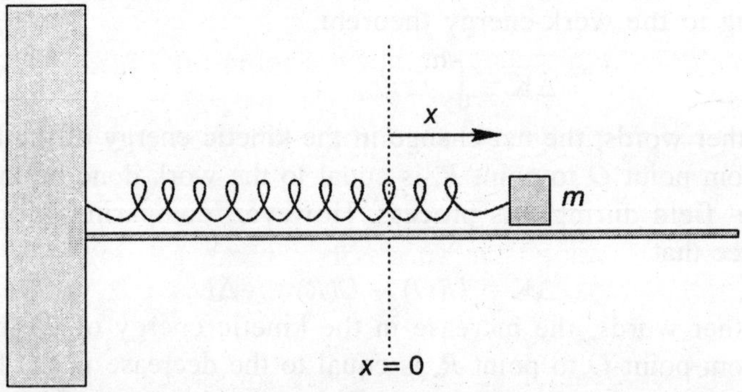

Fig. Mass on a spring

Let x be the extension of the spring: *i.e.*, the difference between the spring's actual length and its unstretched length. Obviously, x can also be used as a coordinate to determine the horizontal displace-ment of the mass. According to Hooke's law, the force f that the spring exerts on the mass is directly proportional to its extension, and always acts to reduce this extension. Hence, we can write

$$f = -kx,$$

where the positive quantity k is called the force constant of the spring. Note that the minus sign in the above equation ensures that the force always acts to reduce the spring's extension: e.g., if the extension is positive then the force acts to the left, so as to shorten the spring.

The work performed by the spring force on the mass as it moves from displacement x_A to x_B is

$$W = \int_{x_A}^{x_B} f(x)dx = -k\int_{z_A}^{x_B} x\,dx = -\left[\frac{1}{2}k\,x_B^2 - \frac{1}{2}k\,x_A^2\right].$$

Note that the right-hand side of the above expression consists of the difference between two factors: the first only depends on the final state of the mass, whereas the second only depends on its initial state. This is a sure sign that it is possible to associate a potential energy with the spring force. Equation, which is the basic definition of potential energy, yields

$$U(x_B) - U(x_A) = -\int_{x_A}^{x_B} f(x)dx = \frac{1}{2}k\,x_B^2 - \frac{1}{2}k\,x_A^2.$$

Hence, the potential energy of the mass takes the form

$$U(x) = \frac{1}{2}k\,x^2.$$

Note that the above potential energy actually represents energy stored by the spring—in the form of mechanical stresses—when it is either stretched or compressed.

Incidentally, this energy must be stored without loss, otherwise the concept of potential energy would be meaningless. It follows that the spring force is another example of a conservative force. It is reasonable to suppose that the form of the spring potential energy is somehow related to the form of the spring force. Let us now explicitly investigate this relationship. If we let $x_B \to x$ and $x_A \to 0$ then equation gives

$$U(x) = \int_0^x f(x')dx'.$$

We can differentiate this expression to obtain

$$f = \frac{dU}{dx}.$$

Thus, in 1-dimension, a conservative force is equal to minus the derivative (with respect to displacement) of its associated potential energy.

This is a quite general result. For the case of a spring force: $U = (1/2)\, kx^2$, so $f = -\, dU/dx = -kx$.

The 3-dimensional equivalent to equation is

$$f = \left(\frac{\partial U}{\partial x}, \frac{\partial U}{\partial y}, \frac{\partial U}{\partial z}\right).$$

For example, we have seen that the gravitational potential energy of a mass m moving above the Earth's surface is $U = mg\, z$, where z measures height off the ground. It follows that the associated gravitational force is

$$f = (0, 0, -\, mg).$$

In other words, the force is of magnitude mg, and is directed vertically downward.

The total energy of the mass shown is the sum of its kinetic and potential energies:

$$E = K + U = K + \frac{1}{2}kx^2.$$

Of course, E remains constant during the mass's motion. Hence, the above expression can be rearranged to give

$$K = E - \frac{1}{2}kx^2.$$

Since it is impossible for a kinetic energy to be negative, the above expression suggests that $|x|$ can never exceed the value

$$x_0 = \sqrt{\frac{2E}{k}}.$$

Here, x_0 is termed the amplitude of the mass's motion. Note that when x attains its maximum value x_0, or its minimum value $-x_0$, the kinetic energy is momentarily zero (*i.e.*, $K = 0$).

MOTION IN A GENERAL 1-DIMENSIONAL POTENTIAL

Suppose that the curve $U(x)$ represents the potential energy of some mass m moving in a 1-dimensional conservative force- field. For instance, $U(x)$ might represent the gravitational potential energy of a cyclist freewheeling in a hilly region. Note that we have set the potential energy at infinity to zero. This is a useful, and quite common, convention (recall that potential energy is undefined to within an arbitrary additive constant). What can we deduce about the motion of the mass in this potential?

Well, we know that the total energy, E—which is the sum of the kinetic energy, K, and the potential energy, U—is a constant of the motion. Hence, we can write

$$K(x) = E - U(x).$$

Now, we also know that a kinetic energy can never be negative, so the above expression tells us that the motion of the mass is restricted to the region (or

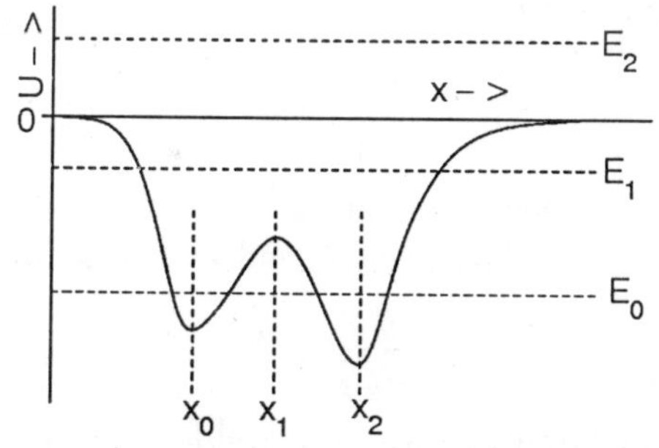

Fig. General 1-dimensional potential

regions) in which the potential energy curve $U(x)$ falls below the value E. Suppose that the total energy of the system is E_0.

The mass is trapped inside one or other of the two dips in the potential—these dips are generally referred to as potential wells. Suppose that we now raise the energy to E_1.

In this case, the mass is free to enter or leave each of the potential wells, but its motion is still bounded to some extent, since it clearly cannot move off to infinity. Finally, let us raise the energy to E_2. Now the mass is unbounded: i.e., it can move off to infinity. In systems in which it makes sense to adopt the convention that the potential energy at infinity is zero, bounded systems are characterized by $E < 0$, whereas unbounded systems are characterized by $E > 0$.

The above discussion suggests that the motion of a mass moving in a potential generally becomes less bounded as the total energy E of the system increases. Conversely, we would expect the motion to become more bounded as E decreases. In fact, if the energy becomes sufficiently small, it appears likely that the system will settle down in some equilibrium state in which the mass is stationary. Let us try to identify any prospective equilibrium states. If the mass remains stationary then it must be subject to zero force (otherwise it would accelerate). Hence, an equilibrium state is characterized by

$$\frac{dU}{dx} = 0.$$

A equilibrium state corresponds to either a maximum or a minimum of the potential energy curve $U(x)$. It can be seen that the U(x) curve shown has three associated equilibrium states: these are located at $x = x_0$, $x = x_1$, and $x = x_2$.

Let us now make a distinction between stable equilibrium points and unstable equilibrium points. When the system is slightly perturbed from a stable equilibrium point then the resultant force f should always be such as to attempt to return the system to this point. In other words, if $x = x_0$ is an equilibrium point, then we require

$$\left.\frac{df}{dx}\right|_{x_0} < 0$$

for stability: i.e., if the system is perturbed to the right, so that $x - x_0 > 0$, then the force must act to the left, so that $f < 0$, and vice versa. Likewise, if

$$\left.\frac{df}{dx}\right|_{x_0} > 0$$

then the equilibrium point $x = x_0$ is unstable. It follows, from equation, that stable equilibrium points are characterized by

$$\frac{d^2U}{dx^2} > 0.$$

In other words, a stable equilibrium point corresponds to a minimum of the po-tential energy curve $U(x)$. Likewise, an unstable equilibrium point corresponds to a maximum of the $U(x)$ curve. Hence, we conclude that $x = x_0$ and $x = x_2$ are stable equilibrium points, whereas $x = x_1$ is an unstable equilibrium point.

Of course, this makes perfect sense if we think of $U(x)$ as a gravitational potential energy curve, in which case U is directly proportional to height. All we are saying is that it is easy to confine a low energy mass at the bottom of a valley, but very difficult to balance the same mass on the top of a hill (since any slight perturbation to the mass will cause it to fall down the hill). Note, finally, that if

$$\frac{dU}{dx} = \frac{d^2U}{dx^2} = 0$$

at any point (or in any region) then we have what is known as a neutral equilib-rium point. We can move the mass slightly off such a point and it will still remain

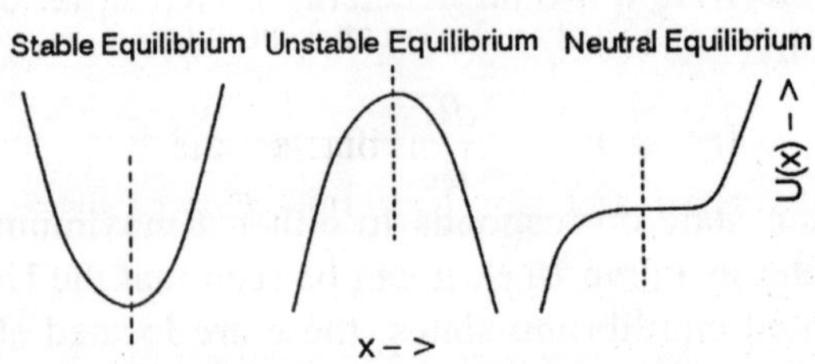

Fig. Different types of Equilibrium

in equilibrium (i.e., it will neither attempt to return to its initial state, nor will it continue to move). A neutral equilibrium point corresponds to a flat spot in a $U(x)$ curve.

POWER

Suppose that an object moves in a general force- field $f(r)$. We now know how to calculate how much energy flows from the force- field to the object as it moves along a given path between two points. Let us now consider the rate at which this energy flows. If dW is the amount of work that the force-field performs on the mass in a time interval dt then the rate of working is given by

$$P = \frac{dW}{dt}.$$

In other words, the rate of working—which is usually referred to as the power—is simply the time derivative of the work performed. Incidentally, the mks unit of power is called the watt (symbol W). In fact, 1 watt equals 1 kilogram metersquared per second-cubed, or 1 joule per second.

Suppose that the object displaces by dr in the time interval dt. By definition, the amount of work done on the object during this time interval is given by

$$dW = f \cdot dr.$$

It follows from equation that

$$P = f \cdot v,$$

where $v = dr/dt$ is the object's instantaneous velocity. Note that power can be positive or negative, depending on the relative directions of the vectors f and v. If these two vectors are mutually perpendicular then the power is zero. For the case of 1-dimensional motion, the above expression reduces to

$$P = fv.$$

In other words, in 1-dimension, power simply equals force times velocity.

BUCKET LIFTED FROM A WELL

Question: A man lifts a 30 kg bucket from a well whose depth is 150 m. Assuming that the man lifts the bucket at a constant rate, how much work does he perform?

Answer: Let m be the mass of the bucket and h the depth of the well. The gravitational force f acting on the bucket is of magnitude mg and is directed vertically downwards. Hence, $f = -mg$ (where upward is defined to be positive).

The net upward displacement of the bucket is h. Hence, the work W' performed by the gravitational force is the product of the (constant) force and the displacement of the bucket along the line of action of that force:

$$W' = f\,h = -\,mgh.$$

Note that W' is negative, which implies that the gravitational field surrounding the bucket gains energy as the bucket is lifted. In order to lift the bucket at a constant rate, the man must exert a force f on the bucket which balances (and very slightly exceeds) the force due to gravity. Hence, $f = -f'$. It follows that the work W done by the man is

$$W = fh = mgh = 30 \times 150 \times 9.81 = 4.415 \times 10^4 \text{ J}.$$

Note that the work is positive, which implies that the man expends energy whilst lifting the bucket. Of course, since $W = -W'$, the energy expended by the man equals the energy gained by the gravitational field.

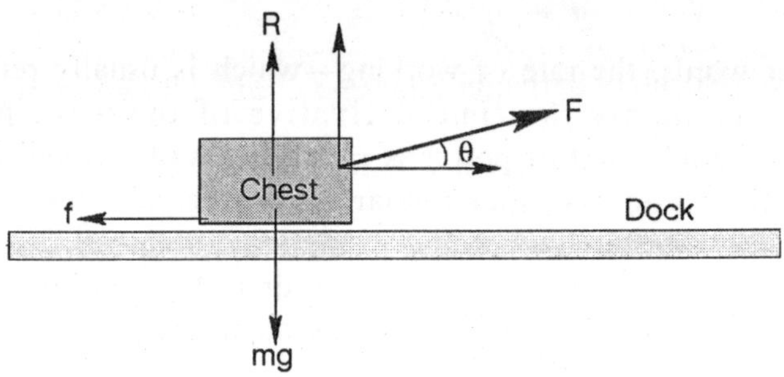

DRAGGING A TREASURE CHEST

Question: A pirate drags a 50 kg treasure chest over the rough surface of a dock by exerting a constant force of 95 N acting at an angle of 15– above the horizontal.

The chest moves 6 m in a straight line, and the coefficient of kinetic friction between the chest and the dock is 0:15. How much work does the pirate perform? How much energy is dissipated as heat via friction? What is the final velocity of the chest?

Answer: Referring to the diagram, the force F exerted by the pirate can be re-solved into a horizontal component $F\cos\theta$ and a vertical component $F\sin\theta$. Since the chest only moves horizontally, the vertical component of F performs zero work.

The work W performed by the horizontal component is simply the magnitude of this component times the horizontal distance x moved by the chest:

$$W = F\cos\theta\, x = 95 \times \cos 15^\circ \times 6 = 550.6 \text{ J}.$$

The chest is subject to the following forces in the vertical direction: the down-ward force $m\,g$ due to gravity, the upward reaction force R due to the dock, and the upward component $F\sin\theta$ of the force exerted by the pirate. Since the chest does not accelerate in the vertical direction, these forces must balance. Hence,

$$R = mg - F\sin\theta = 50 \times 9.81 - 95 \times \sin 15^\circ = 465.9 \text{ N}.$$

The frictional force f is the product of the coefficient of kinetic friction μ_k and the normal reaction R, so

$$f = \mu_k R = 0.15 \times 465.9 = 69.89 \text{ N}.$$

The work W' done by the frictional force is

$$W' = -f x = -69.89 \times 6 = -419.3 \text{ J}.$$

Note that there is a minus sign in front of the f because the displacement of the chest is in the opposite direction to the frictional force. The fact that W' is negative indicates a loss of energy by the chest: this energy is dissipated as heat via friction. Hence, the dissipated energy is 419:3 J. The final kinetic energy K of the chest (assuming that it is initially at rest) is the difference between the work W done by the pirate and the energy $-W'$ dissipated as heat. Hence,

$$K = W + W' = 550.6 - 419.3 = 131.3 \text{ J}.$$

Since $K = (1/2)\, mv^2$, the final velocity of the chest is

$$v = \sqrt{\frac{2K}{m}} = \sqrt{\frac{2\times 131.3}{50}} = 2.29 m/s.$$

STRETCHING A SPRING

Question: The force required to slowly stretch a spring varies from 0 N to 105 N as the spring is extended by 13 cm from its unstressed length. What is the force constant of the spring? What work is done in stretching the spring? Assume that the spring obeys Hooke's law.

Answer: The force f that the spring exerts on whatever is stretching it is $f = -kx$, where k is the force constant, and x is the extension of the spring. The minus sign indicates that the force acts in the opposite direction to the extension.

Since the spring is stretched slowly, the force f' which must be exerted on it is (almost) equal and opposite to f. Hence, $f' = -f = kx$. We are told that $f' = 105$ N when $x = 0.13m$. It follows that

$$k = \frac{105}{0.13} = 807.7 N/m.$$

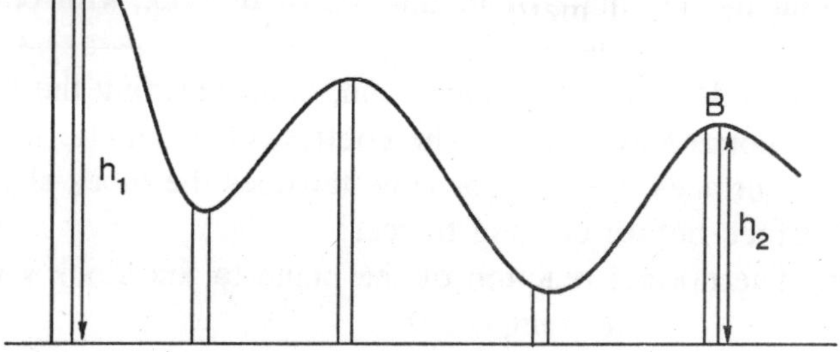

The work *W*A done by the external force in extending the spring from 0 to x is

$$W' = \int_0^x f' dx = k\int_0^x x\,dx = \frac{1}{2}kx^2.$$

Hence,

$$W' = 0.5 \times 807.7 \times 0.13^2 = 6.83\text{ J}.$$

ROLLER COASTER RIDE

Question: A roller coaster cart of mass $m = 300$ kg starts at rest at point A, whose height off the ground is $h_1 = 25\ m$, and a little while later reaches point B, whose height off the ground is $h_2 = 7\ m$. What is the potential energy of the cart relative to the ground at point A? What is the speed of the cart at point B, neglecting the effect of friction?

Answer: The gravitational potential energy of the cart with respect to the ground at point A is

$$U_A = mg\,h_1 = 300 \times 9.81 \times 25 = 7.36 \times 10^4\text{ J}.$$

Likewise, the potential energy of the cart at point B is

$$U_B = mg\,h_2 = 300 \times 9.81 \times 7 = 2.06 \times 10^4\text{ J}.$$

Hence, the change in the cart's potential energy in moving from point A to point B is

$$\Delta U = U_B - U_A = 2.06 \times 10^4 - 7.36 \times 10^4 = -5.30 \times 10^4\text{ J}.$$

By energy conservation, $\Delta K = -\Delta U$, where K represents kinetic energy. However, since the initial kinetic energy is zero, the change in kinetic energy ΔK is equivalent to the final kinetic energy K_B. Thus,

$$K_B = -\Delta U = 5.30 \times 10^4\text{ J}.$$

Now, $K_B = (1/2)\,mv^2_B$, where v_B is the final speed. Hence,

$$v_B = \sqrt{\frac{2K_B}{m}} = \sqrt{\frac{2\times 5.30\times 10^4}{300}} = 18.8 m/s.$$

SLIDING DOWN A PLANE

Question: A block of mass $m = 3$ kg starts at rest at a height of $h = 43$ cm on a plane that has an angle of inclination of $\theta = 35^\circ$ with respect to the horizontal.

The block slides down the plane, and, upon reaching the bottom, then slides along a horizontal surface. The coefficient of kinetic friction of the block on both surfaces is $\mu = 0.25$. How far does the block slide along the horizontal surface before coming to rest?

Answer: The normal reaction of the plane to the block's weight is

$$R = mg\cos\theta.$$

Hence, the frictional force acting on the block when it is sliding down the plane is

$$f = \mu R = 0.25 \times 3 \times 9.81 \times \cos 35^\circ = 6.03 \text{ N}.$$

The change in gravitational potential energy of the block as it slides down the plane is

$$\Delta U = -\,mgh = -\,3 \times 9.81 \times 0.43 = -12.65 \text{ J}.$$

The work W done on the block by the frictional force during this process is

$$W = -\,fx,$$

where $x = h/\sin\theta$ is the distance the block slides. The minus sign indicates that f acts in the opposite direction to the displacement of the block. Hence,

$$W = -\,\frac{6.03 \times 0.43}{\sin 35^\circ} = -4.52J.$$

Now, by energy conservation, the kinetic energy K of the block at the bottom of the plane equals the decrease in the block's potential energy plus the amount of work done on the block:

$$\text{K} = -\,\Delta U + W = 12.65 - 4.52 = 8.13 \text{ J}.$$

The frictional force acting on the block when it slides over the horizontal surface is

$$f' = \mu\, mg = 0.25 \times 3 \times 9.81 = 7.36 \text{ N}.$$

The work done on the block as it slides a distance y over this surface is

$$W' = -\,f\,y.$$

By energy conservation, the block comes to rest when the action of the frictional force has drained all of the kinetic energy from the block: i.e., when $W' = -\,K$. It follows that

$$y = \frac{K}{f'} = \frac{8.13}{7.36} = 1.10m.$$

DRIVING UP AN INCLINE

Question: A car of weight 3000 N possesses an engine whose maximum power output is 160 kW.

The maximum speed of this car on a level road is 35 m/s. Assuming that the resistive force (due to a combination of friction and air re-sistance) remains constant, what is the car's maximum speed on an incline of 1 in 20 (i.e., if θ is the angle of the incline with respect to the horizontal, then $\sin\theta = 1/20$)?

Answer: When the car is traveling on a level road at its maximum speed, v, then all of the power output, *P*, of its engine is used to overcome the power dissipated by the resistive force, f. Hence,

$$P = fv$$

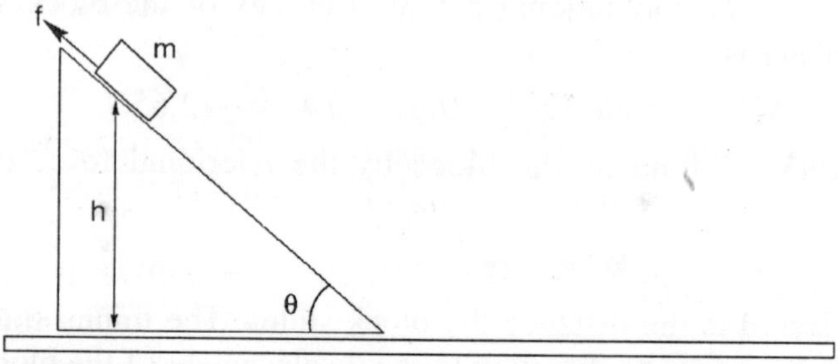

where the left-hand side is the power output of the engine, and the right-hand side is the power dissipated by the resistive force (i.e., minus the rate at which this force does work on the car). It follows that

$$f = \frac{P}{v} = \frac{160\times10^3}{35} = 4.57\times10^3 N.$$

When the car, whose weight is *W*, is traveling up an incline, whose angle with respect to the horizontal is θ, it is subject to the additional force $f = W\sin\theta$, which acts to impede its motion. Of course, this force is just the component of the car's weight acting down the incline. Thus, the new power balance equation is written

$$P = fv' + W\sin\theta\, v',$$

where v' is the maximum velocity of the car up the incline. Here, the left-hand side represents the power output of the car, whereas the right-hand side represents the sum of the power dissipated by the resistive force and the power expended to overcome the component of the car's weight acting down the incline. It follows that

$$v' = \frac{P}{f + W\sin\theta} = \frac{160\times10^3}{1.57\times10^3 + 3000/20} = 33.90 m/s.$$

CONSERVATION OF MOMENTUM

The behaviour of dynamical systems which consist of single point masses (i.e., objects whose spatial extent is either negligible or plays no role in their motion) or arrangements of point masses which are constrained to move together because they are connected via inextensible cables.

Let us now broaden our approach somewhat in order to take into account systems of point masses which exert forces on one another, but are not necessarily constrained to move together. The classic example of such a multi-component point mass system is one in which two (or more) freely moving masses collide with one another. The physical concept which plays the central role in the dynamics of multi-component point mass systems is the conservation of momentum.

TWO-COMPONENT SYSTEMS

The simplest imaginable multi-component dynamical system consists of two point mass objects which are both constrained to move along the same straight-line. Let x_1 be the displacement of the first object, whose mass is m_1.

Likewise, let x_2 be the displacement of the second object, whose mass is m_2. Suppose that the first object exerts a force f_{21} on the second object, whereas the second object exerts a force f_{12} on the first. From Newton's third law of motion, we have

$$f_{12} = -f_{21}.$$

Suppose, finally, that the first object is subject to an external force (i.e., a force which originates outside the system) F_1, whilst the second object is subject to an external force F_2.

Applying Newton's second law of motion to each object in turn, we obtain

$$m_1\,\ddot{x}_1 = f_{12} + F_1,$$
$$m_2\,\ddot{x}_2 = f_{21} + F_2.$$

Here, is a convenient shorthand for d/dt. Likewise, means d^2/dt^2.

At this point, it is helpful to introduce the concept of the centre of mass. The centre of mass is an imaginary point whose displacement x_{cm} is defined to be the mass weighted average of the displacements of the two objects which constitute the system. In other words,

$$x_{cm} = \frac{m_1x_1 + m_2x_2}{m_1 + m_2}$$

Thus, if the two masses are equal then the centre of mass lies half way between them, if the second mass is three times larger than the first then the centre of mass lies three-quarters of the way along the line linking the first and second masses, respectively, if the second mass is much larger than the first then the centre of mass is almost coincident with the second mass, and so on.

Summing Eqs. and, and then making use of equations. and, we obtain

$$m_1\,\ddot{x}_1 + m_2\,\ddot{x}_2 = (m_1 + m_2)\,\ddot{x}_{cm} = F_1 + F_2.$$

Note that the internal forces, f_{12} and f_{21}, have canceled out. The physical significance of this equation becomes clearer if we write it in the following form.

$$M\,\ddot{x}_{cm} = F,$$

where $M = m_1 + m_2$ is the total mass of the system, and $F = F_1 + F_2$ is the net external force acting on the system. Thus, the motion of the centre of mass is equivalent to that which would occur if all the mass contained in the system were collected at the centre of mass, and this conglomerate mass were then acted upon by the net external force. In general, this suggests that the motion of the centre of mass is simpler than the motions of the component masses, m_1 and m_2.

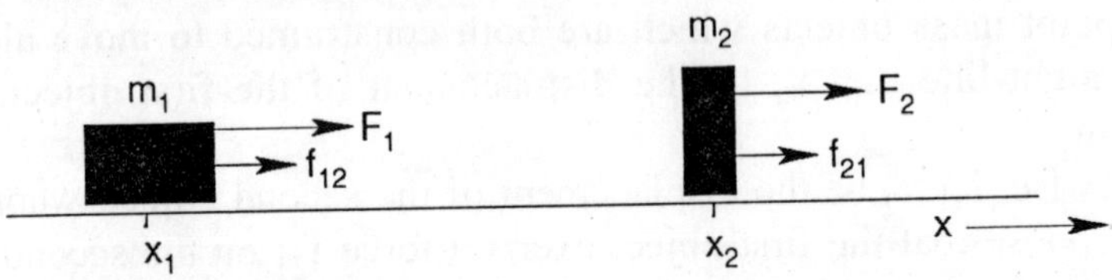

Fig. A 1-dimensional Dynamical System Consisting of two point mass Objects

This is particularly the case if the internal forces, f_{12} and f_{21}, are complicated in nature.

Suppose that there are no external forces acting on the system (i.e., $F_1 = F_2 = 0$), or, equivalently, suppose that the sum of all the external forces is zero (i.e., $F = F_1 + F_2 = 0$). The motion of the centre of mass is governed by Newton's first law of motion. i.e., it consists of uniform motion in a straight-line. Hence, in the absence of a net external force, the motion of the centre of mass is almost certainly far simpler than that of the component masses.

Now, the velocity of the centre of mass is written

$$v_{cm} = \dot{x}_{cm} = \frac{m_1\dot{x}_1 + m_2\dot{x}_2}{m_1 + m_2}.$$

We have seen that in the absence of external forces v_{cm} is a constant of the motion (i.e., the centre of mass does not accelerate). It follows that, in this case,

$$m_1 \dot{x}_1 + \dot{x}_2 \dot{x}_2$$

is also a constant of the motion. Momentum is defined as the product of mass and velocity. Hence, the momentum of the first mass is written $p_1 = m_1 \dot{x}_1$, whereas the momentum of the second mass takes the form $p_2 = m_2 \dot{x}_1$. It follows that the above expression corresponds to the total momentum of the system.

$$P = p_1 + p_2.$$

Thus, the total momentum is a conserved quantity—provided there is no net external force acting on the system. This is true irrespective of the nature of the internal forces. More generally, equation can be written

$$\frac{dP}{dt} = F.$$

In other words, the time derivative of the total momentum is equal to the net external force acting on the system—this is just Newton's second law of motion applied to the system as a whole.

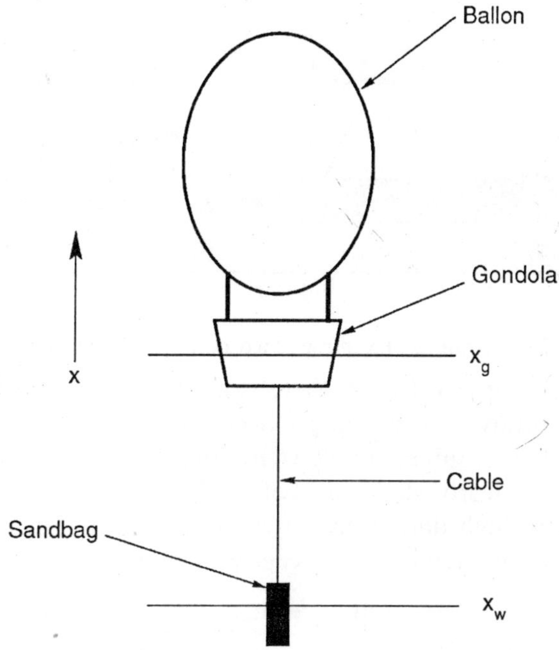

Fig. An Example Two-component System

Let us now try to apply some of the concepts discussed above. Consider the simple two-component system shown. A gondola of mass m_g hangs from a hot-air balloon whose mass is negligible compared to that of the gondola. A sandbag of mass m_w is suspended from the gondola by means of a light inextensible cable.

The system is in equilibrium. Suppose, for the sake of consistency with our other examples, that the x-axis runs vertically upwards. Let x_g be the height of the gondola, and x_w the height of the sandbag. Suppose that the upper end of the cable is attached to a winch inside the gondola, and that this winch is used to slowly shorten the cable, so that the sandbag is lifted upwards a distance Δx_w. The question is this. does the height of the gondola also change as the cable is reeled in? If so, by how much?

Let us identify all of the forces acting on the system. The internal forces are the upward force exerted by the gondola on the sandbag, and the downward force exerted by the sandbag on the gondola.

These forces are transmitted via the cable, and are equal and opposite (by Newton's third law of motion).

The external forces are the net downward force due to the combined weight of the gondola and the sandbag, and the upward force due to the buoyancy of the balloon. Since the system is in equilibrium, these forces are equal and opposite (it is assumed that the cable is reeled in sufficiently slowly that the

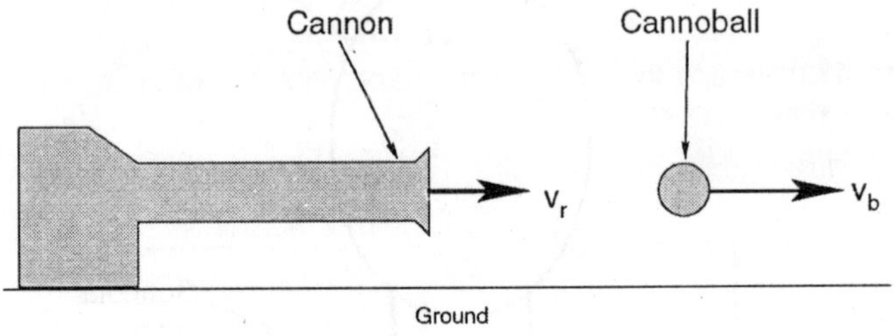

Fig. Another Example Two-component System

equilibrium is not upset). Hence, there is zero net external force acting on the system. It follows, from the previous discussion, that the centre of mass of the system is subject to Newton's first law. In particular, since the centre of mass is clearly stationary before the winch is turned on, it must remain stationary both during and after the period in which the winch is operated. Hence, the height of the centre of mass,

$$x_{cm} = \frac{m_g x_g + m_w x_w}{m_g + m_w},$$

is a conserved quantity.

Suppose that the operation of the winch causes the height of the sandbag to change by Δx_w, and that of the gondola to simultaneously change by Δx_g. If x_{cm} is a conserved quantity, then we must have

$$0 = m_g\ \Delta x_g + m_w\ \Delta x_w,$$

or

$$\Delta x_g = \frac{m_w}{m_g}\Delta x_w.$$

Thus, if the winch is used to raise the sandbag a distance Δx_w then the gondola is simultaneously pulled downwards a distance (m_w=m_g) Δx_w. It is clear that we could use a suspended sandbag as a mechanism for adjusting a hot-air balloon's altitude. the balloon descends as the sandbag is raised, and ascends as it is lowered.

Suppose that a cannon of mass M propels a cannonball of mass m horizontally with velocity v_b. What is the recoil velocity v_r of the cannon? Let us first identify all of the forces acting on the system. The internal forces are the force exerted by the cannon on the cannonball, as the cannon is Δred, and the equal and opposite force exerted by the cannonball on the cannon. These forces are extremely large, but only last for a short instance in time. in physics, we call these impulsive forces. There are no external forces acting in the horizontal direction (which is the only direction that we are considering in this example). It follows that the total (horizontal) momentum P of the system is a conserved quantity. Prior to the Firing of the cannon, the total momentum is zero (since momentum is mass times velocity, and nothing is initially moving). After the cannon is Δred, the total momentum of the system takes the form

$$P = mv_b + Mv_r.$$

Since P is a conserved quantity, we can set P = 0. Hence,

$$v_r = -\frac{m}{M} v_b .$$

Thus, the recoil velocity of the cannon is in the opposite direction to the velocity of the cannonball (hence, the minus sign in the above equation), and is of magnitude (m=M) v_b. Of course, if the cannon is far more massive that the cannonball (i.e., M >> m), which is usually the case, then the recoil velocity of the cannon is far smaller in magnitude than the velocity of the cannonball.

Note, however, that the momentum of the cannon is equal in magnitude to that of the cannonball. It follows that it takes the same effort (i.e., force applied for a certain period of time) to slow down and stop the cannon as it does to slow down and stop the cannonball.

MULTI-COMPONENT SYSTEMS

Consider a system of N mutually interacting point mass objects which move in 3-dimensions. Let the ith object, whose mass is m_i, be located at vector displacement r_i. Suppose that this object exerts a force f_{ji} on the jth object. By Newton's third law of motion, the force f_{ij} exerted by the jth object on the ith is given by

$$f_{ij} = -f_{ji}.$$

Finally, suppose that the ith object is subject to an external force F_i. Newton's second law of motion applied to the ith object yields

$$m_i \ddot{r}_i = \sum_{j=1,N}^{j\neq i} f_{ij} + F_i$$

Note that the summation on the right-hand side of the above equation excludes the case j = i, since the ith object cannot exert a force on itself. Let us now take the above equation and sum it over all objects. We obtain

$$\sum_{i=1,N} m_i \ddot{r}_i \sum_{i=1,N}^{j\neq i} f_{ij} + \sum_{i=1,N} F_i .$$

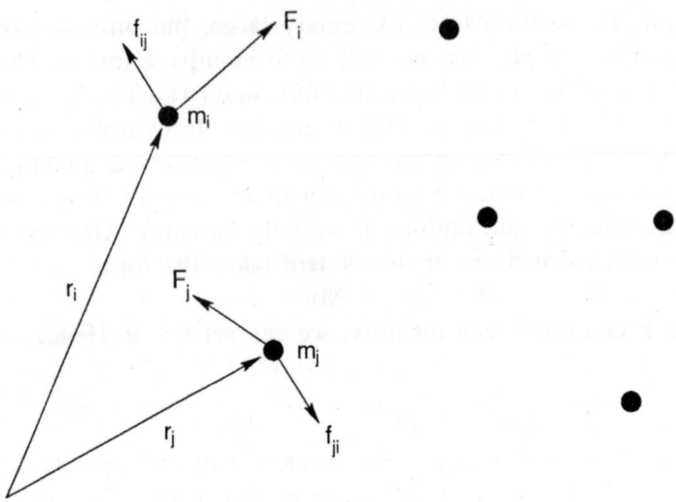

Fig. A 3-dimensional Dynamical System Consisting of Many Point Mass Objects.

Consider the sum over all internal forces. i.e., the first term on the right-hand side.

Each element of this sum—f_{ij}, say—can be paired with another element— f_{ji}, in this case—which is equal and opposite. In other words, the elements of the sum all cancel out in pairs. Thus, the net value of the sum is zero. It follows that the above equation can be written

$$M\,\ddot{r}_{cm} = F,$$

where $M = \sum_{i=1}^{N} m_i$ is the total mass, and $F = \sum_{i=1}^{N} F_i$ is the net external force.

The quantity r_{cm} is the vector displacement of the centre of mass. The centre of mass is an imaginary point whose coordinates are the mass weighted averages of the coordinates of the objects which constitute the system. Thus,

$$r_{cm} = \frac{\sum_{i=1}^{N} m_i r_i}{\sum_{i=1}^{N} m_i}$$

The motion of the centre of mass is equivalent to that which would be obtained if all the mass contained in the system were collected at the centre of mass, and this conglomerate mass were then acted upon by the net external force. The motion of the centre of mass is likely to be far simpler than the motions of the component masses.

Suppose that there is zero net external force acting on the system, so that F = 0. Equation implies that the centre of mass moves with uniform velocity in a straight-line. In other words, the velocity of the centre of mass,

$$\dot{r}_{cm} = \frac{\sum_{i=1}^{N} m_i \dot{r}_i}{\sum_{i=1}^{N} m_i},$$

is a constant of the motion. Now, the momentum of the ith object takes the form $p_i = m_i \dot{r}_i$. Hence, the total momentum of the system is written

$$P = \sum_{i=1}^{N} m_i \dot{r}_i$$

A comparison of Eqs. and suggests that P is also a constant of the motion when zero net external force acts on the system. Finally, equation can be rewritten

$$\frac{dP}{dt} = F.$$

In other words, the time derivative of the total momentum is equal to the net external force acting on the system.

It is clear, from the above discussion, that most of the important results obtained in the previous section, for the case of a two-component system moving in 1-dimension, also apply to a multi-component system moving in 3-dimensions.

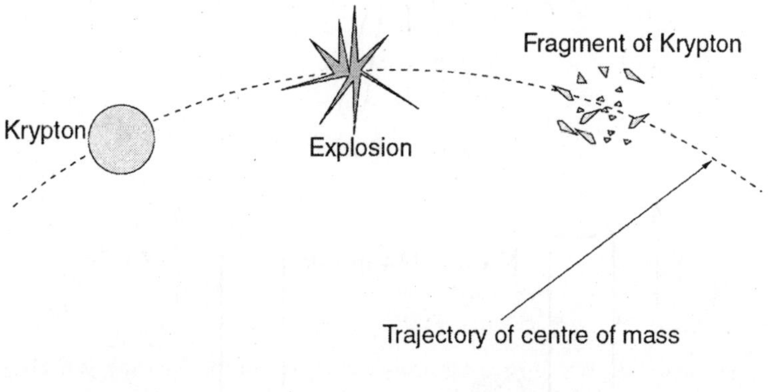

Fig. The Unfortunate History of the Planet Krypton

Let us con-sider the unfortunate history of the planet Krypton. Krypton—Superman's home planet—eventually exploded. Note, however, that before, during, and after this explosion the net external force acting on Krypton, or the fragments of Krypton—namely, the gravitational attraction to Krypton's sun—remained the same. In other words, the forces responsible for the explo-sion can be thought of as large, transitory, internal forces. We conclude that the motion of the centre of mass of Krypton, or the fragments of Krypton, was unaffected by the explosion. This follows, since the motion of the centre of mass is independent of internal forces.

Before the explosion, the planet Krypton presumably executed a standard elliptical orbit around Krypton's sun. We conclude that, after the explosion, the fragments of Krypton (or, to be more exact, the centre of mass of these fragments) continued to execute exactly the same orbit.

ROCKET SCIENCE

A rocket engine is the only type of propulsion device that operates effectively in outer space. A rocket works by ejecting a propellant at high velocity from its rear end. The rocket exerts a backward force on the propellant, in order to eject it, and, by Newton's third law, the propellant exerts an equal and opposite force on the rocket, which propels it forward.

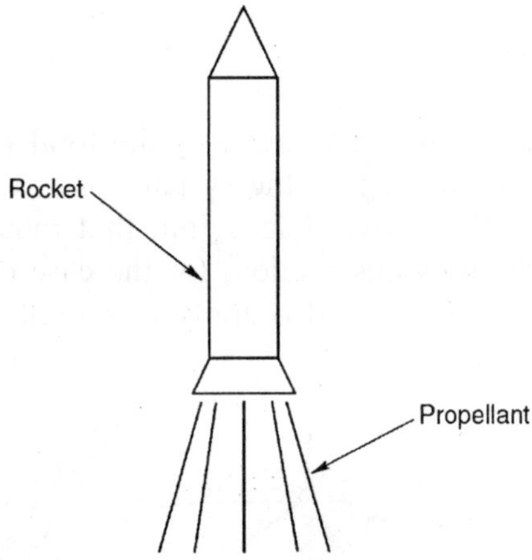

Fig. A Rocket.

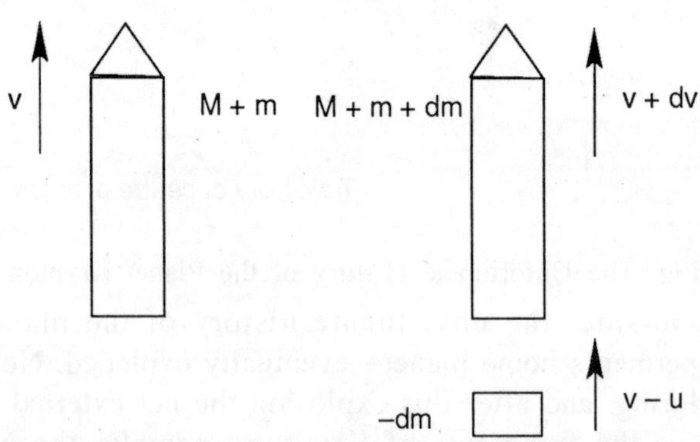

Fig. Derivation of the Rocket Equation.

Let us attempt to find the equation of motion of a rocket. Let M be the fixed mass of the rocket engine and the payload, and m(t) the total mass of the propellant contained in the rocket's fuel tanks at time t. Suppose that the rocket engine ejects the propellant at some fixed velocity u relative to the rocket.

Let us examine the rocket at two closely spaced instances in time. Suppose that at time t the rocket and propellant, whose total mass is M + m, are traveling with instantaneous velocity v.

Suppose, further, that between times t and t + dt the rocket ejects a quantity of propellant of mass –dm (n.b., dm is understood to be negative, so this represents a positive mass) which travels with velocity v – u (i.e., velocity -u in the instantaneous rest frame of the rocket). As a result of the fuel ejection, the velocity of the rocket at time t + dt is boosted to v + dv, and its total mass becomes M + m + dm.

Now, there is zero external force acting on the system, since the rocket is assumed to be in outer space. It follows that the total momentum of the system is a constant of the motion. Hence, we can equate the momenta evaluated at times t and t + dt.

$$(M + m)v = (M + m + dm)(v + dv) + (-dm)(v - u).$$

Neglecting second order quantities (i.e., dm dv), the above expression yields

$$0 = (M + m)dv + udm.$$

Rearranging, we obtain

$$\frac{dv}{u} = \frac{dm}{M+m}.$$

Let us integrate the above equation between an initial time at which the rocket is fully fueled—i.e., $m = m_p$, where m_p is the maximum mass of propellant that the rocket can carry—but stationary, and a final time at which the mass of the fuel is m and the velocity of the rocket is v. Hence,

$$\int_0^v \frac{dv}{u} = -\int_{m_p}^{m} \frac{dm}{M+m}.$$

It follows that

$$\left[\frac{v}{u}\right]_{v=0}^{v=v} = -\left[\ln(M+m)\right]_{m=m_p}^{m=m},$$

which yields

$$v = u\ln\left(\frac{M+m_p}{M+m}\right).$$

The final velocity of the rocket (i.e., the velocity attained by the time the rocket has exhausted its fuel, so that m = 0) is

$$v_f = u \ln\left(1 + \frac{m_p}{M}\right)$$

Note that, unless the initial mass of the fuel exceeds the fixed mass of the rocket by many orders of magnitude (which is highly unlikely), the final velocity v_f of the rocket is similar to the velocity u with which fuel is ejected from the rear of the rocket in its instantaneous rest frame. This follows because $\ln x \sim O(1)$, unless x becomes extremely large.

Let us now consider the factors which might influence the design of a rocket for use in interplanetary or interstellar travel. Since the distances involved in such travel are vast, it is important that the rocket's final velocity be made as large as possible, otherwise the journey is going to take an unacceptably long time.

However, as we have just seen, the factor which essentially determines the final velocity v_f of a rocket is the speed of ejection u of the propellant relative to the rocket. Broadly speaking, v_f can never significantly exceed u. It follows that a rocket suitable for interplanetary or interstellar travel should have as high an ejection speed as practically possible.

Now, ordinary chemical rockets (the kind which powered the Apollo moon programme) can develop enormous thrusts, but are limited to ejection velocities below about 5000 m/s. Such rockets are ideal for lifting payloads out of the Earth's gravitational field, but their relatively low ejection velocities render them unsuitable for long distance space travel.

A new type of rocket engine, called an ion thruster, is currently under development. ion thrusters operate by accelerating ions electrostatically to great velocities, and then ejecting them.

Although ion thrusters only generate very small thrusts, compared to chemical rockets, their much larger ejection velocities (up to 100 times those of chemical rockets) makes them far more suitable for interplanetary or interstellar space travel. The first spacecraft to employ an ion thruster was the Deep Space 1 probe, which was launched from Cape Canaveral on October 24, 1998. this probe successfully encountered the asteroid 9969 Braille in July, 1999.

IMPULSES

Suppose that a ball of mass m and speed u_i strikes an immovable wall normally and rebounds with speed u_f. Clearly, the momentum of the ball is changed by the collision with the wall, since the direction of the ball's velocity is reversed. It follows that the wall must exert a force on the ball, since force is the rate of change of momentum.

This force is generally very large, but is only exerted for the short instance in time during which the ball is in physical contact with the wall. Physicists generally refer to such a force as an impulsive force.

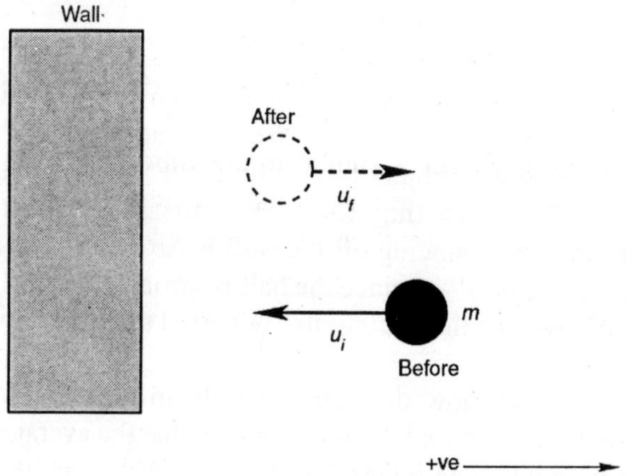

Fig. A Ball Bouncing off a Wall

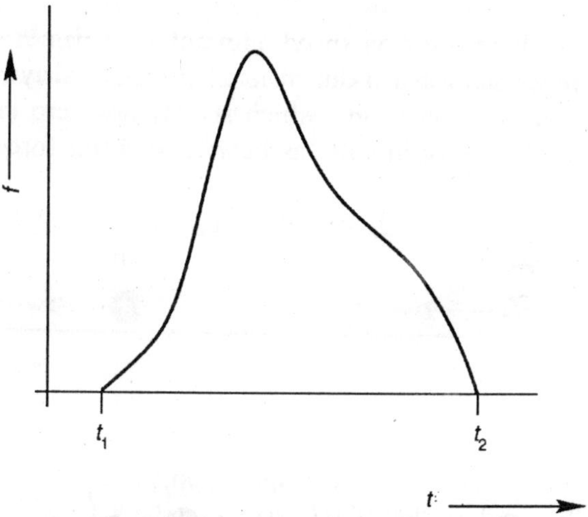

Fig. An Impulsive Force

The typical time history of an impulsive force, f(t). It can be seen that the force is only non-zero in the short time interval t_1 to t_2. It is helpful to define a quantity known as the net impulse, I, associated with f(t).

$$I = \int_{t_1}^{t_2} f(t)dt$$

In other words, I is the total area under the f(t) curve. Consider a object subject to the impulsive force. Newton's second law of motion yields

$$\frac{dp}{dt} = f,$$

where p is the momentum of the object. Integrating the above equation, making use of the definition, we obtain

$$\Delta p = I.$$

Here, $\Delta p = p_f - p_i$, where p_i is the momentum before the impulse, and p_f is the momentum after the impulse. We conclude that the net change in momentum of an object subject to an impulsive force is equal to the total impulse associated with that force. For instance, the net change in momentum of the ball bouncing off the wall is $\Delta p = m\,u_f - m\,(-u_i) = m\,(u_f + u_i)$. The initial velocity is $-u_i$, since the ball is initially moving in the negative direction. It follows that the net impulse imparted to the ball by the wall is $I = m\,(u_f + u_i)$.

Suppose that we know the ball was only in physical contact with the wall for the short time interval Δt. We conclude that the average force f exerted on the ball during this time interval was

$$\bar{f} = \frac{I}{\Delta t}$$

The above discussion is only relevant to 1-dimensional motion. However, the generalization to 3-dimensional motion is fairly straight-forward. Consider an impulsive force $f(t)$, which is only non-zero in the short time interval t_1 to t_2. The vector impulse associated with this force is simply

$$I = \int_{t_1}^{t_2} f(t)\,dt\,.$$

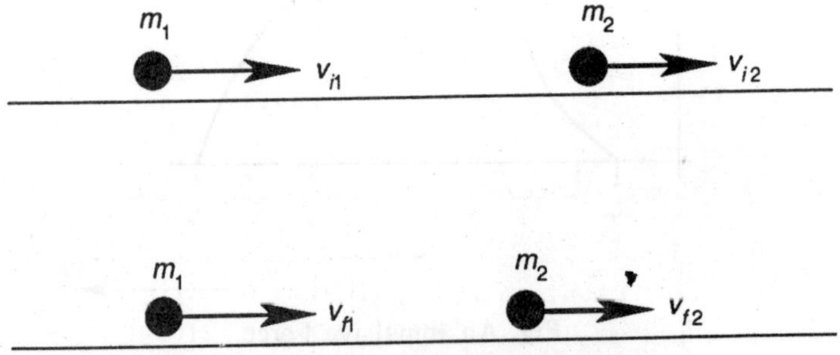

Fig. A Collision in 1-dimension.

The net change in momentum of an object subject to $f(t)$ is

$$\Delta p = I.$$

Finally, if $t_2 - t_1 = \Delta t$, then the average force experienced by the object in the time interval t_1 to t_2 is

$$\bar{f} = \frac{I}{\Delta t}.$$

COLLISIONS IN 1-DIMENSION

Consider two objects of mass m_1 and m_2, respectively, which are free to move in 1-dimension. Suppose that these two objects collide. Suppose, further, that both objects are subject to zero net force when they are not in contact with one another.

Both before and after the collision, the two objects move with constant velocity. Let v_{i1} and v_{i2} be the velocities of the first and second objects, respectively, before the collision.

Likewise, let v_{f1} and v_{f2} be the velocities of the first and second objects, respectively, after the collision. During the collision itself, the first object exerts a large transitory force f_{21} on the second, whereas the second object exerts an equal and opposite force $f_{12} = -f_{21}$ on the first. In fact, we can model the collision as equal and opposite impulses given to the two objects at the instant in time when they come together.

We are clearly considering a system in which there is zero net external force (the forces associated with the collision are internal in nature). Hence, the total momentum of the system is a conserved quantity. Equating the total momenta before and after the collision, we obtain

$$m_1\ vi_1 + m_2\ vi_2 = m_1\ vf_1 + m_2\ vf_2.$$

This equation is valid for any 1-dimensional collision, irrespective its nature. Note that, assuming we know the masses of the colliding objects, the above equation only fully describes the collision if we are given the initial velocities of both objects, and the final velocity of at least one of the objects. (Alternatively, we could be given both final velocities and only one of the initial velocities.)

There are many different types of collision. An elastic collision is one in which the total kinetic energy of the two colliding objects is the same before and after the collision. Thus, for an elastic collision we can write

$$\frac{1}{2}m_1 v_{i1}^2 + \frac{1}{2}m_2 v_{i2}^2 = \frac{1}{2}m_1 v_{f1}^2 + \frac{1}{2}m_2 v_{f2}^2,$$

in addition to equation. Hence, the collision is fully specified once we are given the two initial velocities of the colliding objects.

The majority of collisions occurring in real life are not elastic in nature. Some fraction of the initial kinetic energy of the colliding objects is usually converted into some other form of energy—generally heat energy, or energy associated with the mechanical deformation of the objects—during the collision.

Such collisions are termed inelastic. For instance, a large fraction of the initial kinetic energy of a typical automobile accident is converted into mechanical energy of deformation of the two vehicles. Inelastic collisions also occur during squash/racquetball/handball games. In each case, the ball becomes warm to the touch after a long game, because some fraction of the ball's kinetic energy of collision with the walls of the court has been converted into heat energy.

Equation remains valid for inelastic collisions—however, equation is invalid. Thus, generally speaking, an inelastic collision is only fully characterized when we are given the initial velocities of both objects, and the final velocity of at least one of the objects.

There is, however, a special case of an inelastic collision—called a totally inelastic collision—which is fully characterized once we are given the initial velocities of the colliding objects. In a totally inelastic collision, the two objects stick together after the collision, so that $v_{f1} = v_{f2}$.

Let us, now, consider elastic collisions in more detail. Suppose that we transform to a frame of reference which co-moves with the centre of mass of the system. The motion of a multi-component system often looks particularly simple when viewed in such a frame. Since the system is subject to zero net external force, the velocity of the centre of mass is invariant, and is given by

$$v_{cm} = \frac{m_1 v_{i1} + m_2 v_{i2}}{m_1 + m_2} = \frac{m_1 v_{f1} + m_2 v_{f2}}{m_1 + m_2}$$

An object which possesses a velocity v in our original frame of reference—henceforth, termed the laboratory frame—possesses a velocity $v' = v - v_{cm}$ mass frame. It is easily demonstrated that

$$v'_{i1} = -\frac{m_2}{m_1 + m_2}(v_{i2} - v_{i1}),$$

$$v'_{i2} = +\frac{m_1}{m_1 + m_2}(v_{i2} - v_{i1})$$

$$v'_{f1} = -\frac{m_2}{m_1 + m_2}(v_{f2} - v_{f1})$$

$$v'_{f2} = +\frac{m_1}{m_1 + m_2}(v_{f2} - v_{f1}).$$

The above equations yield

$$-p'_{i1} = p'_{i2} = \mu\,(v_{i2} - v_{i1}),$$

$$-p'_{f1} = p'_{f2} = \mu(v_{f2} - v_{f1}),$$

where $\mu = m_1\ m_2{=}(m_1 + m_2)$ is the so-called reduced mass, and $-p'_{i1} = m_1 v'_{i1}$ is the initial momentum of the first object in the centre of mass frame, etc. In other words, when viewed in the centre of mass frame, the two objects approach one another with equal and opposite momenta before the collision, and diverge from one another with equal and opposite momenta after the collision. Thus, the centre of mass momentum conservation equation,

$$p'_{i1} + p'_{i2} = p'_{f1} + p'_{f2},$$

is trivially satisfied, because both the left- and right-hand sides are zero. Incidentally, this result is valid for both elastic and inelastic collisions.

The centre of mass kinetic energy conservation equation takes the form

$$\frac{p'^2_{i1}}{2m_1} + \frac{p'^2_{i2}}{2m_2} = \frac{p'^2_{f1}}{2m_1} + \frac{p'^2_{f2}}{2m_2}.$$

Note, incidentally, that if energy and momentum are conserved in the laboratory frame then they must also be conserved in the centre of mass frame. A comparison of Eqs.,, and yields

$$(v_{i2} - v_{i1}) = -(v_{f2} - v_{f1}).$$

In other words, the relative velocities of the colliding objects are equal and opposite before and after the collision. This is true in all frames of reference, since relative velocities are frame invariant. Note, however, that this result only applies to fully elastic collisions.

Equations and can be combined to give the following pair of equations which fully specify the final velocities (in the laboratory frame) of two objects which collide elastically, given their initial velocities.

$$v_{f1} = \frac{(m_1 - m_2)}{m_1 + m_2} v_{i1} + \frac{2m_2}{m_1 + m_2} v_{i2},$$

$$v_{f2} = \frac{2m_1}{m_1 + m_2} v_{i1} - \frac{(m_1 - m_2)}{m_1 + m_2} v_{i2}.$$

Let us, now, consider some special cases. Suppose that two equal mass objects collide elastically. If $m_1 = m_2$ then Eqs. and yield

$$v_{f1} = v_{i2},$$

$$v_{f2} = v_{i1}.$$

In other words, the two objects simply exchange velocities when they collide. For instance, if the second object is stationary and the first object strikes it head-on with velocity v then the first object is brought to a halt whereas the second object moves off with velocity v.

It is possible to reproduce this effect in pool by striking the cue ball with great force in such a manner that it slides, rather that rolls, over the table—in this case, when the cue ball strikes another ball head-on it comes to a complete halt, and the other ball is propelled forward very rapidly. Incidentally, it is necessary to prevent the cue ball from rolling, because rolling motion is not taken into account in our analysis, and actually changes the answer. Suppose that the second object is much more massive than the first (i.e., $m_2 \gg m_1$) and is initially at rest (i.e., $v_{i2} = 0$). In this case, Eqs. and yield

$$v_{f1} \simeq -v_{i1},$$
$$v_{f2} \simeq 0.$$

In other words, the velocity of the light object is effectively reversed during the collision, whereas the massive object remains approximately at rest. Indeed, this is the sort of behaviour we expect when an object collides elastically with an immovable obstacle. e.g., when an elastic ball bounces off a brick wall. Suppose, finally, that the second object is much lighter than the first (i.e., $m_2 \ll m_1$) and is initially at rest (i.e., $v_{i2} = 0$). In this case, Eqs. and yield

$$v_{f1} \simeq v_{i1},$$
$$v_{f2} \simeq 2\, v_{i1}.$$

In other words, the motion of the massive object is essentially unaffected by the collision, whereas the light object ends up going twice as fast as the massive one. Let us, now, consider totally inelastic collisions in more detail. In a totally inelastic collision the two objects stick together after colliding, so they end up moving with the same final velocity $v_f = v_{f1} = v_{f2}$. Equation reduces to

$$v_f = \frac{m_1 v_{i1} + m_2 v_{i2}}{m_1 + m_2} = v_{cm}\,.$$

In other words, the common final velocity of the two objects is equal to the centre of mass velocity of the system. This is hardly a surprising result. We have already seen that in the centre of mass frame the two objects must diverge with equal and opposite momenta after the collision. However, in a totally inelastic collision these

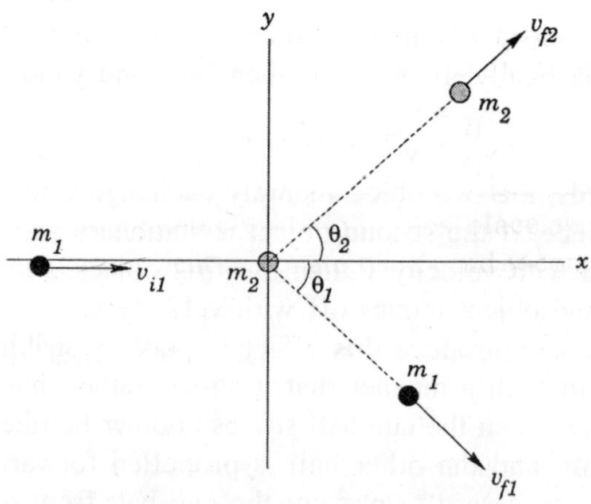

Fig. A Collision in 2-dimensions.

two momenta must also be equal (since the two objects stick together). The only way in which this is possible is if the two objects remain stationary in the centre of mass frame after the collision. Hence, the two objects move with the centre of mass velocity in the laboratory frame.

Suppose that the second object is initially at rest (i.e., $v_{i2} = 0$). In this special case, the common final velocity of the two objects is

$$v_f = \frac{m_1}{m_1 + m_2} v_{i1}.$$

Note that the first object is slowed down by the collision. The fractional loss in kinetic energy of the system due to the collision is given by

$$f = \frac{K_i - K_f}{K_i} = \frac{m_1 v_{i1}^2 - (m_1 + m_2) v_f^2}{m_1 v_{i1}^2} = \frac{m_2}{m_1 + m_2}.$$

The loss in kinetic energy is small if the (initially) stationary object is much lighter than the moving object (i.e., if $m_2 \ll m_1$), and almost 100% if the moving object is much lighter than the stationary one (i.e., if $m_2 \gg m_1$). Of course, the lost kinetic energy of the system is converted into some other form of energy. e.g., heat energy.

COLLISIONS IN 2-DIMENSIONS

Suppose that an object of mass m_1, moving with initial speed v_{i1}, strikes a second object, of mass m_2, which is initially at rest. Suppose, further, that the collision is not head-on, so that after the collision the first object moves off at an angle θ_1 to its initial direction of motion, whereas the second object moves off at an angle θ_2 to this direction. Let the final speeds of the two objects be v_{f1} and v_{f2}, respectively.

We are again considering a system in which there is zero net external force (the forces associated with the collision are internal in nature). It follows that the total momentum of the system is a conserved quantity. However, unlike before, we must now treat the total momentum as a vector quantity, since we are no longer dealing with 1-dimensional motion. Note that if the collision takes place wholly within the x-y plane, then it is sufficient to equate the x- and y-components of the total momentum before and after the collision.

Consider the x-component of the system's total momentum. Before the collision, the total x-momentum is simply $m_1 v_{i1}$, since the second object is initially stationary, and the first object is initially moving along the x-axis with speed v_{i1}.

After the collision, the x-momentum of the first object is m1 $v_{f1} \cos \theta_1$. i.e., m1 times the x-component of the first object's final velocity.

Likewise, the final xmomentum of the second object is $m_2 v_{f2} \cos \theta_2$. Hence, momentum conservation in the x-direction yields

$$m_1 v_{i1} = m_1 v_{f1} \cos \theta_1 + m_2 v_{f2} \cos \theta_2.$$

Consider the y-component of the system's total momentum. Before the collision, the total y-momentum is zero, since there is initially no motion along the y-axis. After the collision, the y-momentum of the first object is $-m_1 v_{f1} \sin \theta_1$. i.e., m_1 times the y-component of the first object's final velocity. Likewise, the final y-momentum of the second object is $m_2 v_{f2} \sin \theta_2$. Hence, momentum conservation in the y-direction yields

$$m_1 v_{f1} \sin \theta_1 = m_2 v_{f2} \sin \theta_2.$$

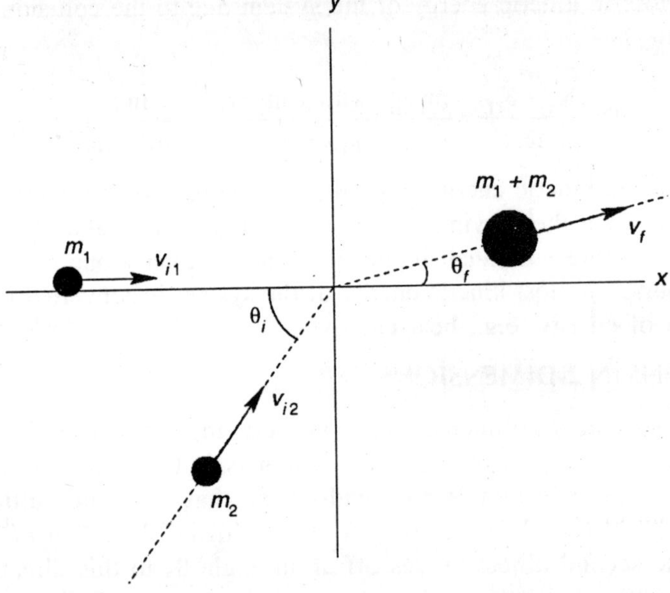

Fig. A Totally Inelastic Collision in 2-dimensions

For the special case of an elastic collision, we can equate the total kinetic energies of the two objects before and after the collision. Hence, we obtain

$$\frac{1}{2} m_1 v_{i1}^2 = \frac{1}{2} m_1 v_{f1}^2 + \frac{1}{2} m_2 v_{f2}^2.$$

Given the initial conditions (i.e., m_1, m_2, and v_{i1}), we have a system of three equations and four unknowns (i.e., θ_1, θ_2, v_{f1}, and v_{f2}). Clearly, we cannot uniquely solve such a system without being given additional information. e.g., the direction of motion or speed of one of the objects after the collision. Figure shows a 2-dimensional totally inelastic collision.

In this case, the first object, mass m_1, initially moves along the x-axis with speed v_{i1}. On the other hand, the second object, mass m_2, initially moves at an angle θ_i to the x-axis with speed v_{i2}.

After the collision, the two objects stick together and move off at an angle θ_f to the x-axis with speed v_f. Momentum conservation along the x-axis yields

$$m_1 v_{i1} + m_2 v_{i2} \cos \theta_i = (m_1 + m_2) v_f \cos \theta_f.$$

Likewise, momentum conservation along the y-axis gives

$$m_2 v_{i2} \sin \theta_i = (m_1 + m_2) v_f \sin \theta_f.$$

Given the initial conditions (i.e., m_1, m_2, v_{i1}, v_{i2}, and θ_i), we have a system of two equations and two unknowns (i.e., v_f and θ_f). Clearly, we should be able to find a unique solution for such a system.

Question: A cannon is bolted to the door of a railway carriage, which is free to move without friction along a straight track. The combined mass of the cannon and the carriage is $M = 1200$ kg. The cannon fires a cannonball, of mass $m = 1.2$ kg, horizontally with velocity $v = 115$ m/s. The cannonball travels the length of the carriage, a distance $L = 85$ m, and then becomes embedded in the carriage's end wall. What is the recoil speed of the carriage right after the cannon is fired? What is the velocity of the carriage after the cannonball strikes the far wall? What net distance, and in what direction, does the carriage move as a result of the firing of the cannon?

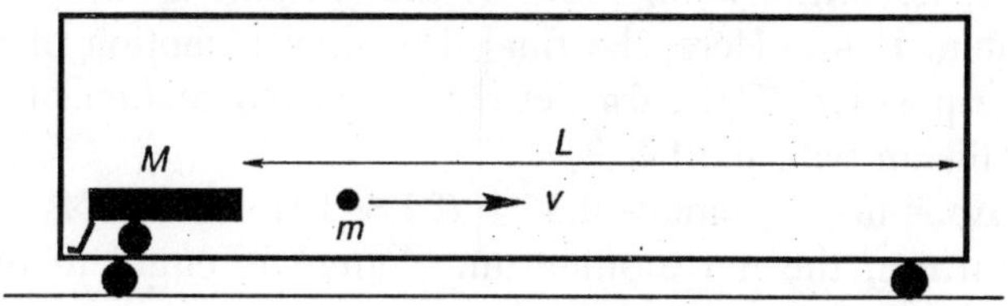

Answer: Conservation of momentum implies that the net horizontal momentum of the system is the same before and after the cannon is fired.

The momentum before the cannon is fired is zero, since nothing is initially moving. Hence, we can also set the momentum after the cannon is fired to zero, giving

$$0 = Mu + mv,$$

where u is the recoil velocity of the carriage. It follows that

$$u = -\frac{m}{M}v = -\frac{1.2 \times 115}{1200} = -0.115 m/s$$

The minus sign indicates that the recoil velocity of the carriage is in the opposite direction to the direction of motion of the cannonball. Hence, the recoil speed of the carriage is $|u| = 0.115$ m/s. Suppose that, after the cannonball strikes the far wall of the carriage, both the cannonball and the carriage move with common velocity w. Conservation of momentum implies that the net horizontal momentum of the system is the same before and after the collision. Hence, we can write

$$Mu + mv = (M + m)w.$$

However, we have already seen that M u + mv = 0. It follows that w = 0. in other words, the carriage is brought to a complete halt when the cannonball strikes its far wall. In the frame of reference of the carriage, the cannonball moves with velocity v – u after the cannon is fired. Hence, the time of Θight of the cannonball is

$$t = \frac{L}{v-u} = \frac{85}{115+0.115} = 0.738s.$$

The distance moved by the carriage in this time interval is

$$d = ut = -0.115 \times 0.738 = -0.0849 \text{ m}.$$

Thus, the carriage moves 8.49 cm in the opposite direction to the direction of motion of the cannonball.

Question: A softball of mass m = 0.35 kg is pitched at a speed of u = 12 m/s. The batter hits the ball directly back to the pitcher at a speed of v = 21 m/s. The bat acts on the ball for t = 0.01 s. What impulse is imparted by the bat to the ball? What average force is exerted by the bat on the ball?

Answer: The initial momentum of the softball is -m u, whereas its final momentum is mv. Here, the final direction of motion of the softball is taken to be positive. Thus, the net change in momentum of the softball due to its collision with the bat is

$$\Delta p = mv - (-)mu = 0.35 \times (21 + 12) = 11.55 \text{ Ns}.$$

By definition, the net momentum change is equal to the impulse imparted by the bat, so

$$I = \Delta p = 11.55 \text{ N s}.$$

The average force exerted by the bat on the ball is simply the net impulse divided by the time interval over which the ball is in contact with the bat. Hence,

Question: A skater of mass M = 120 kg is skating across a pond with uniform velocity v = 8 m=s. One of the skater's friends, who is standing at the edge of the pond, throws a medicine ball of mass m = 20 kg with velocity u = 3 m=s to the skater, who catches it.

The direction of motion of the ball is perpendicular to the initial direction of motion of the skater. What is the final speed of the skater? What is the final direction of motion of the skater relative to his/her initial direction of motion? Assume that the skater moves without friction.

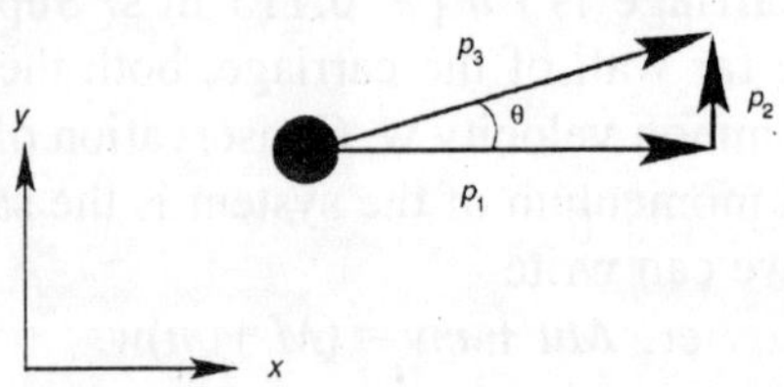

Answer: Suppose that the skater is initially moving along the x-axis, whereas the initial direction of motion of the medicine ball is along the y-axis. The skater's initial momentum is

$$p_1 = (M\ v,\ 0) = (120 \times 8,\ 0) = (960,\ 0)\ \text{N s}.$$

Likewise, the initial momentum of the medicine ball is

$$p_2 = (0,\ m\ u) = (0,\ 20 \times 3) = (0,\ 60)\ \text{N s}.$$

After the skater catches the ball, the combined momentum of the skater and the ball is

$$p_3 = p_1 + p_2 = (960,\ 60)\ \text{Ns}.$$

This follows from momentum conservation. The final speed of the skater (and the ball) is

$$v' \frac{|p_3|}{M+m} = \frac{\sqrt{960^2+60^2}}{120+20} = 6.87\ \text{m/s}.$$

The final direction of motion of the skater is parameterized by the angle θ, where

$$\theta = \tan^{-1}\left(\frac{|p_2|}{|p_1|}\right) = \tan^{-1}\left(\frac{60}{960}\right) = 3.58°$$

Question: A bullet of mass m = 12 g strikes a stationary wooden block of mass M = 5.2 kg standing on a frictionless surface. The block, with the bullet embedded in it, acquires a velocity of v = 1.7 m=s.

What was the velocity of the bullet before it struck the block? What fraction of the bullet's initial kinetic energy is lost (i.e., dissipated) due to the collision with the block?

Answer: Let u be the initial velocity of the bullet. Momentum conservation requires the total horizontal momentum of the system to be the same before and after the bullet strikes the block.

The initial momentum of the system is simply mu, since the block is initially at rest.

The final momentum is (M + m) v, since both the block and the bullet end up moving with velocity v. Hence,

$$mu = (M + m)v,$$

giving

$$u = \frac{M+m}{m} v = \frac{(0.012+5.2)\times 1.7}{0.012} = 738.4\text{m/s}.$$

The initial kinetic energy of the bullet is

$$K_i = \frac{1}{2}mu^2 = 0.5\times 0.012\times 738.4^2 = 3.2714\text{kJ}.$$

The final kinetic energy of the system is

$$K_i = \frac{1}{2}(M+m)v^2 = 0.5\times(0.012+5.2)\times1.7^2 = 7.53\text{J}$$

Hence, the fraction of the initial kinetic energy which is dissipated is

$$f = \frac{K_i - K_f}{K_i} = \frac{3.2714\times10^3 - 7.53}{3.2714\times10^3} = 0.9977.$$

Question: An object of mass m_1 = 2 kg, moving with velocity v_{i1} = 12 m=s, collides head-on with a stationary object whose mass is m_2 = 6 kg. Given that the collision is elastic, what are the final velocities of the two objects. Neglect friction.

Answer: Momentum conservation yields

$$m_1 v_{i1} = m_1 v_{f1} + m_2 v_{f2},$$

where v_{f1} and v_{f2} are the final velocities of the first and second objects, respectively. Since the collision is elastic, the total kinetic energy must be the same before and after the collision. Hence,

$$\frac{1}{2}m_1v_{i1}^2 + \frac{1}{2}m_1v_{f1}^2 + \frac{1}{2}m_2v_{f1}^2 .$$

Let $x = v_{f1}/v_{i1}$ and $y = v_{f2}/v_{i1}$. Noting that $m_2/m_1 = 3$, the above two equations reduce to

$$1 = x + 3\,y,$$

and

$$1 = x^2 + 3\,y^2.$$

Eliminating x between the previous two expressions, we obtain

$$1 = (1 - 3\,y)^2 + 3\,y^2,$$

or

$$6\,y\,(2\,y - 1) = 0,$$

which has the non-trivial solution $y = 1/2$. The corresponding solution for x is $x = (1 - 3y) = -1/2$.

It follows that the final velocity of the first object is

$$v_{f1} = x\,v_{i1} = -0.5 \times 12 = -6 \text{ m/s}.$$

The minus sign indicates that this object reverses direction as a result of the collision.

Likewise, the final velocity of the second object is

$$v_{f2} = y\,v_{i1} = 0.5 \times 12 = 6 \text{ m/s}.$$

Question. Two objects slide over a frictionless horizontal surface. The first object, mass m_1 = 5 kg, is propelled with speed v_{i1} = 4.5 m=s toward the second object, mass m_2 = 2.5 kg, which is initially at rest. After the collision, both objects have velocities which are directed $\theta = 30°$ on either side of the original line of motion of the first object. What are the final speeds of the two objects? Is the collision elastic or inelastic?

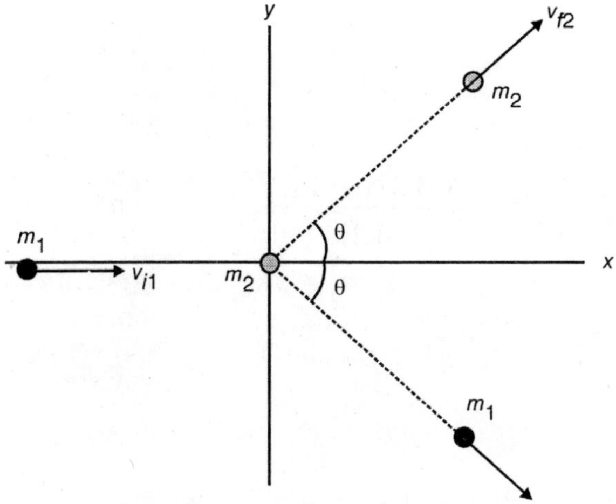

Answer: Let us adopt the coordinate system shown in the diagram. Conservation of momentum along the x-axis yields

$$m_1\, v_{i1} = m_1\, v_{f1} \cos\theta + m_2\, v_{f2} \cos\theta.$$

Likewise, conservation of momentum along the y-axis yields

$$m_1\, v_{f1} \sin\theta = m_2\, v_{f2} \sin\theta.$$

The above pair of equations can be combined to give

$$v_{f1} = \frac{v_{i1}}{2\cos\theta} = \frac{4.5}{2\times\cos 30^\circ} = 2.5981\,\text{m/s},$$

and

$$v_{f2} = \frac{m_1}{m_2} v_{f1} = \frac{5\times 5.981}{2.5} = 5.1962\,\text{m/s}.$$

The initial kinetic energy of the system is

$$K_i = \frac{1}{2} m_1 v_{i1}^2 = 0.5\times 5\times 4\times 4.5^2 = 50.63\text{J}.$$

The final kinetic energy of the system is

$$K_f = \frac{1}{2} m_1 v_{f1}^2 + \frac{1}{2} m_2 v_{f2}^2$$

$$= 0.5 \times 5\ \text{Q}\ 2.5981^2 + 0.5 \times 2.5 \times 5.1962^2 = 50.63\ \text{J}.$$

Since $K_i = K_f$, the collision is elastic.

[illegible]

[illegible]

[illegible]

$= 0.5 \times 5 \times$ [illegible] $\times 0.5 \times 2.5 \times$ [illegible] $= 50.64$ J.

